Global Change, Energy Issues
and Regulation Policies

INTEGRATED SCIENCE & TECHNOLOGY PROGRAM

fondation maison des sciences de l'homme

Volume 2

Aims and Scope

A hub for a large number of national and international scientific and academic networks, the Fondation Maison des Sciences de l'Homme/FMSH (http://www.msh-paris.fr/) helps forge relations between institutions, researchers, and research teams worldwide in order to facilitate research across disciplines. No one discipline is given priority and there are no territorial limitations. The FMSH is committed to its role as a place for experimenting with and fine-tuning ideas, new theories, and types of organization. Founded upon the conviction that interdisciplinarity is essential for understanding the complex questions facing society in a global setting in order to make sense of diverse cultures, the FMSH's mission is to act as "a facilitating, interdisciplinary, and international incubator."

Integrated Science & Technology Program (ISTP) Book Series

Today, the FMSH continues to occupy a unique place in research and to provide a uniquely stimulating environment for intellectual meetings and exchanges. The creation of the *Integrated Science & Technology Programs* (ISTP) facilitates this interdisciplinary exchange and pursues three main goals:

- to highlight the success of integrated approaches in science
- to examine national and international policy with regard to research in science and technology
- to address the cultural, economic, political and social impact of scientific and technological development

The ISTP Book Series includes volumes of selected invited papers and is intended for researchers, practitioners, undergraduate and graduate students in all disciplines who are dealing with complex problems and looking for cutting-edge research as well as methodological tools to set up truly transversal science and technology projects.

Themes

The work generated by exchanges and cooperation between researchers from all countries has helped to delineate research themes that often touch upon the major questions facing contemporary societies. For the Foundation, this means taking a risk on "unconventional" projects that correspond to its mission of "incubator." Through its broad-reaching programs, which address a wide array of cultures and civilizations while maintaining a clear focus on historical depth, international comparisons, and interdisciplinarity, the FMSH assists interdisciplinary projects that seek to improve our understanding of key problems linked to the reshaping of contemporary societies. Examples of themes include:

- Environmental Issues and Sustainable Development
- Earth Sciences and Space Sciences
- Science, Technology and Society
- Cognitive Sciences and Technologies
- Information Technology and Communication
- Health and Life Sciences
- Economics and Finance

Springer is pleased to collaborate with the Fondation Maison des Sciences de l'Homme in publishing this Book Series.

For further volumes:
http://www.springer.com/series/8815

Jean Bernard Saulnier • Marcelo D. Varella
Editors

Global Change, Energy Issues and Regulation Policies

Springer

Editors
Jean Bernard Saulnier
Pprime/ENSMA
Chasseneuil, France

Marcelo D. Varella
Campus do UniCEUB
University Center of Brasilia
Brasilia, DF, Brazil

ISSN 2211-2375 ISSN 2211-2383 (electronic)
ISBN 978-94-007-6660-0 ISBN 978-94-007-6661-7 (eBook)
DOI 10.1007/978-94-007-6661-7
Springer Dordrecht Heidelberg New York London

Library of Congress Control Number: 2013940422

Printed on acid-free paper

Springer is part of Springer Science+Business Media (www.springer.com)

Foreword

A world in change—hasn't it always been in change? Or have we changed our perspective? Questions that challenge global change research efforts are highlighted in this book. I am not sure whether it is our guilty conscience or just new knowledge that drives humanity forward to consider the world is in change. Drastic changes are recognized easily such as population growth and increased lifespan, faster and cheaper transportation and globalization, the Internet and smartphones, but also the political changes and revolutions in various regions. Yes we have changed our lifestyle, our consumption patterns, and our expectations. But who is "we"? Still, millions of people worldwide have no or rare access to fresh water, to sufficient nutrition, to education, and to many other attainments of a modern civilization. They suffer from sickness, war, and crime. It is quite certain they do not have the same perspective as European or American citizens. But even within the so-called industrially developed countries, there is a wide span between those who struggle to survive daily and others who fill biofuels in their big SUV's.[1]

A world in change also means changing societies. Poverty, hunger, or keeping a nation in bondage have always delivered reasonable motives for revolutions. A scarcity of resources such as water, land, ore, coal, and oil has always been a trigger for fighting and war. Understanding this cause–effect relationship therefore should become the strongest driver for another change: our change in mind, our common sense. But is it realistic to expect that the different stakeholders in a multitude of nations follow one global paradigm? A paradigm of sustainability, or even a paradigm of a joint global action policy for a sustainable energy supply or against climate change: can it really become our tour guide for this twenty-first century?

Energy is one of the dominating factors of basic human needs: nutrition, housing, family settlement, health care, and welfare, all depend on a sufficient, reliable, and achievable energy supply. We are used to transforming basic sources of energy such as coal, uranium, solar power, and hydropower into other forms of energy that are better suited for our daily life: electricity, chemical fuels, or heat. Our current

[1] Sport utility vehicle.

consumption of energy is unfair globally, is unfair within our societies, and is unfair between generations. On a long-term perspective fossil fuels are limited and if all coal, oil, and natural gas resources are consumed a huge amount of carbon will be released into the atmosphere. Do we now change our perspective?

It is logical to study various scenarios of future developments. A sustainable energy supply needs to be based on "unlimited" resources and has to minimize the impact on the environment. It has to be achievable for all citizens at an affordable price. We are in a transition phase from using limited fossil resources towards these new sources of energy. Not only wind and solar power but also fusion technology as well as geothermal sources, marine power, bioenergy, and some more exotic ways of energy "harvesting" are worth being investigated.

Researchers have to provide not only technological roadmaps but also socio-economic scenarios. The tallest hurdles for new energy technologies are quite often public opinion, fear, and individual impairments. The well-known NIMBY syndrome—not in my backyard—can be observed very clearly in Germany where politics and public pressure in 2011 forced a sudden change of the whole energy system after the Fukushima nuclear accident. Everybody seem to like the renewable energies but if the rotor blades of big wind power facilities rotate near the village and the tall stands are placed in a beautiful landscape there will immediately be an opposing movement established. No one likes the expansion of the electrical grid if the power lines come to close to his or her house. All would like to have the carbon-free emissions of power plants but most people refuse to consider CCS[2] technology.

The largest challenges of designing and building a future sustainable energy system may be grouped in technology, political boundary conditions, economic incentives, management, and behavior. Changing our consumption pattern is certainly one of the most difficult tasks and the consumer's willingness to pay higher prices as well as to accept impairments might be a Herculean effort. Compared to that I strongly believe in our creativity, the innovative power of mankind, and human intelligence is the only real unlimited resource on earth to be used for the further development of technical solutions.

The conference on energy and environment where I was invited in June 2010 to the small island of Porquerolles in the Mediterranean Sea has been a remarkable milestone on our common way to achieve these future energy solutions. The Entre-Sciences Program at Fondation Maison des Sciences de l'Homme and the European Science Foundation did a wonderful job in organizing such an event. I enjoyed the inspiring atmosphere, our discussions between researchers and industry people, among students, young scientists, and senior experts, in all energy-related fields. The most important was to learn about the different perspectives of participants from developing countries as well as from supersaturated societies. The conference gave me the freedom to focus some days on energy, environment, and societal

[2]Carbon capture and storage.

development. Only 2 years later the energy issue has become a priority of its own, at least in Europe, and at least in Germany.

This book summarizes the conference results of Porquerolles but thanks to the strong power of Jean Bernard Saulnier and his colleagues and coauthors it has been enriched by quite recent contributions and I am very convinced that this book will become a milestone too. I wish this publication will increase appropriate awareness for an energy perspective on global changes and I believe the energy community will be delighted about it.

Chief Science Officer
Karlsruhe Institute of Technology KIT, Germany
Chairman, KIC InnoEnergy SE, Europe

Karl-Friedrich Ziegahn

Acknowledgments

We would like first to thank the following institutions that organized, on the island of Porquerolles (French Riviera), the Global Change Research II Conference, the ESF (European Science Foundation), the FMSH (Foundation Maison des Sciences de l'homme, Paris), the FMSH "Entre-Sciences Program," the UPCAM (University Paul Cézanne Aix-Marseille III, now Aix-Marseille Université), and those who provided financial support:

ESF, FMSH, UPCAM, and also
ADEME (French Environment and Energy Management Agency),
EDF (Electricité de France),
and CNRS (French National Center for Scientific Research).

We also thank all the authors who contributed to the success of the conference itself, and those who accepted the writing of different chapters of this book.

A special acknowledgment is given to Professor Karl Friedrich Ziegahn, co-chair of the conference, who brought particularly dynamic and relevant contributions to the discussions at Porquerolles and wrote an original preface to the book.

Angela Procoli from FMSH and François Rochet from the "Entre-Sciences Program" were particularly active, both in preparing the conference and in revising the manuscript and the editors are sincerely grateful for their support.

We are also grateful to Betty van Herk, from Springer, for her efficient advice in preparing the manuscript, and to Shanthy Gounasegarane, from SPi Content solutions, who helped a lot during the last steps before the printing of the book.

J.B. Saulnier would like to thank all his colleagues involved in the CNRS Interdisciplinary Programme Energy, with a special reference to Bernard Spinner, Edouard Favre, Monique Lallemand, Patrick Le Quéré, Yvan Faure-Miller, Hassan Peerhossaini, and Alain Dollet. And a last but not least thanks to Elisabeth Giacobino who helped so much in the development of CNRS interdisciplinary research programs.

M.D. Varella would like to thank all his colleagues and students involved in the research on law and sciences who participated on this project, especially Priscila Andrade, Carina Costa Oliveira, and Gabriela Batista de Lima.

Contents

Chapter 1
Global Change Research II: Some Keys to the Climate/Energy Crisis

Jean Bernard Saulnier and Marcelo D. Varella

Abstract This introductory chapter analyzes the deep interaction between the environmental crisis (climate change, urbanization/land use, exhaustion of resources, and degradation of ecosystems), energy production, conversion and use, and global regulation policies. It first recalls the main conclusions of the June 2010 Global Change Research II, Porquerolles Conference (environmental degradation related to energy production, links between energy and human needs, energy and environment, interface among technologies, science, and society). It explains the architecture of the book, which fairly faithfully follows the conference plan, including new contributions that were not presented at the conference. It brings some particular comments about climate change and exhaustion of resources, the relationship between basic science and the development of sustainable energy technologies, between global and local environmental policies: technologies, economy, law. The conclusions emphasize five technological keys (and their main issues) to the solution of the energy/environmental crisis: improvement of energy efficiency and savings, green electricity production (if new storage technologies are available), nuclear energy (if its sustainability is increased), carbon management, energy vector use optimization (biofuels if the planet alimentation is not threatened, emergence of the hydrogen society, smartgrids). Finally, last but not least, a sixth key is to be found in the domain of humanities and social sciences, law, politics (negotiations at international level, financial rules, etc.). It also emphasizes the need for basic science research for providing breakthroughs in the field of energy.

J.B. Saulnier (✉)
Pprime/ENSMA, BP 40109, 86961 Chasseneuil, France
e-mail: Jean-bernard.saulnier@let.ensma.fr; jbsaulnier@wanadoo.fr

M.D. Varella
Programa de Mestrado e Doutorado em Direito, University Center of Brasilia,
SEPN 707/907, Campus do UniCEUB, Bloco III, Térreo, Brasília, DF 70790-075, Brazil
e-mail: marcelodvarella@gmail.com

J.B. Saulnier and M.D. Varella (eds.), *Global Change, Energy Issues and Regulation Policies*, Integrated Science & Technology Program 2, DOI 10.1007/978-94-007-6661-7_1,

Keywords CO_2 capture and storage • Renewable electricity and storage • Sustainable nuclear energy • Biofuels • Hydrogen • Financial rules

List of Acronyms

ADEME	Agence de l'environnement et de la maîtrise de l'énergie (France), French Environment and Energy Management Agency
ALPF	Australia Low Pollution Future
AR4	IPCC Fourth Assessment Report (2007)
BAU	Business As Usual
BES	Basic Energy Sciences
BESAC	Basic Energy Sciences Advisory Committee (US DOE)
CC	Climate Change
CCR	Carbon (CO_2) Capture and Recycling
CCS	Carbon (or CO_2) Capture and Storage
CDM	Clean Development Mechanism
CM/JI	CDM Joint Implementation
CNG	Compressed Natural Gas
CNRS	French National Center for Scientific Research
CSP	Concentrated Solar Power
DAFC	Direct Alcohol Fuel Cells
DC	Developing Countries
EC	Emerging Countries
EDF	Electricité de France
EERA	European Energy Research Alliance
EPO	European Patent Office
ESF	European Science Fundation
FMSH	Fondation Maison des Sciences de l'Homme (House of the Sciences of Man Foundation)
GCR	Global Change Research
GDP	Gross Domestic Product
GdR	Group of Research (CNRS tool for creating a consortium of laboratories)
GHG	Greenhouse Gases
Gt	Gigaton (= 1 million ton)
GT3A	Groupe de Travail sur les 3A (CNRS), CNRS Working Group on 3A (Agriculture – Alimentation needs – Agrofuels)
GteqC	Gigaton of Carbon Equivalent
Gteq CO_2	Gigaton of CO_2 Equivalent
HEQ	High Environmental Quality
IC	Industrialised Countries
IEA	International Energy Agency
IPCC	Intergovernmental Panel on Climate Change

LDC	Least Developed Countries
ITER	International Thermonuclear Experimental Reactor
LED	Light Emitting Diodes
LULUCF	Land Use, Land Use Change, and Forestry
Mtoe	Megaton of Oil Equivalent
NAMA	Nationally Appropriate Mitigation Action
NSSSEF	New Science for a Secure and Sustainable Energy Future
OECD	Organisation for Economic Co-operation and Development
OFCE	Observatoire français des conjonctures économiques—The French Economic Observatory
PACTE	CNRS GdR Piles à Combustible Tout Electrolyte (All Electrolyte Fuel Cell) http://www.gdr-pacte.cnrs.fr/
PATSTAT	EPO/OECD World Patent Statistical Database
ppm	Part per Million
PCFC	Proton Ceramic Fuel Cells
PEMFC	Proton Exchange Membrane Fuel Cell
PV	Photovoltaics
RFC	Reasons For Concern
SAMFC	Solid Alkaline Membrane Fuel Cells
SD-PAM	Sustainable Development—Policies And Measures
SOFC	Solid Oxide Fuel Cell
TAR	IPCC Third Assessment Report (2001)
Three-ME	Multisector Macroeconomic Model for the Evaluation of Environmental and Energy policy
UNFCCC	United Nations Framework Convention on Climate Change
UPCAM	University Paul Cézanne Aix Marseille III
WHO	World Health Organization

1.1 Introduction

Further to the first 2008 conference (Global Change Research I 2008), focused on climate modeling and risk/impact assessment, the second conference of the series, "Global Change Research II" (GCR II),[1] tried to examine the deep interaction between the environmental crisis (climate change, urbanization/land use, exhaustion

[1]This GCR II 2010 Conference was organized by the ESF (European Science Foundation), the FMSH (Foundation Maison des Sciences de l'homme, Paris), the FMSH "Entre-Sciences Program," the UPCAM (University Paul Cézanne Aix-Marseille III, now Aix-Marseille Université), and supported by ESF, FMSH, UPCAM, ADEME (French Environment and Energy Management Agency), EDF (Electricité de France), and CNRS (French National Center for Scientific Research).

of resources, and degradation of ecosystems), energy production, conversion and use, and the global regulation policies. It took place on the Island of Porquerolles (French Riviera, near Toulon) from June 11th to 16th, 2010.

The conference was divided into three parts that are summarized as follows.

The relationship between the use of carbon-based energy and global warming which seems to be currently well recognized, was discussed and more general aspects of "environmental degradation related to energy production," in particular impacts on the biosphere and health were also addressed in the first part of this conference. For instance, because in the next 25 years almost two billion more people will move into cities, sustainability and vulnerability analysis of urban areas is an issue of increasing concern due to global change. Urban infrastructure is a vital component of a city including utility and transport networks, water and flood management structures, and underground networks, for which energy needs become now a challenge.

The second part of the conference analyzed the links between "energy and human needs" on the one hand, and "energy and environment." It provided a panorama of research efforts and technological implementations in the field of energy efficiency and environmental performance, for today's energies (nuclear, transport, and buildings, etc.), emerging technologies (biofuels, wind turbines, etc.), and longer-term opportunities (hydrogen and fuel cells, CO_2 capture, storage, recycling and its eventual conversion back to fuel or even to trees!).

The last part of the conference focused on the interface among technologies, science, and society. The implementation of good practices and the diffusion of new energy technologies worldwide strongly depend on global regulation policies, international agreements, and global/local governance. Emphasis was laid on the environmental aspects of international trading and the laws related to the production and transportation of energy (environmental damage liability, environmental catastrophes, etc.). The choice of adapted robust technologies in the less-developed countries (LDC) was illustrated by the presentation of new concepts of improved cooking stoves for poor villagers in Nepal, leading to an improvement in deforestation and to the reduction of lung diseases. Finally, the political, economic, and social effects of energy policies both at the national and international levels were illustrated.

Before discussing the content of the book, we found it interesting first to evoke the main conclusions of the conference (see Sect. 1.2), because, even though we provided a selection of the best contributions in the conference program, good ideas emerged in many presentations. In addition, we complemented these best lectures with new specific contributions that, unfortunately, due to time, it was impossible to include during the five days spent at Porquerolles, and we summarize the final content of the book (see Sect. 1.3).

A first package of connected chapters first deals with the essential analysis of the future of fossils, including a discussion of the many nonconventional resources, which refers to the necessity of developing operational processes dealing with CO_2 mitigation as soon as possible: the capture and sequestration of carbon (CCS) or its capture and recycling (CCR). A general overview on greenhouse gases (GHG) and climatic change brings an up-to-date vision of the knowledge of the complex phenomena involved. Finally, as an example of the social impact of climate change (CC), we have a demonstration of possible urban physical infrastructure adaptation to CC.

A second package presents some examples demonstrating the way science and technology can try to answer the crisis detailed above; two first complementary aspects dealing with CO_2 mitigation, the three other ones with the emergence of uncarbonized production of energy (nuclear) and vectors (H_2 and biofuels), which translate into:

- Two different paths that might contribute to CO_2 mitigation, by capture, and then sequestration (CCS) or recycling (CCR)
- The potential of biofuels and the emergence of advanced technologies based on molecular biology
- The motivation for a hydrogen civilization including fuel cell technologies, with a comparison of well-to-wheel CO_2 emissions between different transport technologies
- The vision of sustainable nuclear energy, and the implications of the Fukushima accident

Concerning nuclear, we have adopted a point of view that refers to a context of peaceful use of nuclear energy, which implies in particular that proliferation danger is under the control of recognized international and national authorities.

A third and last package focuses on the analysis of governance and policies at the international level.

- Two chapters study the main results of the Copenhagen Conference, which, clearly, just as the Durban one, was in no way completely successful. They try to identify what can still be considered as a source of progress, with the evolution of the international climate arena, including geopolitical shifts, new issues on the agenda, and a new cartography of the main actors. The "adoption" of the Copenhagen Accord (2009) did not stop the negotiation process, and the chapters conclude that the Cancun Conference (2010) revived a process that had almost come to a standstill.
- After the adaptation of urban infrastructures to climate change examined in the first package, a second example involving the cities and climate change interaction is presented here with the governance of urban infrastructures in developing cities and the role of carbon finance.
- Finally, an interesting complement to the scientific study of biofuels technologies given in the second package provides a more worldwide and geopolitical analysis of the first-generation biofuels potential and risks (environmental impacts, energy efficiency, hunger in the world, etc.), and attempts to answer the following challenge:

> How relevant is it to build up agrofuel production in terms of a partial solution to energy self-sufficiency, in terms of the reduction of greenhouse gas emissions, in terms of its impact on meeting humanity's alimentation needs?

Then three discussions (Sects. 1.4, 1.5, and 1.6) try to expand a bit more on the main issues of the book.

1.2 The MAIN Conclusions of the 2010 Porquerolles Conference

Let us recall that the conference was organized in seven sessions based on the following classification.

- Climate change, energy and human needs versus energy and environment (two sessions)
- Basic science and the development of sustainable energy technology (two sessions)
- The interface among technology, science, and society (three sessions)

The introduction first gave the basis for green growth (Matarasso 2010), a general overview of the European Energy Research Alliance (EERA) objectives and missions (Ayache 2010), and then offered a view of the necessary compromise between visionary ideas and realistic options to guide research for a sustainable energy future (Ziegahn 2010).

The first two sessions provided a general frame for the introduction of the conference and started with the presentation (Philibert 2010) of the recent work performed by the International Energy Agency. The main figures of 450 Scenario (the objective of which is to try to maintain the CO_2 concentration in the atmosphere at roughly the present value of 450 ppm; see Sect. 1.4 and also Chap. 2) helped to identify the general trends that can be proposed for the next 20 years, concerning the production of electricity, the possible role of renewables, and the content of the energy mix, at the world level as well as at the regional level. The world abatement of energy-related CO_2 emissions and the main related roadmaps were also presented and discussed (Photovoltaic –PV-, Concentrated Solar Power-CSP-. . .). An important point should be kept in mind: these exercises use modeling and the level of confidence of the results does, of course, decrease over prediction time: no one is able today to put reduced values on the uncertainty of these predictions which are all but prognosis. Complementary presentations (Bobylev 2010, see also Chap. 4; Mocanu 2010) introduced some new concepts for the city of the future, demonstrating in particular how they could help urban infrastructures to adapt to climate change.

As regards the role of basic science in the advancement of sustainable energy development, a view of new energy sources using solar energy was first presented (Perathoner 2010; see also Chap. 7). It included rather classical aspects (solar thermolysis, thermochemical cycles, solar cracking, solar reforming, solar gasification, algae and cyanobacteria, etc.) associated with a new way of recycling CO_2 using electrophoto catalysis which might clearly lead to an interesting breakthrough: taking advantage of fuel cell developments, using both photocatalysis and electrocatalysis on two separate electrodes, including a proton membrane. This new process is able to transform CO_2 (entering the cathode) and water (entering the anode, with light) into higher hydrocarbons and alcohols.

Other lectures of interest provided a very comprehensive overview of hydrogen and fuel cells, particularly adapted to the audience (Jones 2010; Skulimowska 2010;

see also Chap. 8). To discuss the problem of the intermittence of some renewable sources of energy (photovoltaic, wind turbines, etc.) and the necessity of energy storage, a practical hydrogen chain operating in Denmark was shown, including production of electricity by a wind turbine, conversion into hydrogen by electrolysis, and reconversion to electricity with a fuel cell.

A clear analysis of the contribution of nuclear power (essentially fission) started from the observation that Europe is clearly a world leader (Abderahim 2010; Ayache 2010; see also Chap. 5), but competition is increasing (Russia, Japan, the United States, India, and China). With 152 reactors (450 at world level) nuclear power is, producing 31 % of electricity in Europe, providing the larger source of low carbon energy, with an excellent safety record. Because fossil and nuclear power plant generation are ageing, there is a need to invest in plant lifetime management and large investments are necessary to build new plants to satisfy the demand. This will be accomplished thanks to scenarios adapted to the transition between the present Generations II and III, and future Generation IV: its characteristics essentially concern an increase in safety and a huge decrease in nuclear waste. From a scientific point of view, the main issues were identified in the field of material sciences and fuel research, simulation, modeling, and experiments for validation. Another important point was mentioned, concerning the need for well-trained and educated specialists in the various fields related to nuclear fission. A long discussion with the audience offered in particular the opportunity to explain clearly the difference between the basic physics mechanisms used in classical plants and in Generation IV.

As an example of detailed possible contribution of renewable energies, the case of wind turbines was analyzed (Therond 2010) and led to the following conclusions.

- Onshore wind is the least expensive after hydro, but limited, as is hydro.
- Solar energy is the most abundant and best distributed in the world, but is still expensive.
- Biomass can be stored, but the resource is not really free.
- Marine energy including offshore wind can provide a massive production of green electricity, but still involves high industrial risk.

So, there is no "technological miracle;" to reach sustainability, the idea of a mix was still recognized, and all renewable energies should be associated with strong energy savings: in this view, wind appears as a necessary step, even if it is not the most efficient way to decrease CO_2 content. In addition, fluctuations in the wind (intermittent) must be managed at system level and seem to be an economic issue as well as a technical one (for instance, best remuneration of hydropower). The main advances still to be accomplished on the wind systems concern:

- Better prediction for different timescales
- Tighter and real-time information exchange with grid operators ("smart grid" approach)
- Storage technologies in extreme cases, such as islanded systems

Let us mention that interesting ideas were also expressed in a series of talks: on the role of the catalyst in physicochemical processes involved in the global warming control (Goswami 2010), or on the consequences of bad concepts of energy production and use on human health. A good example is the case of a project of compressed natural gas conversion of motor vehicles in Dhaka (Wadud 2010; see also Chap. 10). A last one (Rajaure 2010) dealt with a study on cooking stoves for poor villagers in Nepal leading both to an improvement in deforestation and to the reduction of lung diseases (new design of stoves, new concepts of briquettes). Two main ideas should be retained from this last work:

- First, the size of the health problem, with nearly two million people dying prematurely every year from illnesses attributable to indoor pollution due to solid fuel use (WHO 2011 – World Health Organization);
- Second, the efficiency of the research work in this field which, with a quite moderate investment, could induce valuable social, economic, and health improvements. The cooking stove efficiency has increased from 5–10 % to more than 20 %, the exposure to pollutants was reduced by 70 % and wood consumption was also reduced by 30 %.

Moving on to the sessions dealing with the interface among technology, science, and society, contributions covered socioeconomics, law, and politics, among others, of energy and the environment, focusing either at national levels or extending towards European and international standpoints.

An interesting association combined two lectures (Mercier 2010; Callonnec 2010), dealing with the implementation of the Grenelle II law in France (help and constraints for building renovation) and, on the other hand, modeling the efficiency of economical measures enforced to reduce CO_2 emission in the residential sector. The model included 216 typical renovation work scenarios, and their energy performance was assessed using the official energy performance diagnosis method. The results provide the impact of energy efficiency measures in the residential sector:

- Energy efficiency tax credit reform
- Thermal performance modification
- Zero interest loan introduction
- Influence of carbon tax

Such an example appeared to be convincing, showing how modeling can help to prepare the law and to process its followthrough.

Two presentations particularly exemplified the socioeconomic role of biofuel development, one dealing with the transition from oil-based to biofuels-based energy supply in Indonesia (Jupesta 2010) and the second with poverty reduction and sustainable development improvement through biofuels in Brazil (Varella 2010).

A new model of the French economy (Reynès et al. 2011) was especially designed to assess the medium and long-term impact of environmental and energy policies at the macroeconomic and sectorial levels (Yeddir-Tamsamami 2010).

At a wider level, the three targets of Europe's energy policy (energy security, economic competitiveness, environmental protection) were evoked (Pradel 2010), and its different tools defined to ensure the spreading of European norms and structures: political dialogue, law, and economic incentives.

The new use of private law instruments as a new trend for international environmental law was illustrated through two examples: the first with sustainable biofuels production and consumption (Pereira de Andrade 2010), and the second with the issue of damage liability (Costa de Oliveira 2010) in the context of environmental catastrophes linked to energy production (the case of the BP oil spill in the Gulf of Mexico: prevention and reparation of international environment damage).

An original contribution (Dechezleprêtre 2010) used the EPO/OECD (European Patent Office/Organisation for Economic Co-operation and Development) world patent statistical database (PATSTAT) to provide a quantitative description of the geographic distribution of inventions in 13 climate mitigation technologies since 1978 and their international diffusion on a global scale. Statistics suggested that innovation was mostly driven by energy prices until 1990 but since then, environmental policies, and climate policies more recently, have accelerated the pace of innovation.

The global environmental policies at world level were examined first through the effectiveness of the Kyoto protocol on greenhouse gas production (Dahan 2010; see also Chap. 11). A second lecture dealt with the analysis of the Copenhagen process and the different ways to consider it as a failure or a success, particularly by revisiting both scientific and political frames of the climate change regime.

The lack of instruments to help the CDM (clean development means) process was clearly identified (Lemoine 2010; Li 2010; see also Chaps. 12 and 13). In fact, because two thirds (7,900 Mtoe in 2006) of world energy are consumed in cities contributing roughly 80 % to global GHG emissions, and because most of the increased urban population and demand for energy in cities will come from developing countries over the next decades, it is important to examine the role of carbon finance in governing urban infrastructures in developing countries. Existing instruments need improvement; so far, there is no systematic and comprehensive approach with appropriate financial and technological support, and the existing financial tools, including carbon finance (CDM), do not allow cities in developing countries (DC) to change substantially the course of development to avoid long-term energy and carbon lock-in.

Just to give a foretaste of the results of intensive discussions and exchanges, we mention here some main and original comments that came with the conclusions of Professor Ziegahn, co-chair of the conference:

- Acceptance of energy technologies by the public has to be prepared without ideology and with fact-based communication.
- Media and politics, researcher and developer, have to provide information without stereotypes.
- Permanent threat scenarios and "last-days-of-the-earth" prophecies will not provide break-through thinking.
- To awaken unrealistic expectations will lead to disappointment and less power for change.
- Creativity and invention capacity are Europe's only unlimited resources. It should not be stopped by required mainstream thinking or banned because of unconventional ideas.

1.3 From the Conference to the Book

The architecture of the book fairly faithfully follows the way the conference was organized, with a plan in three parts, including new contributions that were not presented in the conference. (Those subjects not addressed at Porquerolles, are in italics.)

1.3.1 Climate Change and Exhaustion of Resources

Chapter 2 presents the essential and up-to-date knowledge on greenhouse gases and climatic changes.

Chapter 3 has collected rich information about *nonconventional fossil resources*.

Chapter 4 discusses the possibility of adaptation of urban physical infrastructure to climate change.

1.3.2 Basic Science and the Development of Sustainable Energy Technologies

Chapter 5 focuses on sustainable nuclear energy and shows how it can help Europe to meet its energy challenges.

Chapter 6 deals with the problem of *CO_2 capture and storage* (*CCS*), which appears to be inescapable, with the use of fossil fuels for several decades to come, as the possibility seems to appear in Chap. 3.

As a complementary point of view, Chap. 7 presents the research paths that could offer new possibilities for CO_2 mitigation: carbon capture and recycling (CCR).

Chapter 8 offers a very complete overview of hydrogen and fuel cells.

Chapter 9 gives the *main scientific issues in the field of biofuels* and emphasizes the role of biology for third generation biofuels (see also the complementary approach in Chap. 14, more oriented towards socio- and geopolitical aspects).

Chapter 10 explains the co-benefits valuation from CNG (compressed natural gas) conversion of motor vehicles for meeting environmental energy and health challenges, with an example in southern countries.

1.3.3 Global and Local Environmental Policies: Technologies, Economy, and Law

Chapter 11 discusses the political results of the Copenhagen Conference and concludes by observing a shift of the focus from the links between science and politics towards the relationships between science and societies.

Chapter 12 also considers the same domain, and tries to identify the drafting of the future international climate regime, through developments from the Copenhagen to the Cancun accords.

Chapter 13 places the institutional implications of using carbon finance to facilitate the development of climate-resilient urban infrastructure in fast-growing cities in developing countries.

Chapter 14 performs the analysis of the *triple A issue* (*Agriculture, Alimentary needs, Agrofuels*) and tries to explicate, quantify, and rationalize the assessment of the stakes concerning human alimentation over the next four decades on a world scale. It also reviews the environmental impact of large-scale farming and its capacity to produce energy quantitatively on a world scale.

1.4 Comments About the Climate Change and Exhaustion of Resources Section

We have no particular comment about the three chapters composing this section, but we emphasize the value of 2 °C which is usually considered a fatal threshold for global warming. This value is also considered in Chaps. 11, 12, and 13.

1.4.1 Why Is the +2 °C Global Warming Considered as a Standard Reference?

This value has become a reference threshold, used in a large number of reports and papers. Many scientists and many governments agree that the world needs to keep an average global temperature rise of 2 °C (by reference to preindustrial times, which means 1.4 °C from now on). We first evoke here the main reasons for the concern of global warming of at least 2 °C. Then we identify the relationship between this value of 2 °C and CO_2 emissions, both in terms of the total mass emitted per year, and of the concentration in the atmosphere.

1.4.2 What May Be the Consequences of a +2 °C Global Warming?

Let us first consider the consequences of a 2 °C global warming. A recent update (IPCC AR4 2007) of previous IPCC studies (IPCC TAR 2001) concludes serious "reasons for concern (RFC)" if global warming exceeds this value of 2 °C, which has now become a kind of reference. This is in fact an average, which means that this does not exclude examples of locally higher increases—between 3 and 5 °C in the Arctic for instance—the effects of which are better known now. A good representation of the different risks we are now examining is given through the so-called "burning embers diagram" (Smith et al. 2008).

1.4.2.1 Risk to Unique and Threatened Systems

It addresses the potential for increased damage to or irreversible loss of unique and threatened systems such as coral reefs (sustained bleaching and widespread demise, if global temperature increases by 1 °C from recent levels), tropical glaciers [Andean Quelacaya: the lives and livelihoods, which are even now difficult under the best conditions, are endangered by the loss of water resources under CC conditions (Thompson et al. 2011), etc.], biodiversity hotspots (polar bears, migrating birds, etc.), small island states, and indigenous communities, among others.

1.4.2.2 Risk of Extreme Weather Events

The second type of RFC includes increases in the frequency, intensity, or consequences (for societies and natural systems) of heat waves, floods, droughts, wildfires (starting at less than 1 °C additional warming), or tropical cyclones and adverse health effects (slightly above 1 °C). We comment, for instance, on the influence of global warming on droughts and precipitation (Trenberth 2011). Increased heating leads to greater evaporation and surface drying, increasing the intensity and duration of drought. However, the water-holding capacity of air increases by about 7 % per 1 °C warming, which induces increased water vapor in the atmosphere. Hence storms, extra tropical rain or snowstorms, or tropical cyclones, supplied with increased moisture produce more intense precipitation events.

1.4.2.3 Impacts Distribution

There is still high confidence that impacts distribution will be uneven and that low-latitude, less-developed areas are generally at the greatest risk due to both higher sensitivity and lower adaptive capacity. However, recent work has shown that vulnerability to climate change is also highly variable within individual countries: there is increasing evidence of greater vulnerability of specific populations, such as the poor and the elderly, to climate variability, and not only in developing countries, but also in developed countries. New studies confirm that Africa is one of the most vulnerable continents, because of the range of the projected impacts, multiple stresses, and low adaptability capacity; really dangerous situations appear between 1 and 2 °C (Costello et al. 2009). We finally mention, as another particular example of uneven impacts, the growing number of coastal dwellers, particularly in areas subject to tropical cyclones, who are facing increasing risks.

1.4.2.4 Aggregate Damages

Impacts distributed across the world can be collected in aggregates, such as monetary damages, lives affected, or lives lost. Some studies reinforce the finding of potential benefits at a very few degrees of warming, followed by damages with more warming (IPCC AR4 2007) but finally impacts find net damages beyond 2 °C. More

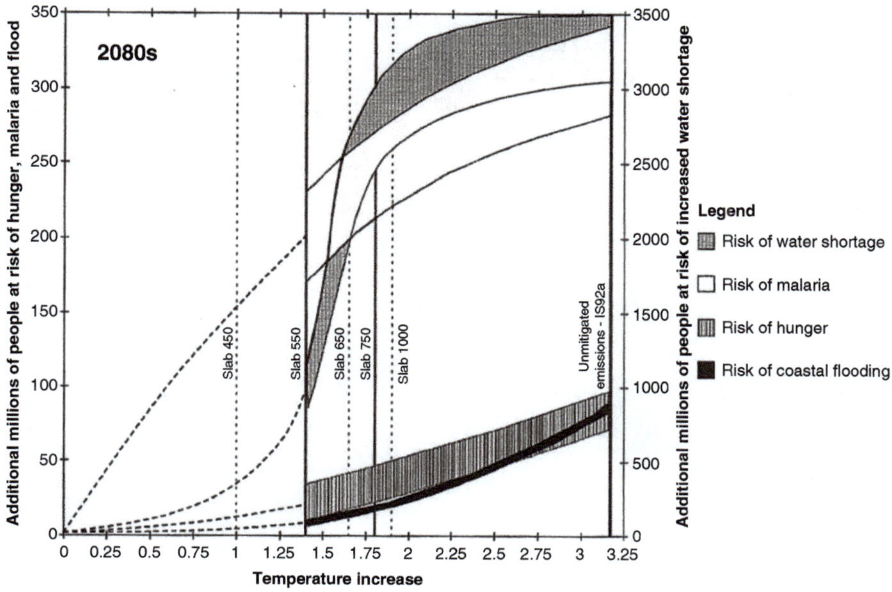

Fig. 1.1 Influence of global warming on people at risk (Parry et al. 2001)

precisely (Smith et al. 2008), hundreds of millions of people might be affected through increased coastal flooding after a 2 °C warming and 0.4–1.7 **billions** of people could be affected by the influence of 1 °C warming on the water supply. Another aggregate concerns the health impact (see Fig. 1.1).

1.4.2.5 Large-Scale Singularities

These risks include the deglaciation (partial or complete) of the West Antarctic or of Greenland ice sheets or a substantial reduction, or even collapse, of the North Atlantic Meridional Overturning Circulation. Complete deglaciation of the Greenland ice sheet would raise the sea level by 7 m, and would occur on a century timescale (Smith et al. 2008). A midpoint of 2.5 °C is considered as a possible trigger for commitment at a large scale of global impacts over a multiple-century timescale.

1.4.3 Relationships Between the +2 °C Increase and CO_2 Emissions

Determining the GHG emissions corresponding to a specified maximum warming is anything but an easy task. This is in fact a kind of inverse problem, for which the effect (global warming) is a target function of time, the cause (CO_2 emissions) a time function to be determined, and the system (the carbon cycle and climate)

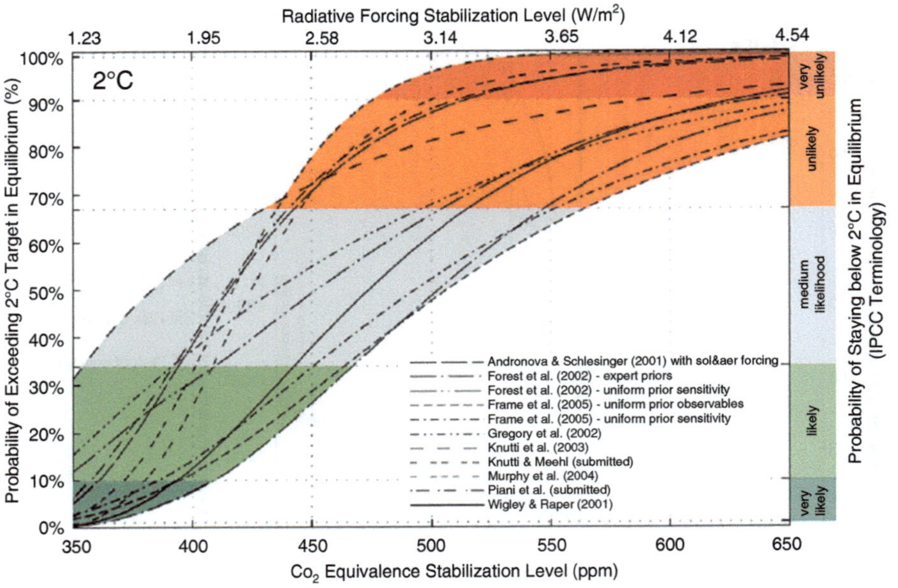

Fig. 1.2 Example of probability for global warming exceeding 2 °C (*Source*: Fig. 28.5, p. 271, from Meinshausen (2006))

represented by quite a complex model, which brings uncertainties into the inverse problem, at both the model structural level and parameter value level (carbon cycle, radiative forcing, influence of sun activity, etc.). A time-consuming, but probably easier way consists in performing sensitivity studies, for instance, to the amount of CO_2 emitted per year, and in observing the responses in terms of CO_2 concentration and CO_2-induced warming. This type of study was performed by introducing a probabilistic approach combining the emissions budget (total mass of CO_2 emitted from a given date) and their implication on the climate (Meinshausen et al. 2009; Allen et al. 2009). As an interesting conclusion, these authors could show, for instance, that the probability of exceeding 2 °C can be limited below 25 % (50 %), by keeping 2000–2049 cumulative CO_2 emissions from fossil sources and land use below 1,000 (1,440) Gt CO_2. Let us observe that we have already emitted 264 Gt CO_2 between 2000 and 2006!

A similar exercise had been performed, to identify the CO_2 concentration stabilized value (which is not the excitation function itself, but another part of the internal response of the system) corresponding to a global warming of 2 °C and similar kinds of answers have been provided as presented in Fig. 1.2 (Meinshausen 2006).

Figure 1.2 shows the results, in terms of the probability of exceeding the 2 °C target (at equilibrium), as a function of CO_2 concentration stabilization level, as predicted by a series of 11 models. It can be described as follows.

There is a 50 % probability for global warming exceeding 2 °C, if the CO_2 stabilized value is 450 ppm in the model of Gregory et al., but, according to the other models, this probability can be situated in fact between 25 and 80 %.

Similarly, the Gregory model predicts that for a 400 ppm value there is a 35 % probability for global warming exceeding 2 °C (or a 65 % probability of stabilization of global warming below +2 °C). But, here, it is better to consider the range of probability between 8 and 57 %.

As a conclusion, we emphasize that because of the different uncertainties attached to the models, there is a necessity not to accept without any restriction a proposal such as "450 ppm is the CO_2 concentration above which the global warming will be over the 2 °C and become uncontrollable." First, these values should be attached to a probabilistic content, and also the transition and all the effects do not start to be amplified suddenly when warming increases above the 2 °C "threshold." Some events might even bring serious concern before reaching this value. This is to be remembered particularly when reading Chaps. 11, 12 and 13.

1.5 Comments About the Basic Science and the Development of Sustainable Energy Technologies Section

We begin by summarizing the analysis of the environmental and energy crisis and include the conclusions of the Stern report (Stern 2006). Then we go on to identify a few broad aspects of the avenues of research to be carried out in an attempt to maintain the main line of CO_2 emissions on the IEA's 450 scenario objective. Finally, we give an example of the implementation of an energy research program, inside which the main chapters of the book may be specified.

1.5.1 From Observing the Crisis to Stern Report Recommendations

Along with water, energy is one of the two most crucial "raw materials" for any human community that wishes to pursue its development. Thanks specifically to the existence of abundant, extremely cheap energy, dramatic economic globalization was able to take place across the planet in the last decades of the twentieth century.

This prosperous period initially enabled industrialized countries (IC) to attain a high standard of living and, secondly, the most important emerging countries (EC) to really get off the ground, from the 1980s onwards. Notwithstanding, this result was obtained at the price of daunting challenges for the years to come: globalization drew indiscriminately from fossil energy resources and showed very little regard as to the environmental impact of their uses.

The consequential perspectives for the planet are:

- First, severe energy shortages even though EC imperatively need increasing quantities of energy to pursue their development and to move closer to Western standards of living. Even if fossil resources are actually more important than they seem (see Chap. 3), they will continue to be more and more expensive, thus in practice, rare.

- Second, dramatic global warming due to the high level of greenhouse gas emissions related to the excessive use of fossil energy.

The current situation is, therefore, characterized by emergency and by constraints. The emergency is to divide the global GHG[2] emissions by a factor of 2 by 2050 (450 Scenario IEA 2009, p. 200) and to divide these emissions by 4 for industrial nations. The constraint is to reorganize our societies with regard to the end of cheap fossil energy and the necessity to seek alternative resources. The extreme emergency aspect of the situation was analyzed in the Stern report,[3] which highlights two major observations:

- The effects, on future climate changes, of actions that we undertake today will be felt after a certain time delay (an action that is started now will have a limited effect on the climate in 50 years' time whereas an action started in 10 or 20 years' time will have a profound effect on the climate in the second half of the century and in the next century).
- The benefits of an intense and fast action on climate change substantially exceed the costs: the costs of climate change could represent 5–20 % of the world gross domestic product (GDP) in 2050 whereas controlling greenhouse gas emissions and stabilizing atmospheric concentrations to 500 or 550 ppmv would only cost 1 % of the world GDP (standing currently around 63,000 billion euros).

1.5.2 What Leads Can We Act On?

Figure 1.3 illustrates the perspectives of change for annual CO_2 emissions (IEA 2009).

Among the large number of scenarios studied, we first recall here that what is called the *reference scenario* does not imply any constraint on the present development of energy use (red color). It should lead (IEA 2009 p.199; ALPF 2008) to an annual rate of emission of 84 Gt CO_2[4]/102 Gteq CO_2[5] in 2050, to 500 ppm CO_2/620 ppm eqCO_2 in 2050 (not stabilized) and to a global warming far above +2 °C.

---

[2]Energy-related CO_2 (29 Gt) emissions represent, in 2009, 67 % of all GHG (43 Gteq CO_2, including N_2O, CH_4, F-gases (with Fluor) and land use, land use change, and forestry, LULUCF) and contribute to 365 ppm for a total of 430 ppm equivalent CO_2 (all GHG being converted into equivalent CO_2).

[3]*Key points of the Stern Report*: All countries will be affected by climate change, but the poorest countries will suffer earliest and most. Warming of 2 °C could leave 15–40 % species facing extinction. Warming of 3 or 4 °C will result in many millions more people being flooded. By the middle of the century 200 million may be permanently displaced due to rising sea levels, heavier floods, and drought. Warming of 4 °C or more is likely to seriously affect global food production. Average temperatures could rise by 5 °C from preindustrial levels if climate change goes unchecked. Before the industrial revolution the level of greenhouse gases in the atmosphere was 280 parts per million (ppm) CO_2 equivalent; the current level is 430 ppm equivalent CO_2. The level should be limited to 450–550 ppm equivalent CO_2.

[4]Energy-related CO_2.

[5]All GHG gases are concerned and the effects of non-CO_2 ones are converted in equivalent quantities of CO_2.

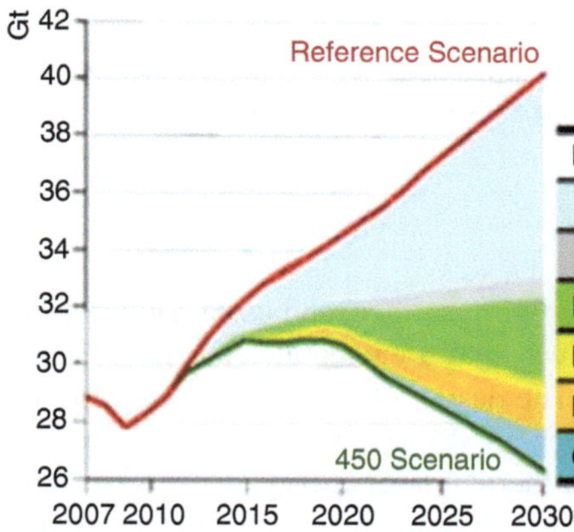

	Abatement ($Mt\ CO_2$)		Investment ($2008 billion)	
	2020	2030	2010-2020	2021-2030
Efficiency	2 517	7 880	1 999	5 586
End-use	*2 284*	*7 145*	*1 933*	*5 551*
Power plants	*233*	*735*	*66*	*35*
Renewables	680	2 741	527	2 260
Biofuels	57	429	27	378
Nuclear	493	1 380	125	491
CCS	102	1 410	56	646

Fig. 1.3 World energy related CO_2 emissions abatement (*Source*: International Energy Agency 2009, fig. 9.2, page 323, with permission)

The *450 Scenario* (green color), aims at maintaining the CO_2 stabilized equivalent concentration at 450 ppm eqCO_2 (Cf. Sect. 1.4), which is slightly above its present value (430 ppm). It should lead in 2050 (IEA 2009, p. 199–200) to 14.5 Gt CO_2 / 21 Gt eqCO_2 emissions and to 410 ppm CO_2/500 ppm eqCO_2, slightly overshooting the stabilized value (projected in 2100).

The actions to be developed concern for instance:

- The increase of the efficiency of energy use (about 50 % of possible mitigation)
- The development of renewables, of biofuels, and of nuclear energy, implying low carbon energy production
- The start of CCS in 2020

1.5.3 An Example of the Energy Research Program at the French CNRS

According to the previous analysis, the main keys for the environment/energy crisis can be found in reinforcing energy efficiency, and producing decarbonized energy. This should be complemented by works on the energy vectors of management optimization (storage, transport, conversion of electricity, heat, cold, etc.) and, of course, by transversal research pertaining to the humanities and social sciences fields (sociology, economy, law, politics, finance, geography, etc.). We also emphasize here that basic sciences are inescapable in order to generate important breakthroughs (NSSSEF 2008; BES 2011), and that research must combine studies both at the phenomena and systems levels. The French CNRS (National Center for Scientific Research) Energy Program has focused its efforts on these major priorities, as described below.

1.5.3.1 Energy Efficiency and Savings

Saving energy largely means improving the efficiency of the energy components and systems, whether this energy stems from classical or renewable resources.

Optimizing Energy Systems

Improving the efficiency of production, transport, storage, and conversion systems means optimizing the components and the way they are integrated into the systems: modeling has a crucial role to play here. The design of heat exchangers (fouling, compactness, multifunctionality, etc.), which are decisive in the energy cascade of machines and of thermal transformers (low thermal-level energy recovery), is to be enhanced as is their optimal performance when they are integrated into the systems. In the field of electrical engineering, 30 % of energy can be spared by introducing control of motor velocity, using power electronic systems, and improving heat losses (new materials with less magnetic or dielectric losses). The lighting sector (10 % of France's electrical consumption, but 20 % at world level) is about to save between 30 and 50 % thanks to the new light emitting diodes (LED) under development (improvement to be achieved in reproducibility, environmental impact, etc.).

Buildings (as an Example)

Limiting consumption in the buildings sector (near 50 % of the energy budget in France) means, in particular, reducing needs and, as such, optimizing energy efficiency (new insulating materials, active glazing, etc.) and integrating renewable sources (PV, geothermal, heat pumps, solar heaters, etc.) so that a zero energy consumption building may be created. Four major areas of research have been identified: atmospheres, the energy efficiency of the envelopes, building environment, and systems and networks adapted to HEQ buildings (high environmental quality standard). This research also takes into account the indispensable socioeconomic analyses (cultural acceptability, behaviors, etc.).

1.5.3.2 Aiming for Energy Production with Reduced CO_2 Emissions

CO_2 Emission-Free Combustion

The combustion of fossil fuels continues to be the main process for producing energy through chemical conversion. The challenge is in developing processes for increasing the efficiency of the overall process and for reducing pollutant emission (nitrogen oxides and sulphur). Working with new combustion regimes and making installations adaptable to changes in fuels today requires better knowledge of basic phenomena (dynamics and stabilization of flames to reduce pollutants, studying chemical kinetics and thermal-acoustic phenomena). Studies are performed to prepare CO_2 capture with good technological and economic conditions (precombustion, chemical looping, etc.; see Chap. 6).

CO_2 Capture and Storage

The program deals with upstream issues (combustion, biomass, process engineering, chemistry, geology, and material sciences) that are capable of leading to a real technological rupture, by assessing new processes that are not yet retained by current technological options (CO_2 capture with adsorption processes, membranes, etc.). In addition, basic studies are performed to help applied research institutions that are more involved in evaluating the storage process at full scale (geosciences, chemical processes, material sciences, chemistry, etc.).

Biomass for Biofuel

The program deals with the fundamental research required for optimizing biomass transformation processes, by taking into account a resource-integrated system approach from the plant, to its thermochemical or biological transformations (specific reactor innovation), according to the application concerned (producing syngas for biofuels or for combustion, oils, tars, charcoal, etc.). The program focuses on second and third generations (specific lignocellulose issues, production using algae, biomimetics).

Clean Electricity Production

Solar energy electrical conversion research deals with PV systems or high-temperature thermodynamic pathways (concentrated solar system, CSP).

In photovoltaics, teams are working, both:

- At the material level, to get new production processes of solar-quality silicon, to improve the efficiency of thin films (silicon, chalcogenides), of organic photovoltaics, or of more advanced (nanostructuration, multiphotons processes, etc.) materials,
- At the system level (reduction of losses, integration of converters).

The basic research in CSP concerns high-temperature materials, themophysical properties (absorption improvement), and heat transfer, both for the collector and the molten salt (heat storage) systems.

Inasmuch as renewable energy production is frequently of an intermittent nature, the issue of storing is important, more for the electricity produced than for heat (see vector management). The highly promising thermoelectricity field is also explored (materials, systems) for energy recovery.

Nuclear

The development of new pathways, fission and fusion, requires highly fundamental research that meets specific scientific challenges (reactor simulation, plasma physics, molten salt chemistry): the CNRS Energy Program works on joint fission/fusion pathway issues that cover the behavior of the materials when irradiated, modeling, physical-chemical issues related to the use of liquid metal and molten salt, fluid flow, and heat transfer. In connection with the Energy Program, the CNRS also works downstream of the electronuclear cycle, as well as on the vast issue of fusion (work carried out within the ITER[6] Program framework).

1.5.3.3 Promoting Energy Vectors

Hydrogen Vector

With regard to H_2 production the CNRS is currently working on two types of bottlenecks:

- In the short term, innovating from existing technologies (reforming hydrocarbons), to meet the demand for embedded, decentralized applications, compact, low-flow reformers, offering high-performance kinetics at a low price
- In the medium to long term, developing new solutions for producing clean hydrogen in large quantities that is low-emission with regard to greenhouse gases (e.g., thermochemical cycles based on concentrated solar resources).

Other more specific research is also being performed:

- Electrolysis (high or low temperature) processes.
- As a more prospective field, bio photo production by algae is also being investigated. Some microalgae, such as *C. reinhardtii*, harbor a hydrogenase interacting with the photosynthetic electron transfer chain and this property allows these organisms to produce hydrogen using light as the sole energy source. (See Chap. 9).

Research in the field of hydrogen storage aims to increase energy density (reservoir compactness) as well as specific energy (reduction in the weight of the reservoir in relation to the mass of hydrogen stored there). Work on solid storage in the form of moderate temperature, low-pressure hydrides is being actively pursued (reversible complex hydrides such as alanates and metallic hydrides, nanoporous matrices, intermetallic compounds, etc.).

---

[6]ITER is an international nuclear fusion research and engineering project currently building the world's largest and most advanced experimental tokamak (a device using a magnetic field to confine a plasma in the shape of a torus) nuclear fusion reactor at Cadarache in the south of France. The ITER project aims to make the long-awaited transition from experimental studies of plasma physics to full-scale electricity-producing fusion power plants.

Fuel Cell

The fuel cell field is extremely well covered by teams integrated into the Group of Research (GdR) PACTE. Basic subjects cover the major technological series, proton membrane cells (PEMFC), and oxide cells (SOFC) dealing with problems of material science, chemistry, electrochemistry, heat transfer, and fluid flow. Complementary approaches are being achieved, both at the core level and the system level. The CNRS' expertise has also been mobilized for specific innovations relative to DAFC (direct alcohol fuel cells) and to SAMFC (solid alkaline membrane fuel cells). In addition, a consensus was recently reached for developing new designs for the PCFC (proton ceramic fuel cells).

Electricity Vector

Research concerning the electricity vector aims at making distributed electricity production, from primary renewable energy sources, highly effective and accessible, at substituting electricity for other energy vectors anywhere the energy or environmental gain may be justified, and at globally consuming less energy. Among the major issues to be solved, we can mention, for example, conversion interfaces (electronic and electromechanical converters, power electronics), architecture optimization, optimal energy management and control, involving lifecycle analysis and operating safety. An important objective is the definition and optimization of functional entities (network-connected energy production and storage units, wind farms, virtual energy plants, etc.), and the notion of smart grids, or even network splitting, which is subject to joint work for different vectors.

Heating and Cooling Vectors

With regard to production, particular attention is given to exploiting solar energy for cooling (buildings, refrigeration, etc.). Transporting heating and cooling vectors is part of the research for long-distance transportation: the threshold at around 10 km and beyond has been identified for new processes based on sorption/desorption systems.

Storage

This research is indispensable in particular as concerns intermittent sources. Electrochemical storage challenges include optimizing existing materials, and finding new materials (anodes, cathodes, electrolytes, connectors, etc.), and authorizing increased energy densities and flows. Let us also mention some quite specific research on the chain PV/electrolysis/gaseous H2 storage and electricity regeneration with a fuel cell (see Chap. 8).

Storing heat means obtaining new materials (latent heat, structural changes, etc.) at various temperatures, improving the stored and drawn-off power, and developing smart sensors that can convey data on storage status.

1.5.3.4 Transversal Research Themes

This research deals in particular with socioeconomic themes and humanities, environmental impact and sustainable development, physics, and materials.

Socioeconomic Themes

Although well-known research subjects exist (prospective scenarios, social acceptability, etc.), new themes are emerging and need to be consolidated, such as the study of institutional, organizational, and social obstacles hindering the transition to a low-profile energy economy, technological, institutional, regulatory, and behavioral innovations, and their diffusion. Energy in human and social development is also a point of interest for this research.

Sustainable Development

The idea of sustainable development includes assessing the environmental impacts of energy choices, at different time and space scales. The program deals here with the example of biomass production, which implies maintaining intensive agricultural practices and needs research on air and water pollution, on the homogenization of the countryside and ground cover, on the decrease in soil fertility, erosion, and biodiversity. In this context, and in order to best clarify the matter, the Program has assigned to a dedicated working group (GT3A, see also Chap. 14) the task of answering the following question:

> How pertinent is it to build up agrofuel production in terms of a partial solution to energy self-sufficiency, in terms of the reduction of GHG emissions, in terms of its impact on meeting humanity's alimentation needs?

Materials

Materials for energy can be basically distinguished by their functional properties: optic, electromagnetic, thermal, and the like, and often have in common the need for environmental resistance properties or adequate mechanical capacities. The issues that involve performance properties under operating conditions, under instantaneous and durable stress give rise to more highly transversal actions.

Another problem concerns the availability of material for energy, involving the necessity:

- Of improving the processes to reduce the quantity used (CNRS teams have reduced by a factor of 40 in 10 years the mass of platinum used by a PEMFC)
- Of developing ad hoc processes for recycling
- Of finding possible substitutions by a less expensive and more easily available material (case of zinc instead of indium for PV)

Physics

Finally, the contribution to energy of teams working in physics can specifically be illustrated with the study of phenomena for identified micro-/nanoscales, in sectors such as heat transfer, thermoelectricity, and photovoltaics (enhancement of the efficiencies). Let us also mention here the breakthrough in the domain of refrigeration, thanks to giant magneto caloric effect materials, leading to new cooling systems, with improved efficiency when compared with classic thermodynamic ones, with less noise and environmental impact.

1.6 Comments About the Section Global and Local Environmental Policies: Technologies, Economy, and Law

The legal and economic aspects are built around two central axes:

- Elements related to the economic, legal, and political difficulties to achieve an effective multilateral system on climate change
- Aspects of urban and agricultural planning related to climate change

This section is therefore within the logic of multilevel governance, which emphasizes some of the international role in negotiations, but also the multiplicity of public and private stakeholders. The challenge is to build a system with sufficient driving forces to establish commitments strong enough to promote changes in the logic of sustainable world development.

1.6.1 International Issues

In fact, the climate change regime, post-2012, asks for an ambitious objective, to meet scientists recommendations, which involves important economic resources and rethinking national and international measures. In the mid-term (2020), it involves commitment to make sharp reductions of GHG emissions to keep emissions trajectory in the long term (2050).

Regarding the difficulties of multilateral negotiations, new problems have presented themselves.

First, there are political and economic difficulties in reaching the +2 °C global warming target. The maintenance of this 2 °C signifies a massive reduction of emissions: to achieve this objective, developed countries should reduce their emissions by 25–45 % below 1990 levels by 2020, and developing countries "need to deviate below their projected baseline emissions" compared with "business as usual" scenarios.

Second, how is differentiation of obligations among different groups of countries to be included?

Third, how can support mechanisms, such as funding or assistance to poorer countries be created?

Binding reduction targets for developing countries makes it is necessary to differentiate between medium–high income developing countries and other poorer countries. China, India, and Brazil are among the main emitters of greenhouse gases. Their commitment to a global regime with reduction targets is essential for the maintenance of the 2 °C objective. In addition, the United States conditioned their entry into a new regime upon the participation of developing countries. In the latest negotiations, Brazil and China have come to accept to assume some obligations, but without setting goals. The question would be what mechanisms would be appropriate and if a new post-Kyoto regime could be built on strengthening the role of nationally appropriate mitigation actions (NAMAs), on what some important developing countries are founding their proposals.

Finally, the African countries are prompting an expansion of the assistance by developed countries up to 5 % of their GDP. It has been a longstanding demand since the first multilateral environmental conventions, which has rarely been implemented. In Copenhagen, developing countries, and in particular African countries, called for a very significant increase in financial assistance. The "fast-start" budget aimed at preparing developing countries for their total participation to further the international climate regime amounts up to USD 30 billion for the 2010–2012 period, expenses being equitably shared between mitigation and adaptation (Copenhagen Accord, Para. 5). However, developed countries have not been able to provide financial resources.

1.6.2 National Issues

At the national level, we highlight the importance of work on the development of urban infrastructure facing the issue of climate change. In this case, the contributions of Li (Chap. 13) and Bobylev (Chap. 4) show that 80 % of global greenhouse gases are emitted in cities. Urban infrastructure could be conceptualized as "a complex set of systems and networks that provide vital services to a city." The urbanization is expected to remain in expansion throughout the world, particularly in developing countries, and global population will grow to nine billion in 2050, of which more than 70 % will be located in urban areas (UN Habitat 2008). In fact, cities are at the forefront of combatting global warming, simply because 80 % of greenhouse gases are emitted in cities. It is projected that cities would account for 73 % of world energy consumption in 2030 in the reference scenario of the International Energy Agency (IEA 2009). Because there is fast and recent urbanization in many developing countries, as in China, it is necessary to rethink ways of development in urban areas and urban construction. But any solution must consider efficiencies, vulnerabilities, interconnectivities, instruments of sustainability, and elements of interdependence and convergence typical of each urban area. Cultural aspects will certainly contribute to increasing this complex scenario.

Many other problems could be identified. One of the most important is the inadequate support for climate-friendly infrastructure in developing cities. It is important to find alternatives and mechanisms to induce conformity with international climate change goals. First, urban projects could become eligible for carbon finance players. However, the financial resources existing today under the UNFCCC (United Nations Framework Convention on Climate Change) are highly ineffective to induce changes.

There is a major waste of resources to build and maintain models of cities, not concerned with the phenomenon of climate warming. Although some countries, including France and Germany, have so far been directing significant resources to improving the energy efficiency of urban centers, we are still far from being able to reverse the main common scenario. Despite some developing countries having specific projects related to climate change concerns in urban areas for building or insulation, most of them still use the same building techniques practiced hundreds of years ago. One example of change is the Brazilian program *Minha Casa, Minha Vida* (my house, my life), one of the major housing programs in the world, which only provides credits to private companies if projects preview the installation of solar heating panels.

Regarding the sectoral policies, it has not taken into account the requirements necessary for the calculation of changes in urban development options. In this case, it is necessary to create mechanisms for economic incentives for reducing emissions in urban areas, either through NAMAs, or by other mechanisms deemed appropriate. Mechanisms could be created according to national priorities. States could accept that urban development projects are eligible in the context of national appropriate mitigation actions (NAMA mechanism) and create specific incentives for sectorial approachs or targets for urban administrators.

Among the new mechanisms that can be stimulated, the authors identified sectoral targets and Non Lose in SD-PAM or Integrating Cities into NAMAs.[7] These common mechanisms could have a very important role to play in shaping sustainable urban development and low carbon trajectory, However, the schemes and institutional aspects must be designed carefully to ensure environmental integrity and economic performance in cities.

---

[7]*Nationally appropriate mitigation action* (NAMA) refers to a set of policies and actions that countries undertake as part of a commitment to reduce greenhouse gas emissions (see Chap. 13). The term recognizes that different countries may take different nationally appropriate actions on the basis of equity and in accordance with common but differentiated responsibilities and respective capabilities. It also emphasizes financial assistance from developed countries to developing countries to reduce emissions.

Project-based carbon finance. This is the "conventional" carbon finance window, where the facility will target LDC project opportunities. These projects are taken through the CDM/JI mechanisms and will have a particular focus on programmatic approaches.

Sectoral crediting and trading mechanisms. This is an "emerging" carbon finance modality, in which financing can be applied across an entire industrial sector (e.g., steel, or cement production, or aviation).The aim of sustainable development policies and measures (SD-PAMs) is to encourage the development of policies that contribute to developing countries' economic and social objectives, with the possibility of lowering GHG emissions at the same time. As indicated in the name Non Lose in SD-PAM, no penalties are incurred in case of failing to meet a target, but emissions reductions achieved beyond the target level earn emissions reduction credit.

1.7 Conclusions

The present status of our knowledge on greenhouse gases and climatic changes (Chap. 2 of this book) can indicate that the +2 °C increase of global warming—with all its probabilistic characteristics—might be a possible trigger for commitment at a large scale of global impacts over a multiple-century timescale (irreversible loss of reefs, of tropical glaciers, and of water resources, risk of extreme weather events, coastal flooding, various heath impacts, etc.).

Because of the climate change threats on the big cities of the future (particularly the intense precipitation events) people have to conceive systems to help urban physical infrastructure adaptation to climate change. Huge underground infrastructures for prevention of such flooding during rain and typhoon seasons have already been built, for instance, in Japan and in Malaysia (G-Cans 2006; SMART Project 2006; Chap. 4).

But, to define more general objectives, on what leads can we act?

According to IEA analysis (see Sect. 1.5.2), the most effective key to moving the trajectory from a business as usual (BAU) scenario to the 450 Scenario is clearly the improvement of energy efficiency and savings (key K1). It involves many mechanisms, among which we can mention the improvement of components, both thermal and electrical (heat exchangers, electrical machines, lighting, etc.) and of systems, including buildings, transport, and industrial processes (see Sect. 1.5.3), or new technologies such as thermoelectricity or improved power electronics.

Key K2 relies on all the Renewables pathways (PV, CSP, wind turbines, etc.) that can contribute to green electricity production, but their frequent intermittent character has to be bypassed, which depends on the technological and economical resolution of the crucial problem of electricity storage on a large scale.

The next one, key K3, may be nuclear energy, (Chap. 5), as far as it able to reinforce its sustainable features (waste, security, availability of fuel and new generations, etc.).

Key K4 deals with carbon management and is probably as crucial as K1. Recall that on a world-scale basis, fossil fuels (Chap. 3) are likely to remain the main sources for electricity generation in the twenty-first century, and many industrial processes that are also large CO_2 emitters, will still be active for many decades. Therefore, carbon capture and storage (CCS, Chap. 6) is generally considered a necessary option for reducing CO_2 emissions to the atmosphere. Because 14 large-scale, integrated CCS projects are in operation or under construction today, with a rough total storage capacity of 33 million tons a year, this option is clearly on the way. As a research perspective for future works carbon capture and recycling (CCR, Chap. 7) should also be kept in mind and the corresponding results carefully observed, with experts claiming that its potential is similar to that of CCS (250–350 Mt/y in the short to medium term, compared to the 1,500–2,000 Mt/y required in 2030 by the 450 Scenario). But, because the industrial deployment of CCS is scheduled no earlier than one decade, preliminary measures should be taken in between, such as the interesting example of conversion of petroleum vehicles to run on compressed natural gas (contribution to CC, premature deaths avoided, savings, and GDP increase; Chap. 10).

Key K5 concerns the energy vectors, and at first, development of biofuels (Chap. 9) in such conditions that with their emergence (first generation essentially today) at a significative scale, a competition between alimentation and agrofuel production will not be threatening. The conclusion of Chap. 14 on this point is clear: there is no reason to dramatize and excessive alarm could well serve only to amplify the instabilities and thus encourage speculation. The use of other vectors is to be carefully optimized: hydrogen and the possible evolution toward an H_2 society (Chap. 8), electricity (storage, smart grids), and the couple heat/cold (storage and long distance transport).

Furthermore, a last key K6 involves transversal research pertaining to the humanities and social sciences fields, with a series of domains covering economics, social issues, law (both at national and international negotiations level), politics, and financial rules (Chaps. 11, 12, and 13). The challenge with K6 is to build a system with driving forces sufficient to establish commitments strong enough to promote changes in the logic of a sustainable world development, based on the breaks made by science and technology.

As a last comment, we would also emphasize the role of basic sciences (nanosciences, molecular biology, catalyst, time-reversed acoustics in geosciences, new X-ray light sources for in situ testing, etc.) without which no effective breakthrough will help for the resolution of this energy/environment crisis.

References

Abderahim H (2010) Perspectives on nuclear global change research II: environmental crisis, energy issues and global regulation policies FMSH. University Paul Cezanne, Aix-en-Provence, 11–18 June 2010

Allen MR, Frame DJ, Hutingford C, Jones CD, Lowe JA, Meinshausen M (2009) Warming caused by cumulative carbon emissions towards the trillionth tonne. Nature 458:1163–1166, April 2009

ALPF (2008) Australia low pollution future – the economics of climate change mitigation – report – Chapter 3: The reference scenario assumptions and projections – the treasury and minister for climate changer and water. http://www.treasury.gov.au/lowpollutionfuture/report/html/03_Chapter3.asp

Ayache C (2010) An overview of the European energy policy and/or European energy research global change research II: environmental crisis, energy issues and global regulation policies FMSH. University Paul Cezanne, Aix-en-Provence, 11–18 June 2010

BES (2011) Basic Energy Sciences 2011 summary report. U.S. Department of Energy Office of Science. http://science.energy.gov/bes/news-and-resources/reports/

Bobylev N (2010) Urban physical infrastructure adaptation to climate change global change research II: environmental crisis, energy issues and global regulation policies FMSH. University Paul Cezanne, Aix-en-Provence, 11–18 June 2010

Callonnec G (2010) Modelling the efficiency of economical measures enforced to reduce CO_2 emission in residential sector global change research II: environmental crisis, energy issues and global regulation policies FMSH. University Paul Cezanne, Aix-en-Provence, 11–18 June 2010

Costa de Oliveira C (2010) International environmental challenges: some solutions brought by private international law global change research II: environmental crisis, energy issues and global regulation policies FMSH. University Paul Cezanne, Aix-en-Provence, 11–18 June 2010

Costello A, Abbas M, Allen A, Ball S, Bell S, Bellamy R, Friel S, Groce N, Johnson A, Kett M, Lee M, Levy C, Maslin M, McCoy D, McGuire B, Montgomery H, Napier D, Pagel C, Patel J, Puppim de Oliveira JA, Redclift N, Rees H, Rogger D, Scott J, Stephenson J, Twigg J, Wolff J,

Patterson C (2009) Managing the health effects of climate change. Lancet 373(9676):1693–1733, May 2009, www.thelancet.com
Dahan A (2010) After Copenhagen, revisiting both scientific and political frames of the climate change regime global change research II: environmental crisis, energy issues and global regulation policies FMSH. University Paul Cezanne, Aix-en-Provence, 11–18 June 2010
Dechezleprêtre A (2010) Invention and transfer of climate change mitigation technologies o global scale: a study drawing on patent data global change research II: environmental crisis, energy issues and global regulation policies FMSH. University Paul Cezanne, Aix-en-Provence, 11–18 June 2010
G-Cans Project (2006) Tokyo. Retrieved 10 Jan 2012 from http://www.g-cans.jp/
Global Change Research I (2008) New methodologies and interdisciplinary approaches. In: Global change research, FMSH, Island of Porquerolles, 5–10 Nov 2008
Goswami VK (2010) The physico-chemical characteristics of the chemical substances to control the global warming by the development of physico-chemical process using selective catalyst beds global change research II: environmental crisis, energy issues and global regulation policies FMSH. University Paul Cezanne, Aix-en-Provence, 11–18 June 2010
IEA (2009) World energy outlook. OECD/International Energy Agency, Paris (France)
IPCC AR4 (2007) *Assessment report 4* Working Group II § 19.4.2.2. International Panel for Climate Change, Geneva
IPCC TAR (2001) Third assessment report. International Panel for Climate Change, Geneva
Jones D (2010) Contribution of hydrogen and fuel cells technologies to a sustainable energy future global change research II: environmental crisis, energy issues and global regulation policies FMSH. University Paul Cezanne, Aix-en-Provence, 11–18 June 2010
Jupesta J (2010) Analysis of the transition energy systems from oil-based into bio-based energy supply: Indonesia case global change research II: environmental crisis, energy issues and global regulation policies FMSH. University Paul Cezanne, Aix-en-Provence, 11–18 June 2010
Lemoine M (2010) Energy issue and the clean development mechanism of the Kyoto protocol global change research II: environmental crisis, energy issues and global regulation policies FMSH. University Paul Cezanne, Aix-en-Provence, 11–18 June 2010
Li J (2010) Governing urban infrastructure in developing cities: the role of carbon finance global change research II: environmental crisis, energy issues and global regulation policies FMSH. University Paul Cezanne, Aix-en-Provence, 11–18 June 2010
Matarasso P (2010) Compatibility between GHG reduction and sustainable growth: basis for a "green growth" global change research II: environmental crisis, energy issues and global regulation policies FMSH. University Paul Cezanne, Aix-en-Provence, 11–18 June 2010
Meinshausen M (2006) What does a 2°C target mean for greenhouse gas concentrations? - a brief analysis based on multi-gas emission pathways and several climate sensitivity uncertainty estimates. In: Schellnhuber JS, Cramer W, Nakicenovic N, Wigley TML, Yohe G (eds) Avoiding dangerous climate change. Cambridge University Press, Cambridge
Meinshausen M, Meinshausen N, Hare W, Raper SCB, Frieler K, Knutti R, Frame DJ, Allen MR (2009) Greenhouse-gas emission targets for limiting global warming. Nature 458:909
Mercier V (2010) The buildings energetic performance global change research II: environmental crisis, energy issues and global regulation policies FMSH. University Paul Cezanne, Aix-en-Provence, 11–18 June 2010
Mocanu C (2010) Wastewater pre-treatment processes numerical simulations in the sewage system global change research II: environmental crisis, energy issues and global regulation policies FMSH. University Paul Cezanne, Aix-en-Provence, 11–18 June 2010
NSSSEF (2008) New science for a secure and sustainable energy future. A report from the Basic Energy Sciences Advisory Committee (BESAC), US DOE, Washington, DC
Parry M, Arnell N, McMichael T, Nicholls R, Martens P, Kovats S, Livermore M, Rosenzweig C, Iglesias A, Fischer G (2001) Millions at risk: defining critical climate change threats and targets. Glob Environ Chang 11:181–183, Elsevier
Perathoner S (2010) New energy sources and CO_2 treatment global change research II: environmental crisis, energy issues and global regulation policies FMSH. University Paul Cezanne, Aix-en-Provence, 11–18 June 2010

Pereira de Andrade P (2010) The use of private law instruments as a new trend of international environmental law: the case of renewable energy promotion global change research II: environmental crisis, energy issues and global regulation policies FMSH. University Paul Cezanne, Aix-en-Provence, 11–18 June 2010

Philibert C (2010) The uses of energy global change research II: environmental crisis, energy issues and global regulation policies FMSH. University Paul Cezanne, Aix-en-Provence, 11–18 June 2010

Pradel N (2010) The environmentally protective dimension of the EU's external energy policy global change research II: environmental crisis, energy issues and global regulation policies FMSH. University Paul Cezanne, Aix-en-Provence, 11–18 June 2010

Rajaure N (2010) Improved cooking stoves and alternative fuel for poor villagers global change research II: environmental crisis, energy issues and global regulation policies FMSH. University Paul Cezanne, Aix-en-Provence, 11–18 June 2010

Reynès F, Yeddir-Tamsamani Y, Callonnec (2011) Presentation of the three-ME model: multisector macroeconomic model for the evaluation of environmental and energy policy. OFCE, Paris, May 2011, 97 p. (Document de travail –Working Document; n° 2011–10). Retrieved 10 Jan 2012 from http://www.ofce.sciences-po.fr/pdf/dtravail/WP2011-10.pdf

Skulimowska A (2010) Hydrogen fuel production by using medium temperature PEM water electrolysis global change research II: environmental crisis, energy issues and global regulation policies FMSH. University Paul Cezanne, Aix-en-Provence, 11–18 June 2010

SMART Project (2006) Stormwater management and road tunnel. Retrieved 10 Jan 2012 from http://www.smarttunnel.com.my/

Smith JB, Schneider SH, Oppenheimer M, Yohe GW, Hare W, Mastrandrea MD, Patwardhan A, Burton I, Corfee-Morlot J, Magadza CHD, Füssel HM, Pittock AB, Rahman A, Suarez A. van Ypersele JP (2008) Assessment dangerous climate change through an update of the Intergovernmental Panel on Climate Change (IPCC) "reasons for concern". In: Proceedings of the US National Academy of Science PNAS Early Edition. www.pnas.org/cgi/doi/10.1073/pnas.0812355106, 5p. (252 Ko)

Stern N (2006) Stern review report on the economics of climate change. Cambridge University Press, Cambridge

Therond PG (2010) Renewable and wind energy: a testimony from an operator global change research II: environmental crisis, energy issues and global regulation policies FMSH. University Paul Cezanne, Aix-en-Provence, 11–18 June 2010

Thompson LG, Mosey-Thompson E, Davis ME, Bretcher HH (2011) Tropical glaciers, recorders and indicators of climate change, are disappearing globally. Ann Glaciol 52(59):23–34

Trenberth KE (2011) Changes in precipitation with climate change. Clim Res 127:123–138

UN-Habitat (2008) Cities and climate change adaptation. UN-Habitat Donors meeting, Seville, 15–16 Oct 2008

Varella MD (2010) Biofuels, poverty reduction and sustainable development global change research II: environmental crisis, energy issues and global regulation policies FMSH. University Paul Cezanne, Aix-en-Provence, 11–18 June 2010

Wadud Z (2010) CNG conversion of motor vehicles in Daka: air quality and climate benefits global change research II: environmental crisis, energy issues and global regulation policies FMSH. University Paul Cezanne, Aix-en-Provence, 11–18 June 2010

WHO (2011) Fact sheets, Sept 2011. World Health Organisation, Geneva

Yeddir-Tamsamami Y (2010) Presentation of the three-ME model: multisector macroeconomic model for the evaluation of environmental and energy policy global change research II: environmental crisis, energy issues and global regulation policies FMSH. University Paul Cezanne, Aix-en-Provence, 11–18 June 2010

Ziegahn KF (2010) Research for a sustainable energy future between visionary ideas and realistic options global change research II: environmental crisis, energy issues and global regulation policies FMSH. University Paul Cezanne, Aix-en-Provence, 11–18 June 2010

Chapter 2
Greenhouse Gases and Climatic Change

Vincent Moron

Abstract Earth climate is determined by the equilibrium between the amount and distribution of incoming radiation absorbed from the sun and the outgoing longwave radiation emitted at the top of the atmosphere. Several atmospheric trace gases, including water vapor, carbon dioxide, methane, and nitrous oxide, absorb far more efficiently the longwave radiation than solar radiation. These so-called greenhouse gases increase the amount of energy available to the earth and keep it much warmer than it would be otherwise. Although water vapor (and clouds that contribute both to the greenhouse effect and cooling through the back reflection of the incoming solar radiation) does not stay in the atmosphere more than ~2 weeks, most of the other greenhouse gases stay far more than 10 years. Anthropogenic use of fossil fuels, cement production, and deforestation already increased the atmospheric concentration of greenhouse gases and human activities also created new synthetic and powerful ones such as chlorofluorocarbon. The corresponding positive radiative already contributed to the ~0.8 °C increase of the global surface temperature since 1850 and will act as the main climate driver for at least the next century. This chapter outlines the bases of the greenhouse effect and its impact on the earth climate from ~1850 to 2100.

Keywords Greenhouse gas • Earth climate • Radiative balance • Climate model • Kyoto protocol

---

V. Moron
Département de Géographie, Pôle SGAE, UFR ALLSH,
Aix-Marseille University, Aix en Provence, France

CEREGE, UMR 7330 CNRS, Aix en Provence, France

Institut Universitaire de France, Paris, France

International Research Institute for Climate and Society,
Columbia University, Palisades, NY, USA
email: moron@cerege.fr

J.B. Saulnier and M.D. Varella (eds.), *Global Change, Energy Issues and Regulation Policies*, Integrated Science & Technology Program 2, DOI 10.1007/978-94-007-6661-7_2,

List of Acronyms

AOGCM	Atmosphere–Ocean General Circulation Model
CDIAC	Carbon Dioxide Information Analysis Center
CEREGE	Centre Européen de Recherche et d'Enseignement des Géosciences de l'Environnement
CFC	Chlorofluorocarbon
CO_2	Carbon dioxide molecule
CH_4	Methane molecule
CRU	Climatic Research Unit
ESRL	Earth System Research Laboratory
GCM	Global Climate Model
GISS	Goddard Institute for Space Studies
GHG	Greenhouse gas
Gt	Gigaton (1 Gt = 10^9 t)
GtC	Gigaton of equivalent carbon
HCFC	Hydro-carbo-fluoro-carbone molecule
HFC	Hydro-fluoro-carbone molecule
H_2O	Water molecule
IPCC	Intergovernmental panel on climate change
IR	Infrared radiation
m a.s.l	Meter above sea level
N_2O	Nitrous oxide molecule
O_3	Ozone molecule
ppm	Parts per million (in volume)
ppt	Parts per trillion (in volume)
SRES	Special Report on Emissions Scenarios
UNEP	United Nations Environment Programme

2.1 Introduction

In 1975, W.S. Broecker wrote, "*If man-made dust is unimportant as a major cause of climatic change, then a strong case can be made that the present cooling trend will, within a decade or so, give way to a pronounced warming induced by carbon dioxide. By analogy with similar events in the past, the natural climatic cooling which, since 1940, has more than compensated for the carbon dioxide effect, will soon bottom out. Once this happens, the exponential rise in the atmospheric carbon dioxide content will tend to become a significant factor and by early in the next century will have driven the mean planetary temperature beyond the limits experienced during the last 1,000 years.*" This visionary prediction follows some earlier calculations (Fourier 1827; Arrhenius 1896) establishing the physical relationship between certain atmospheric traces (i.e., the greenhouse gases, GHG hereafter) and the earth temperature variations. There is an increase in GHG atmospheric concentration[1] because human activities (fossil fuel burning, land clearing, deforestation, etc.) release it at a rate[2] surpassing the natural capacity of the earth's system to remove it from the atmosphere. Some of the current knowledge about the earth's radiative balance and GHG atmospheric concentration as well as its link with temperature variations from 1850 to 2100 and the basic mechanisms of the climatic response to current and near-future GHG atmospheric concentrations are reviewed here.

2.2 The Radiative Balance and the Greenhouse Gases

The climate system is a thermodynamical engine fueled by solar radiation (Trenberth et al. 2009). The energy gained from inner earth through volcanism and geothermal sources is considered to be negligible at global scale. At equilibrium, the total absorbed solar radiation at the outer limit of the climate system is counterbalanced by the same amount of emitted radiation.[3] All objects above 0 K emit electromagnetic radiation. Planck's, Stefan–Boltzmann's and Wien's laws define the relationship between the amount and spectrum of emitted radiation and the surface temperature of the emitter.[4] The radiative equilibrium between absorbed solar radiation and outgoing emitted infrared radiation defines a radiative equilibrium temperature, which is approximately 255 K with the assumption that emissivity of the climate system is close to 1.[5] A mean temperature of 255 K is not observed at the earth surface (the mean observed temperature at the earth surface is actually close to 288 K) but rather close to an altitude of 5 km in mean. The fact that observed surface

[1] See, for example, http://www.esrl.noaa.gov/gmd/ccgg/trends/.

[2] Recent estimates show that global fossil and cement emissions equal 9.5 +/– 0.5 GtC/year (that is +54 % from 1990) while global land-use change emissions equal 0.9 +/– 0.5 GtC/year in 2011 (Le Quéré et al. 2012).

[3] This quantity equals the solar "constant" (1,366 W/m^2 = solar energy intercepted by the earth disk) divided by 4, for geometry constraint, multiplied by 1 minus albedo (= 0.3), the albedo being the fraction of radiation reflected by the earth's system, that is, 1366/4 × 0.7~239 W/m^2. The earth's surface absorbs 50 % of total solar radiation and the atmosphere absorbs 20 % of it.

[4] The amount of emitted radiant energy is proportional to the emissivity and fourth power of the surface temperature of the emitter and its spectral peak is inversely proportional to its surface temperature. The emissivity is the ability of a material to emit energy by radiation. This ability is relative to an idealized physical body at the same temperature, called a *black body*, that absorbs all incident electromagnetic radiation and is also the best possible emitter of thermal radiation. The solar surface, roughly near 5,750 K, emits roughly 160,000 times more radiation than the earth surface per unit of surface, mostly in the ultraviolet (spectral power < 0.4 μm), "visible" (i.e., light a human eye could see, within 0.4 and 0.7 μm) and infrared bands (IR, spectral band > 0.7 μm), whereas earth surface and atmosphere (roughly between 180 and 340 K) emit IR only.

[5] The emissivity of land, ocean surface and thicker clouds than cirrus is close to 1, that is, the one of a black body. The clear-sky atmosphere has an emissivity of 0.4–0.8, and cirrus clouds have a typical emissivity of 0.2.

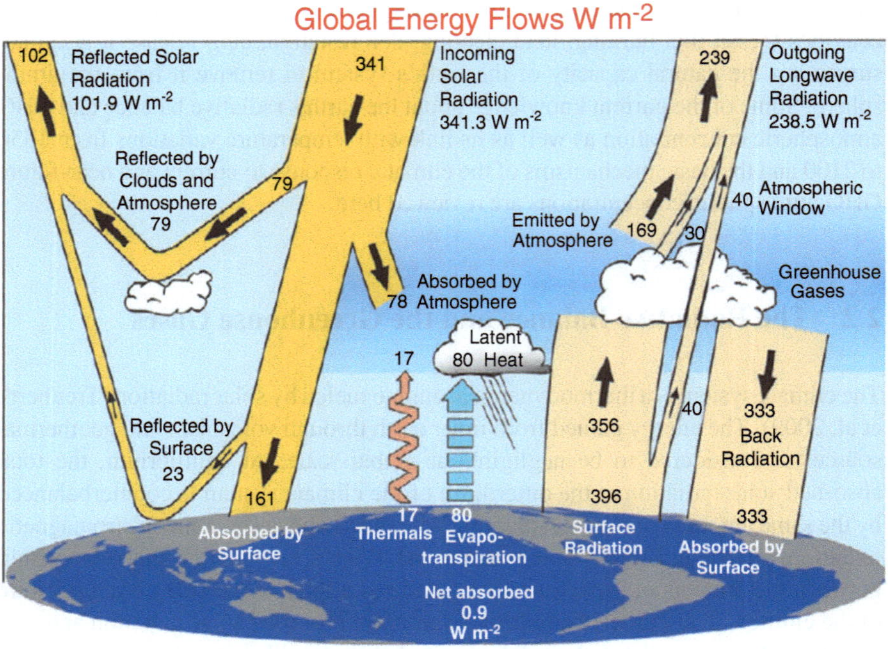

Fig. 2.1 Global annual mean Earth's energy budget (W/m^2) for the period of March 2000 to May 2004 (from Trenberth et al. 2009). The *broad arrows* indicate the schematic flows of energy in proportion to their importance (Reprinted with permission of the American Meteorological Society)

temperature is higher than radiative mean temperature is explained by the greenhouse effect.[6]

Atmosphere is composed of a mixture of various gases. The most abundant gases are di-atomic nitrogen and oxygen accounting for ~98 % in volume of dry air. The GHGs, accounting for less than 1 % of the atmospheric volume, have the property to absorb infrared radiation efficiently, whereas the atmosphere overall absorbs only 20 % of solar radiation (Fig. 2.1). The most important "natural" GHG is water vapor (H_2O), then carbon dioxide (CO_2), methane (CH_4), nitrous oxide (N_2O), and ozone (O_3). Human activities (burning of fossil fuels in industry and agriculture, heat production and transport, deforestation, cement production) add some of these GHGs, but also create synthetic GHGs, as chlorofluorocarbons (CFCs). These "natural" and "synthetic" GHGs (except H_2O and O_3) remain at least 12 years in the atmosphere and thus affect the planetary radiative balance independently on the location of their emission into the atmosphere because the time to mix the whole troposphere is faster than 1 year. Any increase of the energetic content, for example, related to increased GHG atmospheric concentration, leads to a

[6]See http://www.realclimate.org/index.php/archives/2010/07/a-simple-recipe-for-ghe/ for a simple explanation of how the greenhouse effect works. For a more comprehensive review, see, for example, Danny Harvey (2000).

temperature increase until a new equilibrium is eventually reached, so that the climatic system emits IR as much as absorbed net radiation at its outer limit (i.e., the top of the atmosphere for the earth).

It should be noted that although gaseous H_2O is the most efficient[7] GHG, it could not be considered as a major driver of temperature variations because, as liquid and solid water, it is unable to remain in the atmosphere for a long time (typically less than 2 weeks). But increased atmospheric temperature could increase the atmospheric concentration of gaseous H_2O leading to a positive feedback, because warmer air could contain more gaseous H_2O and an higher atmospheric concentration of H_2O increases the greenhouse effect. In other words, despite its largest contribution to the current greenhouse effect, H_2O is more a passive actor of temperature variations whereas other GHGs are active drivers (i.e., they are able to physically drive temperature variations) of temperature variations in current conditions[8] due to their capacity to accumulate in the atmosphere coupled with their physical properties.

2.3 Temporal Variations of the Greenhouse Gases

Many human activities release carbon into the atmosphere. Figure 2.2 displays the monthly mean mole fraction of CO_2 (in parts-per-million—ppm—of volume) measured at three distant stations: Mauna Loa (Hawaii), Barrow (Alaska), and South Pole. These three locations are far away from any large industrial local source of anthropogenic carbon. The long-term positive trend due to global anthropogenic carbon consumption is superimposed to an annual cycle with maximum/minimum recorded at the end of the boreal winter and summer periods at Barrow and Mauna Loa. This annual cycle is due to the vegetation cycle in the northern hemisphere inasmuch as photosynthesis exceeds respiration in spring and summer whereas the opposite, that is, a net CO_2 release by the vegetation, occurs in autumn and winter.

[7] One of the last estimates of the relative contribution of atmospheric long-wave absorbers to the current-day greenhouse effect is: 50 % for water vapor, 25 % for clouds, and 20 % for CO_2. (Schmidt et al. 2010; available at http://pubs.giss.nasa.gov/docs/2010/2010_Schmidt_etal_1.pdf).

[8] At glacial–interglacial scale (i.e., between 10,000 and 100,000 years), CO_2 variations tend to follow temperature variations in Antarctica by ~200–1,000 years at the glacial termination. At this scale, temperature variations are mostly driven by orbital changes (Milankovitch theory). The time lag between CO_2 and temperature seems at least partly due to the adjustment of the ocean deep circulation that releases some CO_2 into the atmosphere when the earth warms. This relationship does not invalidate the current one because the increased concentration of atmospheric CO_2 released by human activities drives the current warming whereas glacial termination was initiated by orbital changes 20,000 years ago. Moreover, it is assumed that GHG variations at glacial–interglacial scale had exerted a positive feedback on temperature variations with other processes as the ice–albedo–temperature feedback, that is, the fact that deglaced areas decrease the mean earth albedo, increasing the amount of absorbed solar radiation (Lorius et al. 1990). Lastly, recent analyses (Shakun et al. 2012) demonstrate that global temperatures mostly lag CO2 variations in Antarctica during the last deglaciation.

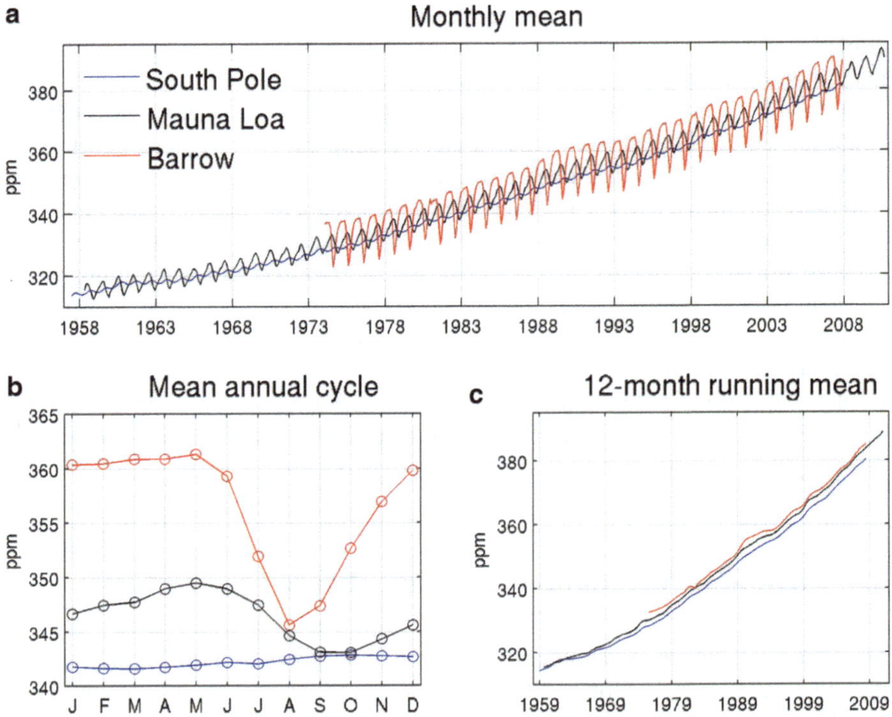

Fig. 2.2 Atmospheric concentration of CO_2 measured at Barrow (Alaska, 71°19′N, 156°36′W, 11 m a.s.l. in *red*), South Pole (89°59′S, 24°48′W, 2810 m a.s.l. in *blue*), and Mauna Loa (19°32′N, 155°35′W, 3397 m a.s.l. in *black*) observatories from September 1957 to July 2010: (**a**) monthly means, (**b**) mean annual cycle, and (**c**) 12-month running mean. The mole fraction of CO_2, expressed as parts per million (ppm), is the number of molecules of CO_2 in every one million molecules of dried air (water vapor removed) (Mauna Loa data come from ESRL, and Barrow and South Pole data are extracted from CDIAC)

The annual cycle is large at Barrow, which is closer to large Eurasiatic and North America continental masses and almost flat at South Pole.

The long-term increase is very consistent among the three records (Fig. 2.2) and not perfectly linear with some steps. On a longer time scale, CO_2 concentration ranges between 270 and 290 ppm in interglacial periods to 190–200 ppm during glacial periods, although concentrations could have been far larger before the quaternary. The atmospheric concentration of other GHG increases due to their use in various human activities already drives the contemporaneous warming of the global surface temperature. Table 2.1 gives the atmospheric concentrations of the most important GHG in October 2008–September 2009 compared to the preindustrial period.

Each GHG has a different lifetime and each molecule has a different radiative forcing. The relative impact of each GHG could also be compared with the integrated radiative forcing from the preindustrial period (IPCC 2007). With that frame, CO_2 is the largest forcing, due to its absolute concentration and its long lifetime

Table 2.1 GHG concentration (in parts-per-million, ppm, and parts-per-trillion, ppt) in 1750 and in 2008/2009 with their lifetime and radiative forcing from preindustrial to current conditions

GHGs	Preindustrial (~1750; ppm)	Current concentrations (2008/2009)	Lifetime (in years)	Radiative forcing (in W/m^2)
CO_2	280	385 ppm	~100	+1.66
CH_4	0.7	1.8 ppm	12	+0.48
N_2O	0.27	0.32 ppm	114	+0.16
Tropospheric O_3	0.025	0.034 ppm	<Few days	+0.35
Synthetic GHG (CFC, HCFC, HFC, SF_6, etc.)	0	Few ppt to 536 ppt	From few years to thousands of years	~+0.30 (most from CFC-11 and CFC-12[a])

Data come from http://cdiac.ornl.gov/pns/current_ghg.html

[a]Some of these synthetic GHGs including CFC-11 and CFC-12 have been banned after the Montreal conference (1987) and its amendments under the UN convention on the stratospheric ozone hole. CFCs are replaced by other components that are not harmful to stratospheric ozone but usually have a huge global warming potential

even if each CO_2 molecule is not the more efficient heat trapper, and synthetic GHGs (CFCs, HCFCs, HFCs, etc.) have a significant impact despite their very small concentrations. The increase of atmospheric GHG increases the energy content of the climate system.[9] The uncertainty on the net anthropogenic forcing is low (IPCC 2007) and as stated before, its extent is global.

2.4 Observed Temperature Variations

Temperature has been recorded for more than 400 years. The longest continuous temperature record is the "Central England Temperature," available since 1659 (Plaut et al. 1995). The temperatures are recorded worldwide in normalized environments so that they can be compared and spatial averages can be computed.[10]

[9]Since 1750 the total radiative forcing related to the increase of atmospheric GHG concentrations due to human activities equals +2.9 W/m^2. The net anthropogenic effect including cooling effect mostly due to sulfur emissions equals +1.6 W/m^2. The direct cooling effect of anthropogenic sulfur associated with the aerosol veil, that increase albedo at a regional scale is complicated by its indirect effect through the modification of the optical properties of clouds. The cooling effect is less certain than the one associated with GHG increase (IPCC 2007).

[10]There is a debate about the sense of a "global" (in the sense of planetary) mean of surface temperature. Everybody could experience very large temperature variations on small time and spatial scales, for example, simply moving from shade to sunlight in a summer day. It seems then unreasonable to compute a spatial mean from a few samples. But, the range of temperature variations strongly decreases when time means (instantaneous record to annual mean) are considered, especially when raw temperatures are scaled to the local mean annual cycle (theoretically estimated with at least 30 years of data). It is because the drivers of temperature variations at this timescale are from regional (e.g., atmospheric Rossby waves, which are giant meanders in the atmosphere.

Sea surface temperature recordings have been generalized since the Brussels maritime workshop in 1853. Several research centers across the world (such as GISS and CRU) have established gridded datasets of surface temperatures freely available on the web. There is considerable work to remove known biases related to urbanization and changes of thermometers. The uncertainty is less than one order of magnitude relatively to long-term increases in these gridded datasets.

Figure 2.3 shows the annual mean of planetary-scale surface temperature computed by the Climatic Research Unit (Brohan et al. 2006) from 1850 to 2009. A nonlinear increasing trend is already visible concentrated in two periods, from 1910s to 1940s and then after 1975 (Fig. 2.3). Other datasets lead to the same conclusions (IPCC 2007).

Beyond GHG variations, there are two major "external" forcings operating at timescales between 1 and ~100 years: (i) the solar "constant" varies and we are now in cycle 24 of an 11-year cycle (the last minimum occurred in 2007–2009; cf. http://www.climate4you.com/Sun.htm#Recent solar irradiance); (ii) explosive volcanism able to disseminate large amounts of sulfur dioxide in the stratosphere cools down the earth's surface by as much as 0.3 °C for 2–3 years, such as after the eruption at Pinatubo in June, 1991. Both forcings are estimated to be minor relative to the GHG variations (Fig. 2.3). Internal climate system interactions also superimpose some variations at these timescales. For example, the warm phase of the El Niño Southern Oscillation (as in 1997–1998, Fig. 2.3) transfers a large amount of heat from the upper levels of the tropical Pacific to the large-scale atmosphere, and thus adds a transient (lasting a few years at maximum) warming signal of a few tenths of degrees Celsius in global surface temperature (Klein et al. 1999). All these factors are combined with GHG increase and drive the temperature variations. Looking at a perfect match between mean temperature variations, that integrate all these causes, with the time evolution of a single forcing is physically wrong, and the natural climate variability is able to generate nonlinear variations even from a monotonic forcing, as the GHG increase. In that respect, numerical simulation is a decisive tool inasmuch as it allows multiple scenarios where the relative impact of each possible forcing could be compared and scaled, beyond basic estimate of their radiative forcing as shown in Table 2.1.

The increase of mean surface temperature during the twentieth century is not disputable. Moreover, it is also corroborated by other planetary-scale climatic variations that could hardly be explained by any alternative plausible factors. Two of

---

There are typically 4–6 such Rossby waves around the globe between subtropical and subpolar latitudes, that are the main factor determining the spatial scale of monthly or seasonal temperature anomalies at the extratropical latitudes) to zonal/near-global (as El Niño Southern Oscillation phenomenon) or even planetary scales (e.g., variations of the solar constant or GHG concentrations, large volcanic eruptions, etc.). The anomalies of temperatures relative to the annual cycle at monthly and moreover annual timescales have thus a far larger spatial coherence and less amplitude than localized records. In that way, it is possible, and physically plausible, because of the link between temperature variations and the change in radiative balance, to compute the spatial mean of surface temperature at continental or even planetary scales.

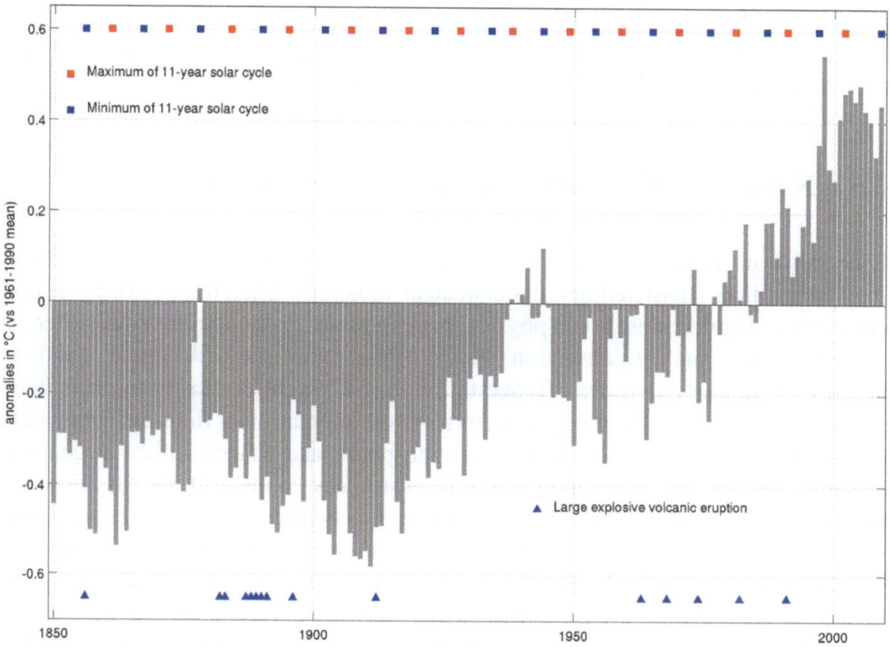

Fig. 2.3 Annual mean surface temperature from 1850 to 2009 expressed as anomalies in °C relatively to the 1961–1990 mean (from the Climatic Research Unit, Norwich, UK). Note that other estimates (such as the one provided by the Goddard Institute Space Studies, New York, USA) are very consistent. The *red* and *blue squares* on the *top* indicate, respectively, the maximum and minimum of the 11-year solar cycle (i.e., sunspot cycle). The solar constant variation is close to ~1 W/m^2 from minimum to maximum of the 11-year cycle, that is, equivalent to a radiative forcing of ~0.17 W/m^2. A slower variation is superimposed on the 11-year solar cycle, but its amplitude is still uncertain with a maximum of ~2 W/m^2 (equivalent to a radiative forcing of ~0.34 W/m^2) between a minimum in the late nineteenth century and a maximum in the second half of the twentieth century. The *blue triangles* at the *bottom* indicate a major volcanic eruption able to inject a massive amount of sulfur into the stratosphere. These volcanic eruptions could have a radiative forcing close to −0.5 to −3.3 W/m^2 in three years from the eruption at maximum. Note that such volcanic eruptions are absent from 1913 to 1962, then from 1992 onwards. For example, the eruption of Eyjallajökull in April 2010 was not explosive enough to inject sulfur into the stratosphere and its planetary-scale forcing is thus negligible

these interrelated variations are (i) sea level rise and (ii) the melting of most of mountain glaciers. The mean sea level rose by ~20 cm since 1870 (+1.7 mm/year during the twentieth century) and the rate of rises has recently increased (Church and White 2006). The recent rate of global mean sea level rise measured by satellite (since 1993) equals +3.3 mm/year with ~30 % due to thermal expansion and ~55 % from mass loss in mountain glaciers and ice sheets (Cazenave and Llovel 2010). Worldwide glaciers have been shrinking significantly with strong retreats in the 1940s, followed by stable or growing conditions around the 1970s and again increasing rates of ice loss from 1985 onward (UNEP 2008). The reaction of a single glacier is not only linked to local temperature variations but also changes in solar radiation as well as cloudiness, amount and annual cycle of precipitation, and so on. But the fact that almost all mountain glaciers retreat at the same time should be considered as a fingerprint of current global warming.

2.5 Numerical Simulation of the Climate System

Global climate models (GCM) are mathematical artifacts of the climate system based on physical laws translated into mathematical equations. The climate system is discretized in 3D gridboxes. Some physical principles are explicitly described whereas some processes (such as convection) should be parameterized with ad hoc equations because of unresolved scales (Trenberth 1993). GCMs differ mostly by their horizontal and vertical resolutions, parameterizations, integration of different submodels (the current minimal set includes ocean, atmosphere, and a land–surface–vegetation scheme), and the way to compute numerical estimates on the grid. GCMs allow the running of numerical experiments with a single or a combination of plausible radiative forcings. It allows separating the cause of variations and the climate system response. This advantage is decisive when there are no known historic or paleoclimatic analogues (as in the case of the current GHG increase). GCMs are in fact not perfect and they remain considerably "simple" compared to the true climate. Nevertheless, the comparison between the first IPCC crude predictions of sea level rise and temperature increase made in 1990 and the observations until 2008 is very encouraging about the ability of current GCMs to simulate a realistic response to well-calibrated radiative forcing as the current GHG increase (Rahmstorf et al. 2007).

Basically, there are three types of uncertainty regarding the near-future numerical simulations of the climate: (i) the uncertainty linked to the simulation of the response to large and regional-scale forcing; (ii) the uncertainty linked to the amount of the future GHG emissions; and (iii) the uncertainty linked to the relative and/or absolute impact of other forcings. These uncertainties are considered in different ways.

The first uncertainty is inherent to any modeling because (i) the climate system integrates a continuum of time and spatial scales that could not be explicitly and fully considered (i.e., we cannot explicitly simulate all air molecules and our knowledge of the whole climate system is far from comprehensive) and (ii) the climate system is chaotic, that is, very sensitive to initial conditions. This means that a single numerical experiment contains one part related to the forcing (i.e., "forced" response) but also another part coming from the initial conditions and which is "free," that is, not reproducible (i.e., another experiment with exactly the same forcing but different initial condition leads to a different output). We need to have a probabilistic approach by running multiple experiments with the same or, better, different GCMs to estimate the intensity and shape of the forced response relative to the free one. The fact that the climate system is chaotic does not forbid probabilistic long-term prediction as soon as the forced response surpasses the free one for a given time and spatial scale. For example, the exact temperature on July 14th or January 1st, 2100 in Paris will be unknown until 10–15 days before, but we can predict that July 2100 will be warmer

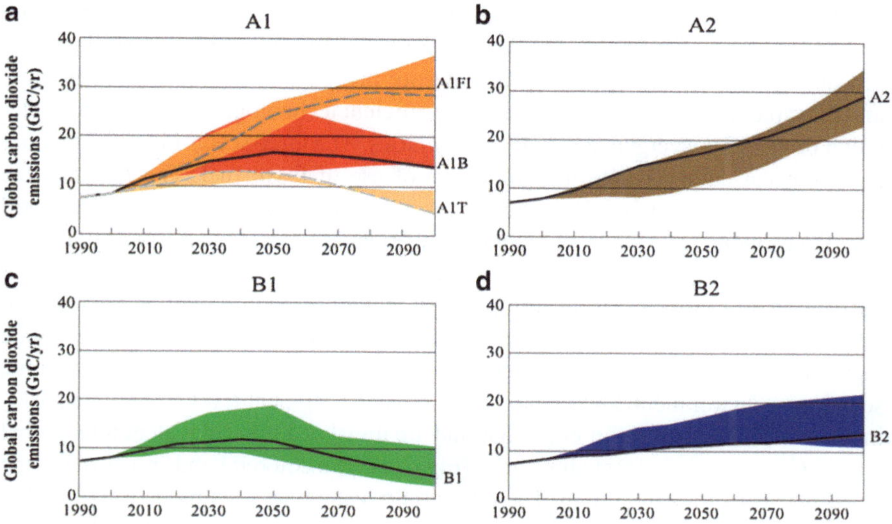

Fig. 2.4 Total global annual CO_2 emissions from all sources from 1990 to 2100 in gigatonnes of carbon per year (GtC/year) for the six scenario groups (panel **a**: A1FI in orange, A1B in red and A1T in light orange; panel **b**: A2 in brown; panel **c**: B1 in green; panel **d**: B2 in blue). The 40 SRES scenarios are presented by the 5 scenario groups with each colored emission band showing the scenario range within each group. The equivalent radiative forcing (in 2100 relatively to 1990) ranges from around +3 W/m^2 (B1) to +7 W/m^2 (A1FI and A2) (IPCC 2007) (*Climate Change 2007: Synthesis Report.* Contribution of Working Groups I, II and III to the Fourth Assessment Report of the Intergovernmental Panel on Climate Change, Figure 3.2. IPCC, Geneva, Switzerland)

in mean than January 2100 in Paris because (i) a month is considered instead of a single day, (ii) the thermal difference between January and July 2100 in Paris (and all northern extratropical and subtropical zones) is primarily forced by the annual cycle of solar radiation between these 2 months, and (iii) the prediction of the polarity and the amplitude of this forcing is almost certain. On the contrary, this does not say anything about the temperature variation between January 1st, 2100 and the next day because this variation relies partly on the precise atmospheric state, which is unpredictable before December 15th to 20th, 2099 at best. Prediction of the climate response to atmospheric GHG increase is a climatic prediction of the same type as the temperature response to the annual cycle of the solar radiation.

The second uncertainty is explicitly taken into account through a whole range of scenarios considering mostly demographic growth, socioeconomic variations and technological efforts. The first scenarios, called IS92, were defined in 1992. The scenarios developed in 1996 for the IPCC third assessment report are broader. They include improved emissions baselines and allow the examination of different rates and trends of socioeconomic and demographic changes throughout the world. They are detailed in a special report of the IPCC.[11] There are a total of 40 scenarios and 6 scenario groups within 4 families (A1, A2, B1, and B2; Fig. 2.4) summarized here: A1FI

[11] See http://www.ipcc.ch/pdf/special-reports/spm/sres-en.pdf

(Fossil fuel Intensive), A1B (Balanced), and A1T (predominantly nonfossil fuel) are embedded in the A1 storyline, that is, a future world of very rapid economic growth, global population that peaks in the mid-twenty-first century and declines thereafter, and the rapid introduction of new and more efficient technologies. There is an expectation of global convergence among the countries. The three scenarios differ by their technological emphasis and the proportion of use of fossil fuel. The A2 storyline combines a continuously increasing global population and a less converging world (i.e., the economic growth is more regionally oriented) than in the A1 family. The B1 storyline describes the same population scenario as A1, but with a rapid transition of the economy toward service and information, including the introduction of clean and resource-efficient technologies. The B2 storyline describes a world with a moderate increase of global population coupled with emphasis on local solutions to economic, social, and environmental sustainability at local and regional levels. These scenarios include GHG but also sulfur emissions and are translated into equivalent total global annual CO_2 emissions from all sources (Fig. 2.4). The colored band shows the range of all scenarios for each group. The only group leading to a net reduction of global CO_2 emissions in 2100 relative to 1990 (~7 GtC/year) is the B1 and A1T group although the increase is moderate in B2 and A1B (roughly 13–15 GtC/year in 2100) and strong for A1FI and A2 (close to 30 GtC/year in 2100). The equivalent radiative forcing ranges from +2 W/m^2 (B1) to +7 W/m^2 (A1FI). Within each group, the scenario explores the differences and uncertainties in the driving forces. These emissions are then converted into GHG atmospheric concentrations. In summary, the 40 scenarios describe different pathways and cover a wide range of possible "futures." Note that the last estimate of the current growth rate of fossil-fuel emission (+3.5 %/year between 2000 and 2007) is above the largest predicted growth rate, that is, +2.7 %/year (AIFI) on 2000–2010.

The last uncertainty is related to the relative impact of GHG increase with other independent forcings operating at similar time scales (i.e., between 10 and 1,000 years). Many studies analyze the possible external forcings of the radiative balance of the earth. If we exclude major changes related to massive meteorites and/or comets falling on earth as the major impact occurring at Chixulub (Yucatan, Mexico) 65×10^6 years ago, there are two other possible forcing on the 10–1,000 years scale. The first one is the solar "constant" which exhibits an almost continuous scale of variation from seconds to billions of years. The 11-year cycle has a net radiative impact of 0.17 W/m^2 although longer (and less regular) 80–200 year cycles could have a net radiative impact of 0.3–0.7 W/m^2. The second natural forcing is related to major volcanic eruptions that could decrease the net radiation by 0.5–3.5 W/m^2 for 2–3 years. We do not know about the future evolution of the solar constant and even more, volcanic activity. But, we can hypothesize at least that variations of the solar constant similar to those experienced during the twentieth century are unable to add a significant signal to the GHG increase postulated by IPCC scenarios. In the same way, if volcanic eruptions keep the same variation as in the twentieth century (few significant eruptions before 1912, then after 1963 with the last one in 1991), its impact would be negligible.

2.6 Climate Projections for the Near Future

Figure 2.5 shows the evolution of mean annual temperatures for a set of scenarios. Each curve is the mean of multiple experiments (23 GCMs) and obviously filters the interannual–decadal variability. Even if the forcing is monotonic, the climatic response should be irregular with near-stationary periods interrupted by more or less abrupt increases. The superposition of additional factors reviewed just above will also add a degree of complexity to the planetary signal. Anyway, the global mean temperature in 2100 is expected to increase from +1.8 °C (B1) to +4 °C (A1FI) with the full range between +1.1 °C and +6.4 °C.[12] This signal is weak at the beginning of the twenty-first century and progressively increases in power as time goes by.

Even if the GHG increase is spatially uniform, the thermal response is heterogeneous with a larger increase over the continents, especially over the subpolar continents of the northern hemisphere, rather than over oceans (especially in the northern

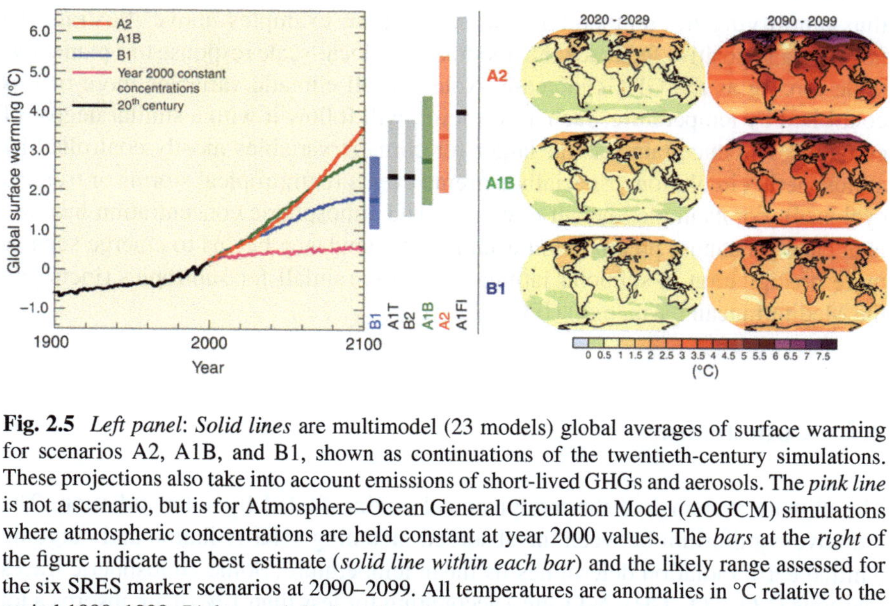

Fig. 2.5 *Left panel*: *Solid lines* are multimodel (23 models) global averages of surface warming for scenarios A2, A1B, and B1, shown as continuations of the twentieth-century simulations. These projections also take into account emissions of short-lived GHGs and aerosols. The *pink line* is not a scenario, but is for Atmosphere–Ocean General Circulation Model (AOGCM) simulations where atmospheric concentrations are held constant at year 2000 values. The *bars* at the *right* of the figure indicate the best estimate (*solid line within each bar*) and the likely range assessed for the six SRES marker scenarios at 2090–2099. All temperatures are anomalies in °C relative to the period 1980–1999. *Right panel*: Projected surface temperature change for the early (2020–2029) and late twenty-first century (2080–2099). The map shows the multi-AOCGM average projection for the A2, A1B, and B1 SRES scenario. Temperatures are anomalies in °C relative to the period 1980–1999 (IPCC 2007) (*IPCC 2000: Special Report on Emissions Scenarios*. Prepared by Working Group III of the Intergovernmental Panel on Climate Change, Figure SPM-3. Cambridge University Press)

[12]Note that the thermal difference between glacial and interglacial periods during the quaternary equals 6–10 °C in Antarctica and Greenland.

Atlantic and around Antarctica; Fig. 2.5). This emphasizes the role of positive and negative feedback that, respectively, amplify and weaken the temperature increase. For example, the response is weaker across the ocean partly because of the thermal vertical structure of the ocean with an efficient vertical mixing of the heat and also the infinite source of water, thus limiting the surface temperature increase through evaporation. On the contrary, the response will be stronger for the continents having a current seasonal snow cover and where its duration will decrease as the temperature increases (Fig. 2.5). In that case, the replacement of the snow cover/sea ice by open water, vegetation, or soils during a certain amount of time will greatly increase the amount of absorbed solar radiation (because snow reflects 80–95 % of incident solar radiation whereas water or vegetation absorbs 85–99 % of incident solar radiation) and thus amplify the response. The amplification could also be due to the vertical structure of the atmosphere (i.e., vertical thermal inversion in the lower troposphere as above subpolar continents in winter and permanent icy surfaces). The response will also be stronger over the continents where soils become drier in consequence of higher temperature, especially in spring and summer (Fig. 2.5). In that case, latent heat will decrease and sensible heat and IR emission will increase, thus amplifying the temperature increase. These examples above illustrate the negative and positive feedback that modulate the local-scale response to a planetary-scale forcing as the GHGs increase. Note that all climatic variables more or less controlled by temperature, such as sea level, will follow it with a similar degree of certainty. The uncertainty is far larger for climatic variables mostly controlled by atmospheric circulation, especially extreme events (extratropical storms or tropical cyclones). For example, the net effect of GHG atmospheric concentration increase is unclear on annual precipitation even if some tendency begins to emerge such as more precipitation in subpolar latitudes and less rainfall for subtropics (including the Mediterranean basin, e.g.) (IPCC 2007).

2.7 Conclusion

The climate system is a thermodynamical engine fueled by solar radiation. The radiative equilibrium between the absorbed incoming solar radiation and outgoing emitted earth radiation determines its mean temperature. Some atmospheric traces (gaseous H_2O, CO_2, CH_4, N_2O, etc.) accounting for less than 1 % of air absorbs a lot of infrared radiation although they are almost transparent to visible light, thus increasing the amount of absorbed energy by the atmosphere and earth surface. The GHG atmospheric concentration increases due to various human activities increases monotonically, with an increased rate from the 1950s and thus forces a warming of the global surface temperature. In fact, the global surface temperature has increased mostly from 1910 to 1940 and from 1975. This increase is fully consistent with global-scale variations as sea-level rise and the melting of most of the mountain glaciers during the twentieth century. Even if the global surface temperature variations are not only controlled by GHG atmospheric concentration, there is a large consensus about the significant role of its current increase, especially during the second half of the twentieth century.

For the next centuries, if large explosive volcanic eruptions injecting a large amount of sulfur above 12–18 km do not occur every 2–3 years, if solar constant variations are similar to those experienced in the last centuries at least, and lastly if we exclude extremely rare events as the collision with a massive meteorite, then the main forcing to temperature variations will be the GHG atmospheric concentration increase due to human activities.[13] Despite the Kyoto Protocol[14] ratified in 1997 and ending in 2012 (currently extended till 2020), the current increase of fossil-fuel emission from 2000 is slightly above the "worst" IPCC scenario (i.e., A1FI). Global warming is thus virtually certain, in response to the increase of heat trapped by the climatic system and the continuous restoration of the radiative balance, but its rate and amplitude depend either on human choices or natural processes such as the oceanic ability to remove some of excess carbon from the atmosphere.[15] The whole scope of the IPCC scenario helps us to consider a wide range of possible future pathways. In that context, the increase of global surface temperature is also virtually certain during the twenty-first century (and beyond) and should be between +1.8 °C and +4 °C in 2100 (+6.4 °C if we follow the A1FI scenario until 2100). Beyond the fact that human choices could slow and delay this increase, it will be modulated in space and in time, even if other forcings are kept constant, due to the intrinsic nature of the climate system and the interplay of physical processes able to amplify/weaken the response at regional and zonal scales. The consequences of GHG increase on variables more or less controlled by temperature are also virtually certain such as the sea-level rise or the decrease of extent and/or of duration of snow cover or sea ice. The consequences are less clear for the hydrological cycle but higher temperatures will increase potential evaporation and will dry soils even if rainfalls are constant.

Acknowledgments I thank B. Hamelin and J. Guiot (CEREGE) for their careful reading.

References

Arrhenius SA (1896) On the influence of carbonic acid in the air upon the temperature of the ground. Philos Mag J Sci Ser 5, 41:237–276

Broecker WS (1975) Climatic change: are we on the brink of a pronounced global warming. Science 189:460–463

---

[13]This hypothesis also excludes engineering solutions able either to remove massive amounts of carbon from the atmosphere or to increase the earth's albedo.

[14]See http://unfccc.int/kyoto_protocol/items/2830.php.

This ability should decrease with time—and perhaps saturate—inasmuch as the ocean acidifies itself as it absorbs more and more carbon. The recent estimates show that sinks of Carbon averaged since 1959 equal respectively: atmosphere (44 % of total Carbon anthropogenic emissions), land (28 %), ocean (28 %) (Le Quéré et al. 2012).

Brohan JJ, Kennedy I, Harris SFB, Tett S, Jones PD (2006) Uncertainty estimates in regional and global observed temperature changes: a new dataset from 1850. J Geophys Res 111:D12106. doi:10.1029/2005JD006548

Cazenave A, Llovel W (2010) Contemporary sea level rise. Annu Rev Mar Sci 2:145–173

Church JA, White NJ (2006) A 20th century acceleration in global sea-level rise. Geophys Res Lett 33:L01602. doi:10.1029/2005GL024826

Danny Harvey LD (2000) Global warming, the hard science. Prentice Hall, New York, 336 pp

Fourier J (1827) Mémoires sur les températures du globe terrestre et des espaces planétaires (Memoir on the temperature of the earth and planetary spaces). Mémoires de l'Académie Royale des Sciences de l'Institut de France, tome VII: 570–604

IPCC (2007) Synthesis report. IPCC, Geneva, 52 pp. Available at: http://www.ipcc.ch/publications_and_data/publications_ipcc_fourth_assessment_report_synthesis_report.htm

Klein SA, Soden BJ, Lau NC (1999) Remote sea surface temperature variations during ENSO: evidence of a tropical atmospheric bridge. J Climate 12:917–932

Le Quéré C, Andres RJ, Boden T, Conway T, Houghton RA, House JL, Marland G, Peters GR, van der Werf G, Ahlström A, Andrew RM, Bopp L, Canadell JG, Ciais P, Doney SG, Enright C, Friedlingstein P, Huntingford C, Jain AK, Jourdain C, Kato E, Keeling RF, Klein GK, Levis S, Levy P, Lomas M, Poulter B, Raupach MR, Schwinger J, Stich S, Stocker BD, Viovy N, Zaehle S, Zeng N (2012) The global carbon budget 1959–2011. Earth Syst Sci Data Dis 5:1107–1157. doi:10.5194/essdd-5-1107-2012

Lorius C, Jouzel J, Raynaud D, Hansen JA, Le Treut H (1990) The ice core: climate sensivity and future greenhouse warming. Science 347:139–145

Plaut G, Vautard R, Ghil M (1995) Interannual and interdecadal variability in 335 years of Central England temperatures. Science 268:710–713

Rahmstorf S, Cazenave A, Church JA, Hansen JA, Keeling RF, Parker DE, Somerville RCJ (2007) Recent climate observations compared to projections. Science 316:709

Schmidt GA, Ruedy RA, Miller RL, Lacis AL (2010) Attribution of the present-day greenhouse effect. J Geophys Res 115:D20106. doi:10.1029/2010JD014287

Shakun JD, Clark PU, He F, Marcott SA, Mix AC, Liu Z, Otto-Bliesner B, Schmitter A, Bard E (2012) Global warming preceded by increasing carbon dioxide concentrations during the last deglaciation. Nature 484:49–54. doi:10.1038/nature10915

Trenberth KE (ed) (1993) Climate system modeling. Cambridge University Press, New York, Xxix + 788 pp

Trenberth KE, Fasullo JK, Kiehl JT (2009) Earth's global energy budget. Bull Am Soc 90:311–323

UNEP (2008) Global glaciers changes: fact and figures. World Glacier Monitoring Service, Zurich, 88 pp. Availabe at: www.grid.unep.ch/glaciers

Chapter 3
New Perspectives for Fossil Fuels: Hydrocarbons in "Unconventional" Settings

Alain-Yves Huc and Roland Vially

Abstract Due to the current concern regarding the remaining reserve of oil and gas, there is a strong incentive for the innovative exploitation of fossil fuels occurring in unconventional situations. The latter include a wide variety of resources, some of them known for a long time, but set aside in deference to the profit of the more conventional oil and gas; others rely on more recent identification or on new production technologies; and some are already in the operational stage, with others in the pilot stage or even only prospective.

In this respect this chapter first briefly reviews the concept of the petroleum system and then tentatively extends this concept to the definition of the different hydrocarbon occurrences associated with "nonconventional" settings. These settings include: coal, oil shale, heavy oils and bitumens, primary and secondary biogenic gas, methane hydrate, shale gas, coal bed methane, tight oil/oil producing shale, tight gas, aquifer gas, and so on.

It does not pretend to address exhaustively all aspects of nonconventional hydrocarbons, but to provide a synthetic overview of the geological meaning of these new players in the energy domain.

Keywords Nonconventional hydrocarbons • Coal • Biogenic gas • Petroleum system • Continuous-type reservoir • Heavy crude oil

List of Acronyms

bcf billion cubic feet (= 10^9 ft^3)
bcm billion cubic meters (= 10^9 m^3) 1 bcm = 35.31 billion cubic feet

A.-Y. Huc (✉) • R. Vially
Geology, Geochemistry, Geophysics Division, IFP Energies Nouvelles,
1 and 4 avenue de Bois Préau, 92852 Rueil-Malamaison, France
e-mail: alainyveshuc@gmail.com; roland.vially@ifpen.fr

J.B. Saulnier and M.D. Varella (eds.), *Global Change, Energy Issues and Regulation Policies*, Integrated Science & Technology Program 2, DOI 10.1007/978-94-007-6661-7_3,

BTU	British Thermal Unit, English unit of energy (1 BTU = 1,055 J)
CBM	Coalbed methane
cP	Centipoise unit of dynamic viscosity (1 cP = 1 mPa s)
CCS	Carbon capture and sequestration
CTL	Coal to liquid
EOR	Enhanced oil recovery
GTL	Gas to liquid
IEA	International Energy Agency
IFPEN	IFP Energies Nouvelles
MBTU	Mega BTU (= 10^6 BTU)
SAGD	Steam-assisted gravity drainage
tcf	Trillion cubic feet (= 10^{12} ft^3)
tcm	Trillion cubic metres (= 10^{12} m^3) 1 tcm = 35.31 trillion cubic feet
Tm^3	Tera m^3 (= 10^{12} m^3 = 1 tcm)
UCG	Underground coal gasification
US LLNL	US Lawrence Livermore National Laboratory
USGS	US Geological Survey

3.1 Introduction

According to the International Energy Agency (IEA 2010) in 2008 fossil fuels met more than 80 % of the world demand for total primary energy. Although it is mandatory that fossil fuel use should be replaced by more sustainable and more environmentally friendly sources of energy, the growing global population and fast economic development of countries such as China, India, and Brazil imply a progressive transitional period during which fossil fuels will continue to be instrumental in the energy supply. Due to the current concern regarding the remaining reserve of oil and gas, there is a definitive need for intensive exploration and improved operational practices in order to increase the recovery of conventional hydrocarbons from the in situ accumulations, and to reduce the environmental impact of the production and transport of these products. In addition there is a strong incentive for the innovative exploitation of fossil fuels occurring in unconventional situations. The latter include a wide variety of resources, some of them known for a long time, but set aside in deference to the profit of the more convenient conventional oil and gas; others rely on more recent identification or on new production technologies; and some are already in the operational stage, with others in the pilot stage or even only prospective.

Whatever the considered nonconventional resource, it belongs to the same carbon cycle that underlies the concept of petroleum system.

In this respect this chapter first briefly reviews the concept of the petroleum system and then tentatively extends this concept to the definition of the different hydrocarbon occurrences associated with nonconventional settings. It does not pretend to address exhaustively all the aspects of nonconventional hydrocarbons, but to provide a synthetic overview of the geological meaning of these new players in the energy domain.

3.2 The Petroleum System Concept

Sedimentary basins are formed within depressions of the earth's upper crust, usually filled in by seas, lakes, or oceanic bodies. They are the result of geodynamic processes associated with tectonic plates. The earth's upper crust is made of crystalline rocks such as granite for the continental crust and basalt for the oceanic crust.[1] Together with metamorphisized sediments[2] they form the base of sedimentary basins. During geological time, the basins are progressively filled with sedimentary rocks, such as shale, sandstone, carbonates, or evaporites.[3]

This filling generally lasts tens to hundreds of millions years, with a sedimentation rate averaging a few millimeters per year. The weight of the deposited sediments associated with tectonic-driven processes may result in the downward deformation of the underlying crust increasing the thickness of the pile of sediments which can reach in extreme situations up to 20 km. This deepening of the basin is called subsidence. The Paris basin is an example of such a sedimentary basin lying on a crystalline basement currently outcropping in the Armorican massif, the Vosges, and the Central massif. The sedimentary filling amounts around 3,000 m of Mesozoic-Tertiary sediments in its deepest parts, underlain by an Upper Paleozoic succession. It is within such sedimentary basins that hydrocarbons are generated and trapped.

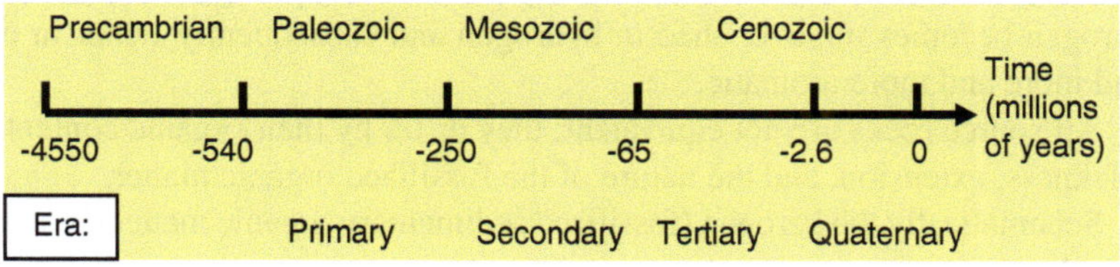

In this context a petroleum system is a sedimentary basin or part of a sedimentary basin including elements (source rock, carrier system, seal, and trap) and on the other hand processes (generation, migration, trapping, and eventual alteration of the fluids). The elements are as follows (Fig. 3.1):

- *Source rock* is a sedimentary unit containing fossilized organic matter, the kerogen, whose alteration (thermal cracking[4] associated with the temperature increase resulting from the progressive burial of the source rock, the local heat flow history, and the thermal conductivity of the sediment pile) generates hydrocarbon

[1]Crust is the outermost solid shell of the earth. Together with the upper mantle they form the lithosphere.

[2]Metamorphisized sediments are sediments that have been transformed into crystalline rocks by high temperature and high pressure.

[3]Evaporites are sediments made of salts deposited following the evaporation of saltwater bodies (i.e., sea water); examples are halite (NaCl) and calcium sulfate (Ca SO_4).

[4]Thermal cracking is the cleavage of chemical bonds as a result of thermal energy action.

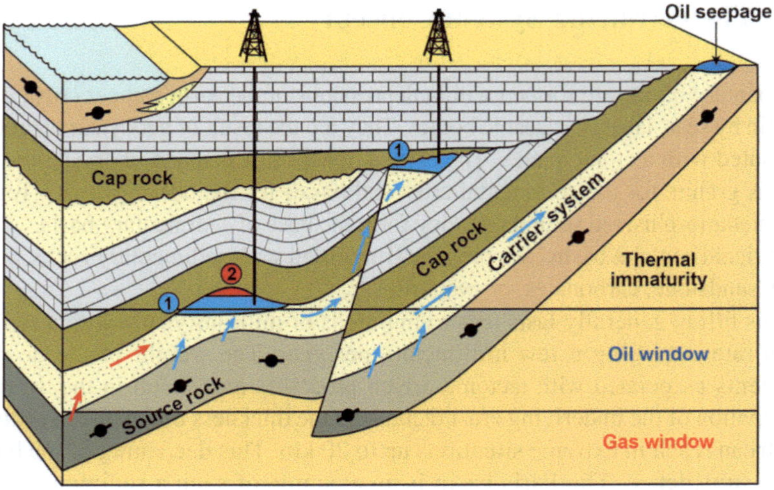

Fig. 3.1 Conventional petroleum system including source rock, carrier system, cap rock (seal), reservoir, oil: *1* and gas: *2* (*Source*: IFPEN, with permission)

fluids, oil and gas, leaving behind an organic residue. In the process of thermal cracking the released fluids are relatively enriched in hydrogen, and the residual kerogen becomes impoverished in hydrogen and subsequently richer in carbon and more and more aromatic.

All source rocks are not equivalent; they differ by their organic content, their thickness, extension, and the nature of the fossilized organic matter.

Schematically the kerogen (fossilized sedimentary organic matter) is classified into three main types.

Type I is mainly derived from the remains of bacteria and primitive algae populating in great amounts closed aquatic bodies such as lakes or isolated lagoon complexes. This type is not widespread but is prone to generate oil and gas inasmuch as 70–80 % of the kerogen mass can be transformed into hydrocarbons under an adequate thermal history. Source rocks of Type I occur, for instance, in the lacustrine sediments within the rifts associated with the Early Cretaceous of the South Atlantic margins, in several Tertiary basins in SouthEast Asia or in intracontinental basins in China.

Type II is by far more common and is associated with marine environments. It derives mainly from the remains of phytoplankton material. The main source rocks of the Arab-Persian Gulf, North Sea, Venezuela, or western Siberia are representative examples. Forty to sixty percent of the mass of this type of kerogen can be transformed into hydrocarbons.

Type III derives from the remains of land plants. This type is often associated with marine or lacustrine deltaic settings. A distinctive state of this type of kerogen is its occurrence as massive organic matter containing a limited amount of mineral sediment and known as "humic coal". The hydrocarbon potential of this organic matter is relatively low with 10–30 % of the mass of the kerogen which can be transformed into oil and gas. However because the involved sediment piles are often very thick, the quantity of generated hydrocarbons can be important. Source rocks of type III are instrumental in the petroleum potential of the Niger delta or of the Mahakam delta in Indonesia

It should be noted that the sedimentary processes leading to the formation of source rocks imply a depositional environment exhibiting very low energy water[5] and consequently the source rocks are not only characterized by a high sedimentary organic content, but also by being fine-grain sediments.

- *Carrier system/reservoir rock*: as a consequence of the fine-grain nature of the source rocks; the displacement of the generated fluids at the basin scale requires first that the petroleum fluids are expulsed from the source rock (primary migration) then that the migration from source to trap (secondary migration) relies on the occurrence of a carrier system involving geological features displaying a permeability which allows a sufficient capacity for fluid flow. This system of drains includes porous and pervious sedimentary rocks, tectonic-induced fractured rocks, faults, and boundary layers surrounding salt diapirs.[6] The latter two have complex behaviors because, depending on the geological circumstances, they can act as a conduit or barrier for the displacement of the hydrocarbons. Moreover, part of the carrier system, porous and permeable sedimentary rocks and fractured massifs, can act as reservoir rocks hosting accumulated fluids.
- *Seal/cap rock*: in order to keep the hydrocarbons confined within the carrier system an impervious sedimentary bed is needed overlying the drains. To be efficient at the basin scale, this seal should correspond to a widespread feature such as regional transgressive shales or massive evaporites intervals. In the absence of such a geological object the fate of the hydrocarbons would be to be dispersed towards the surface where biological and chemical processes can actively proceed to destroy them. At the field scale this seal is identified as the cap rock whose petrophysic properties should be able to overcome the injection pressure generated by the low-density oil and gas column and mechanical properties able to avoid breaching.
- *Structural or stratigraphic traps* occur when, according to the rules of physics, the secondary migration, controlled by the buoyancy, the ultimate fate of the moving fluids is to reach the surface of the basin where they will seep as "hydrocarbon shows" and be oxidized by biological and chemical processes, unless they meet a trapping situation when traveling along the carrier system. The traps can be structural, resulting from geometrical features induced by tectonic

---

[5] *Low energy water* is the water energy controlled by water agitation and movement induced by currents, waves, storms.

[6] *Salt diapirs* are the salt structures formed as a consequence of the high buoyancy of salt when compared to other sediments. As a result, when buried the salt flows upward intruding on sediments and forming bodies of different shapes such as domes, pillars, and sheets, called diapirs.

activity (anticline, fault, deformation due to salt tectonics, etc.) or stratigraphic, resulting from the lateral change in petrophysic properties of the carrier system (isolated sedimentary bodies, change of depositional environments, local diagenesis, unconformities, etc.). It should be noted that at the geological timescale and according to their respective efficiencies and geological histories the trapped accumulations undergo different fates and should be considered as transitional, but with a wide range of possibilities in terms of life span.

In addition to the previously listed elements, which are all mandatory, the petroleum system concept includes the occurrence and timing of processes that control the scenario leading to the possible emplacement of oil and gas accumulation in sedimentary basins:

- The *generation of hydrocarbons* is the result of the thermal alteration of the kerogen as previously described. According to a progressive increasing thermal history a given source rock goes from:
  - An immature state (also termed organic *diagenesis* step) when the provided thermal energy is not high enough to generate hydrocarbons by cracking of chemical bonds within the kerogen
  - To an oil and gas generation stage (*catagenesis*), starting generally with the formation of a viscous and heavy "immature" oil and followed by the generation of increasingly lighter oil and of a growing relative quantity of gas (oil window then wet gas window)
  - And finally to an only gas generation stage (*metagenesis*, dry gas window)

The process of oil and gas generation is usually mathematically formalized as a set of first-order kinetic reactions which is used as a basis for the algorithms developed in basin models designed to simulate the thermal transformation of the sedimentary organic matter.

- The *expulsion* out of the source rock of a part of the generated hydrocarbons is a phenomenon that needs to overcome the very low permeability of the source rock (in the order of nanoDarcys) which implies the in situ generation of a natural injection pressure on the order of megaPascals. It is largely controlled by the huge difference of the pore pressure that is built up between the source rock and the carrier system due to:
  - A difference in the "compressibility" (difference in the effective strength) of the two geological objects
  - The increase of the internal pressure within the source rock itself as a result of the thermal cracking of the kerogen promoting a substantial rise of the "molar volume" of the contained organics

However, it should be noted that a fraction of the hydrocarbons (oil and gas, ultimately only gas) remains trapped within the source rock (this is reflected by the observed occurrence of the measured residual hydrocarbons that routinely allows analytical assessment of the "expulsion efficiency" of a given source rock). The driving mechanism for the primary migration being the difference of pressure between the source rock and a drainage system, the direction of expulsion is independent of gravity (buoyancy is not implied) and can proceed upward or downward, feeding carriers located above and/or below the source rock.

- The *secondary migration* conveys the hydrocarbons along the carrier system. Thanks to a sufficient porosity and permeability of the drains the driving force needed for the displacement of the oil and gas is the buoyancy,[7] eventually modulated, when active, by "positive" or "negative" hydrodynamics flow.[8]

Subsequently the secondary migration proceeds only towards the surface. Depending on the geological characteristics of the basin this migration can be vertical (usually along faults and diapir bodies) or lateral (usually following pervious sedimentary features), with the possibility to transport the hydrocarbons very long distances (eventually up to hundreds of km). Because the hydrocarbons are mainly migrating as separate phases within sediments initially saturated with water, the rate of displacement is described according to Darcy's law extended to polyphasic flow, involving the concept of relative permeability controlled by the hydrocarbon saturation. The ultimate reach of the secondary migration, if not hindered by the occurrence of traps in the followed pathways, is the soil surface displaying oil and gas seeps.

- The alteration of hydrocarbons in the reservoir is a rather common situation. Biological, chemical, or physical processes are instrumental in these alteration phenomena. They can result in a change in the composition of the hydrocarbons, and even in their partial or total destruction. For instance, these alteration processes can be illustrated by
  - The biodegradation of oil which is mediated by microbes belonging to the "deep biosphere" and can lead to the transformation of conventional oil into a viscous heavy crude oil.
  - The thermal sulfate reduction occurring when the hydrocarbons are in contact with sulfate rocks such as anhydrite ($Ca\ SO_4$) at high temperature. This is an oxido–reduction reaction involving the reduction of sulfates and the oxidation of the hydrocarbons resulting in the formation of H_2S and CO_2 (the latter being eventually mineralized into secondary calcite: $Ca\ CO_3$).
  - Gas washing involves the sweeping of a light fraction of oil from an accumulation by a circulating hydrocarbon gas, leaving behind heavier oil. The striped light fraction can eventually condensate at a shallower depth when the loaded gas meets lower temperature and pressure conditions.

[7] *Buoyancy* is the upward force that a fluid exerts on a body (here a stringer of hydrocarbon) with a lower volumetric mass than itself.

[8] *Hydrodynamics* is the flow of the formation water that can be "positive" when moving in the direction of the hydrocarbon movement and consequently facilitating the migration, for instance, water derived from the compaction of sediments, or negative when moving in the opposite direction and consequently interfering with the migration, for instance, meteoric water.

3.3 The Unconventional Fossils Fuels in an "Extended Petroleum System" Perspective

Herein the purpose is to provide a "petroleum system" rationale to the sources of energy appearing in the arena of fossil fuels and referred to as "unconventional hydrocarbons," coal-derived fluid fuels, oil shale, heavy oil and bitumen, primary and secondary biogenic gas, methane hydrate, shale gas, coal bed methane, tight oil/oil producing shale, tight gas, aquifer gas, and the like.

As a point of fact all these fossil fuels do not depart from the petroleum system concept but correspond rather to an extension of our vision of the latter, or at least to specific niches within the system that were previously (or are still) disregarded assuming a lack of technical or economical productivity (Fig. 3.2).

3.3.1 Coal

In the framework of the petroleum system concept, humic coal[9] corresponds to a massive expression of the sedimentary organic matter (kerogen) derived from land plants and belonging to Type III (Fig. 3.2). The different types of humic coals (peat,

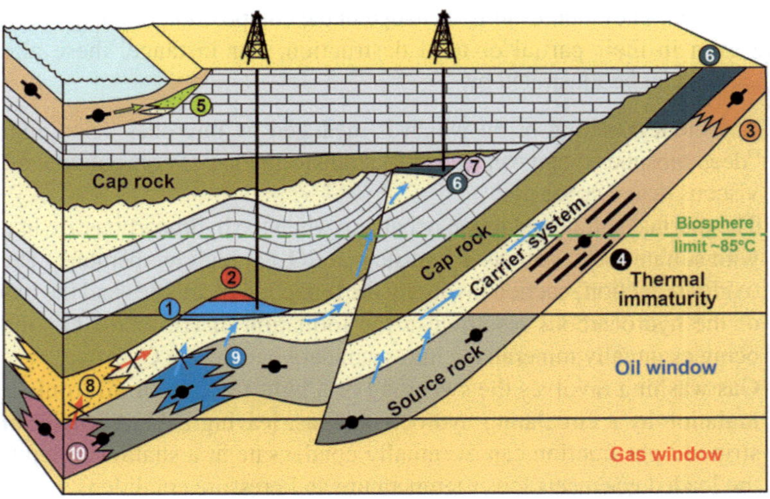

Fig. 3.2 Extended petroleum system: *1* conventional oil, *2* thermogenic gas, *3* oil shale (immature organic-rich source rock, type I or II), *4* coal seams (type III organic matter) and coalbed methane, *5* primary biogenic gas, *6* heavy/extra heavy oils, bitumen, *7* secondary biogenic gas, *8* tight gas in a basin-centered situation, *9* tight oil/oil from mature source rock, *10* shale gas (*Source*: IFPEN, with permission)

[9]*Humic coal:* Coal refers to sediment made almost entirely of organic matter; humic means that this organic matter derives from land plants. Other coal exists, less abundant, composed of, for example, algal remains: algal coal.

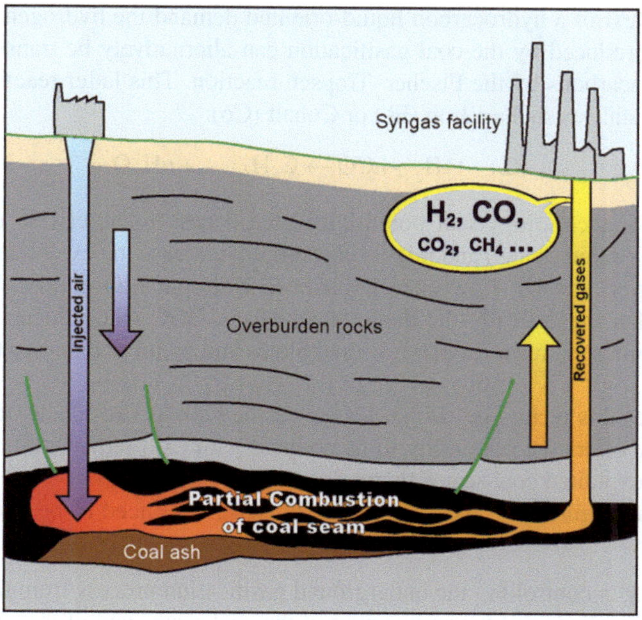

Fig. 3.3 Underground coal gasification (*Source*: Modified from http://groundtruthtrekking.org/Graphics/UndergroundCoalGasification.html)

lignite, brown coal, bituminous coal, anthracite, etc.) only reflect the incremental increase in the thermal maturity of this organic matter.

Burned in power plants and steam-powered vehicles, humic coal which was the main driver of the nineteenth-century industrial revolution, today is again a major focus for energy seekers thanks to its near 10 trillion metric tons of proved reserves.

Technologies are being developed that aim to reduce the environmental impact of coal energy generation ("clean coal"). They include improved processing techniques to be applied ex situ, after mining and associated with Carbon Capture and Sequestration (CCS), but also underground operations such as in situ gasification.

Underground coal gasification (UCG) relies on the in situ partial combustion of coal seams thanks to the controlled injection of air; the major recovered gases include CO and H_2,

$$3C(\text{Coal}) + O_2 + H_2O \rightarrow 3CO + H_2$$

associated with CO_2, CH_4 and H_2O (Fig. 3.3). The mixture (syngas) can be burned directly and used as a fuel source for power plant facilities.

However, in the case of an economy system requiring a high supply of hydrogen, additional hydrogen can be generated from CO by adding steam in order to carry out the "water gas shift" reaction:

$$CO + H_2O \rightarrow CO_2 + H_2$$

In a context of a hydrocarbon liquid-oriented demand the hydrogen and carbon monoxide produced by the coal gasification can alternatively be transformed into liquid hydrocarbons by the Fischer–Tropsch reaction. This latter reaction requires the use of catalysts such as Iron (Fe) or Cobalt (Co):

$$(2n+1)H_2 + nCO \rightarrow C_nH_{(2n+2)} + nH_2O$$

The UCG technology can potentially give access to otherwise unmineable coals (deep or thin seams) and might substantially increase the exploitable reserves. In this respect the US Lawrence Livermore National Laboratory (US LLNL) estimates that theoretically the deployment of the UGC would increase by more than three times the reserves in the United States and by more than two the reserves of India.

Moreover this technique, which leaves gasification ash residuals underground, additionally offers the possibility to be coupled with CO_2 sequestration within the previously exploited coal seams themselves.

However, engineering and environmental issues still need to be addressed, for instance:

- Difficulty in controlling the underground gasification process from the surface
- Reliable prediction of the architecture of the coal seams and of their internal heterogeneities, including lateral and sequential variability of the organics properties
- Protection of the surrounding aquifers from contamination by gasification products
- Managing the possibility of terrain subsidence

3.3.2 Oil Shales

In a petroleum system framework, oil shales correspond to immature source rocks that can be retorted (pyrolyzed)[10] in order to recover the oil resulting from the industrial cracking of the contained kerogen (Fig. 3.2).

As a consequence of the kinetics aspect of hydrocarbon generation, the temperature required on an industrial timescale, for ex situ retorting of an oil shale, is very high (~350 °C).

In this respect, and in order to be economical, the process can be applied only to source rocks with a very high petroleum potential (usually organic-rich Type I or II source rocks), and located at very shallow depths (Fig. 3.2). Actually the qualification as oil shale is related to the economical payoff of the industrial operations, including production and processing. In any case, the thermodynamic definition of an oil shale is that the retorting process provides at least as much energy (in the form of shale oil, recently tentatively renamed kerogen oil) than is used for the retorting

[10] *Retorting* or *pyrolyzing* oil shale consists in heating the rock in the absence of oxygen (this is not combustion). Heat is the only driver of the reaction.

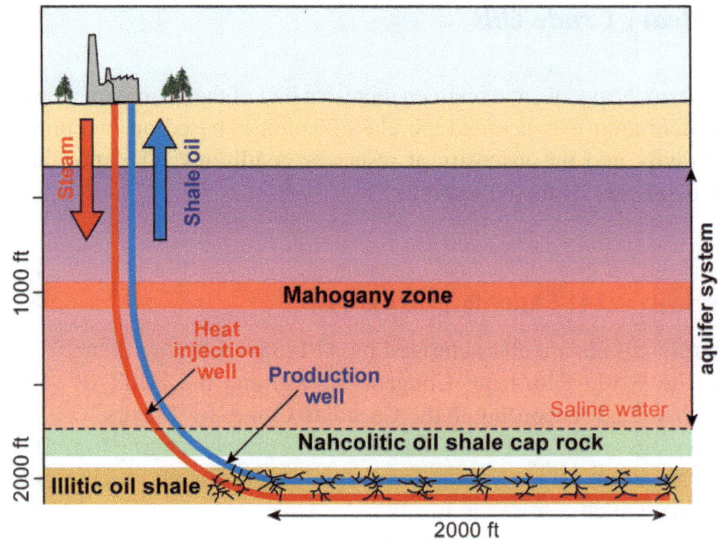

Fig. 3.4 Example of in situ process for oil shale retorting (*Source*: Modified from: www.Amso.net http://www.amso.net/Our-Concept/Our-Process.aspx, with permisssion)

operation. In this respect the energy equilibrium is roughly reached for a production of around 25 l/t of rock. However, in order to be competitive with conventional oil, an indicative figure for the required production of shale oil is ~40 l/t (Dyni 2005).

If above-ground retorting operations are still in progress, mainly as in Estonia (Ordovician Kukersite) and in Brazil (Permian Irati Shale) or in the development phase in Australia (Eocene Stuart oil shale), such ex situ exploitation generates important environmental impacts that need to be carefully managed. This situation led, in spite of very large potential reserves, to a dramatic decrease in the production of fuels from oil shales by the end of the twentieth century. New technologies, which are at the pilot stage and involve different approaches to perform an in situ retorting are currently being developed. They usually imply the coupling of down hole pyrolysis, horizontal drilling, and engineered fracturing. These projects, which mainly target the Eocene Green River Shale in the United States, a formation which is by far the most important resource in the world, rely on a wide range of techniques including the use of electric heaters, hot gas injection inducing fracturing and retorting (eventually associated with horizontal drilling), hydraulic fracturing with electrically conductive material, radio frequency and microwaves (Fig. 3.4). Taking advantage of the kinetic nature of the kerogen alteration the operators try to optimize the provided energy and the timespan of treatment (lasting up to several years).

As with all underground exploitation operations, environmental risks need to be assessed, monitored, and mastered. The environmental footprint is limited when compared with ex situ processing. However, depending on the geological situation and depth these might involve prevention of ground subsidence, protection of aquifers, and control on emitted effluents.

3.3.3 Heavy Crude Oils

Heavy oil, extra heavy oil, and bitumen represent half of the petroleum world resource. As far as definition is concerned the classification is based on two properties: the specific gravity and the viscosity at reservoir conditions. The specific gravity is routinely expressed by the API gravity.

API gravity = (141.5/specific gravity) − 131.5

Conventional oils are characterized by API gravity higher than 20° according to the World Petroleum Congress (1980) and the UNITAR (1982), or higher than 22.3° according to the Canadian Centre for Energy.

Heavy oils are defined as oils having an API gravity ranging from 20°/22.3° to 10°. Note that 10° corresponds to oil having the same specific gravity as pure water, which is given as 1.

Extra heavy oils and natural bitumens display an API gravity lower than 10.

The difference between extra heavy oils and bitumen is based on the viscosity at reservoir conditions: lower than 10,000 mPa·s (cP) for extra heavy oils and higher than 10,000 mPa·s (cP) for natural bitumens.

Their high viscosity and detrimental composition (rich in heteroatomic heavy "molecular" weight moieties (entities) such as asphaltenes and resins, depleted in hydrogen, rich in sulphur, rich in metals, often acidic) are a great challenge in terms of production, transport, and upgrading.

Some of these oils are the result of an early generation from specific kerogens; however, the vast majority is the remaining part of previously conventional oil that has been altered by the action of microorganisms (biodegradation). In term of the petroleum system, they correspond to the ultimate fate of oil on its travel following the carrier system towards the surface where it meets thermal conditions allowing the activity of microorganisms using the oil as a source of carbon and energy (Figs. 3.2 and 3.5). As shown in the following, a by-product of the biodegradation activity is the formation of gases such as methane and carbon dioxide.

These gases, when still present within the oil and upon depressurization generate the formation of lower viscosity foam which can allow "cold production" of the heavy crude oils.

Other technologies mainly rely on the use of heat, which strongly affects the state of aggregation of asphaltenes and resins, and substantially reduces the oil viscosity. The processes involve the injection of steam into the reservoir. A recently developed concept is steam-assisted gravity drainage (SAGD), which combines the use of steam with horizontal well technology. In the case of bitumen, such as the oil sand deposits in Alberta (Canada), production relies on mining techniques followed

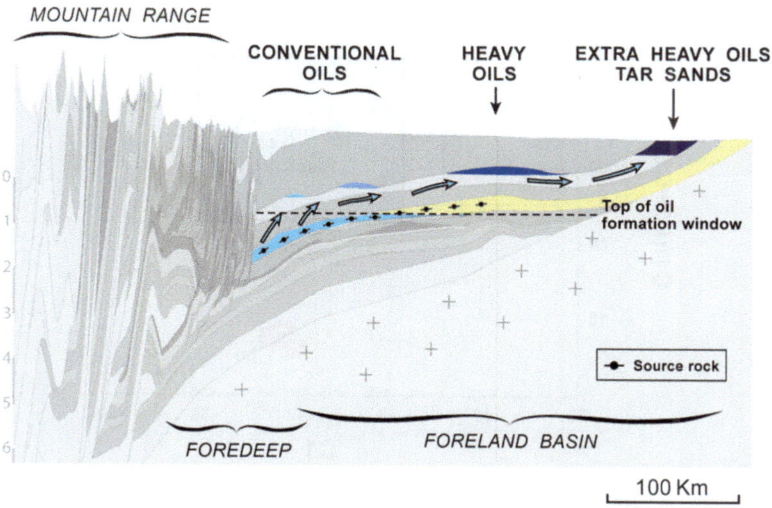

Fig. 3.5 Schematic distribution of conventional oil, heavy oil, extra heavy oil, and tar sands in a foreland basin situation (Eschard and Huc 2008) (*Source*: IFPEN, with permission)

by an ex situ process designed to separate the bituminous material from the mineral. Basically this is performed by subjecting a slurry (mixture of oil, sand, and water) to air bubbling under specific temperature conditions to promote the formation of a froth (foam made of air and bitumen) separated from the settling sand. The exploitation of these resources implies complying with rules and procedures designed to mitigate environmental concerns, such as land reclamation, water and tailing management, and CO_2 emission (Huc 2011).

3.3.4 Geological Biogas

3.3.4.1 Primary Biogenic Gas

The role of microbiology in geological processes has been largely underestimated up to the turn of the twenty-first century. The discovery of the deep biosphere, a huge microbial biomass living underground (with a low metabolitic rate) down to a few hundred (or even thousand) meters drastically changed the perspective (Parkes et al. 2000).

During the early stage of burial (early *diagenesis*) the sedimentary organic matter undergoing fossilization is subjected to the action of microorganisms involving complex processes leading at depth to the generation of methane.

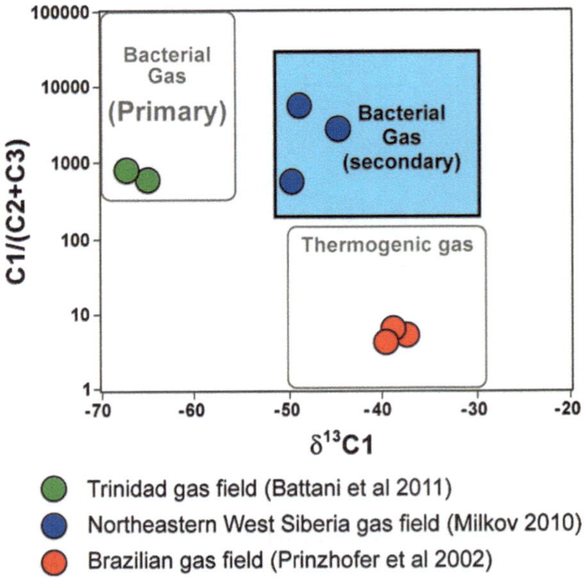

Fig. 3.6 Carbon isotope signature of primary biogenic gas, secondary biogenic gas, and thermogenic gases in a modified Bernard's diagram (Data: Battani et al. 2011; Milkov 2010b; Prinzhofer and Battani 2002). The *y*-axis displays the ratio between the methane (C1) and the sum ethane+propane (C2+C3). The *x*-axis displays the isotope signature of the carbon of the methane. It is expressed in the difference per mil between the ratio $^{12}C/^{13}C$ of the sample and the same ratio of an international standard (*Source*: IFPEN, with permission)

The Biochemical Process

Microbial communities rely on electron acceptors. The latter are consumed in a sequential manner: schematically, first they use oxygen, then when oxygen is exhausted, they use sulfates.

In organic-rich sediments, due to the high demand for oxygen, the pore environment rapidly becomes anoxic. In marine systems the occurrence of sulfates allows sulfate-reducing bacteria to oxidize the organic matter with the generation of H_2S. Sulfate reducers keep the partial pressure of H_2 too low for methanogens to be active. But if the sedimentation rate is sufficient to prevent the renewal of sulfates by diffusion from sea water, microbial activity will rapidly be driven by fermentatives and methanogens. Methanogenesis in a freshwater environment is dominated by acetate fermentation whereas in marine systems (after sulfate exhaustion) the main pathway is CO_2 reduction.

In any instance this biologic activity leads to the generation of a very dry gas that consists almost entirely of methane, exhibiting a very light carbon isotope signature (Fig. 3.6).

The accumulation of this biologic gas, coined *primary biogenic gas*, in commercial quantities requires relatively unusual geologic conditions (Fig. 3.2). Actually it can only migrate as a separate phase if the quantity formed is enough to overcome its saturation in water in order to be available for trapping if the geological conditions allow, for example, the occurrence of adequate trap and seal in a shallow context (Lin et al. 2004). The importance of this primary biogenic gas occurring in conventional traps is then a matter of debate (Rice and Claypool 1981; Clayton 1991, Katz 1995). A figure of 20 % of the global gas reserves has been proposed by Katz (1995) and references herein, and this is without mentioning its occurrence associated with unconventional situations.

3.3.4.2 Secondary Biogenic Gas

Biodegradation of oil is mainly taken as responsible for the formation of heavy crude oils (see Sect. 3.3.3). However this process, which is mediated by anaerobic microorganisms, results in the generation of gas products: H_2S if sulfate is present, but mainly CH_4 and CO_2.

The Biochemical Process

The metabolitic processes are not fully deciphered but they apparently involve the fermentation of hydrocarbons to acetate and hydrogen, syntrophic oxidation of acetate to carbon dioxide and hydrogen, and finally reduction of CO_2 mediated by methanogens. In terms of the petroleum system concept and similarly to the previously mentioned heavy crude oil, the secondary biogenic gas corresponds to a facet of the ultimate fate of the oil on its travels following the carrier system towards the surface where it meets thermal conditions allowing the activity of microorganisms using the oil as a source of carbon and energy (Fig. 3.2).

When considering the worldwide volume of biodegraded oil, this corresponds potentially to a huge amount of produced gas. More specifically, this secondary biogenic gas might account for a large part of the gas fields and gas dissolved within oil, in petroleum provinces where biodegradation is an active and significant process.

A tentative figure of around 2,350 tcm has been proposed (Milkov 2010a). A situation of secondary biogenic gas generation has been proposed for the Devonian Antrim Shale on the northern margin of the Michigan basin where reservoir rocks are actually fractured organic-rich black shale that also serve as source rocks (Shurr and Ridgey 2002). A recent regional study would assign a secondary biogenic origin to the extensive gas reserves of the northern part of the Western Siberia basin (Milkov 2010b). Thanks to the different pathways leading to the generation of methane, their respective impact on the chemical composition of the resulting gases and on the isotope fractionation provides discriminating signatures (Figs. 3.6 and 3.7).

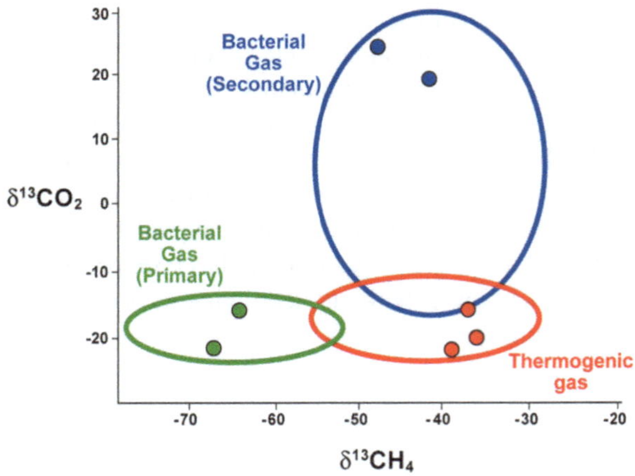

Fig. 3.7 Carbon isotope signature of primary and secondary biogenic gases and thermogenic gases; diagram inspired by Milkov (2010a) (*Data*: Battani et al. 2011; Milkov 2010b; Prinzhofer and Battani 2002). The *y*-axis is the isotope signature of the CO_2 carbon and the *x*-axis is the isotope signature of the methane carbon (*Source*: IFPEN, with permission)

These signatures can be exploited in order to identify the processes at the origin of the gases (thermogenic, primary, and secondary biogenic gas), assuming that there is no bias associated with complex mixing or contamination from external gas input. For instance, diagnostic methods are proposed by combining the gas wetness (methane vs. higher molecular weight homologues) and the isotope signature of the methane, or the carbon isotope signature of the methane and carbon dioxide (Bernard et al. 1977; Prinzhofer and Battani 2003; Milkov 2010a) (Figs. 3.6 and 3.7).

3.3.5 Continuous-Type Reservoirs

Several fossil fuel accumulations do not respond to the strict definition of the classical trapping processes including a reservoir rock, cap-rock, and trap (structural or stratigraphic).

They actually correspond to a physical trapping within a sedimentary layer (or a set of sedimentary layers). The trapping process can occur in the source rock system itself (coal bed methane, shale gas, tight oil/oil producing shale) or in a carrier rock exhibiting a reduced pervasive capability (tight gas in basin-centered gas systems) (Fig. 3.2).

In continuous type reservoirs the production procedure relies on overcoming the physical trapping (i.e., desorption, hydraulic fracturing, etc.) and by expanding the volume of exploitation (i.e., horizontal drilling, hydraulic fracturing, etc.; Fig. 3.8).

Similarly to other underground exploitation operations, environmental risks need to be assessed, monitored, and mastered. According to the geological context, depth and engineering technology, risk management might involve, for example, prevention

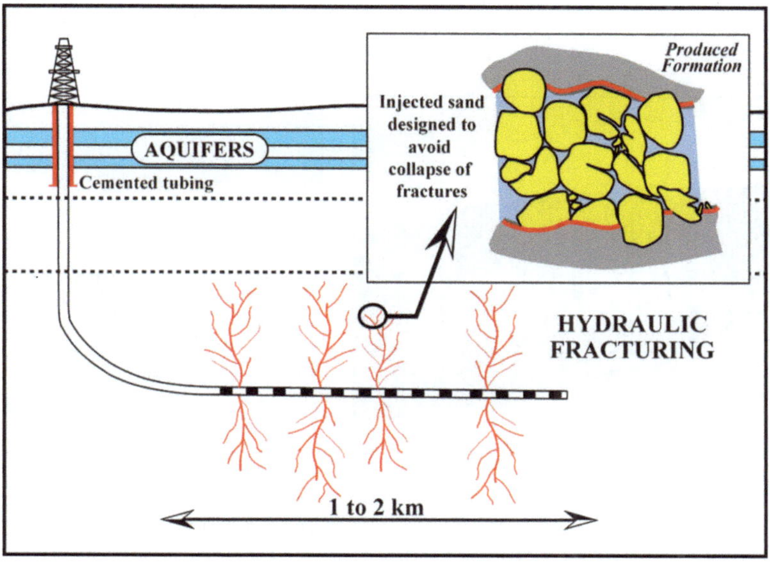

Fig. 3.8 Schematic view of the coupling of horizontal drilling and engineered hydraulic fracturing (*Source*: IFPEN, with permission)

of ground subsidence, protection of aquifers, control of injected chemicals, processing of emitted effluents, and water management.

3.3.5.1 Coalbed Methane (CBM)

Coal can be schematically described as made of a matrix with low permeability and a network of cracks (the cleats) affecting the whole volume of a coal seam (Fig. 3.9).

Methane is mainly trapped by adsorption within the organic matrix. In order to be produced, the methane needs to be desorbed by depressurization, and subsequently driven from the matrix into the cleats. This requires first pumping out the water filling the cleats and water occurring within the matrix, in order to obtain a sufficient relative permeability for the gas. If, in term of the petroleum system, the coalbed methane is clearly associated with a source rock (the coal seam; Fig. 3.2), according to each specific geological setting the recovered gas can be the result of early diagenesis (primary biogenic gas), of the generation of gas by thermal cracking (catagenesis and metagenesis), or by the latter alteration of the coal by microbial alteration (equivalent to secondary biogenic gas). The adsorption capacity of the coal is related to intrinsic properties such as maceral composition, coal maturity, gas composition, and ash content, as well as moisture content and is strongly dependent on external conditions such as temperature and pressure.

For a given coal composition, increasing pressure promotes adsorption of gas until the amount of stored gas approaches the maximum capacity of the coal; on the other hand, increasing temperature progressively hinders adsorption of gas on

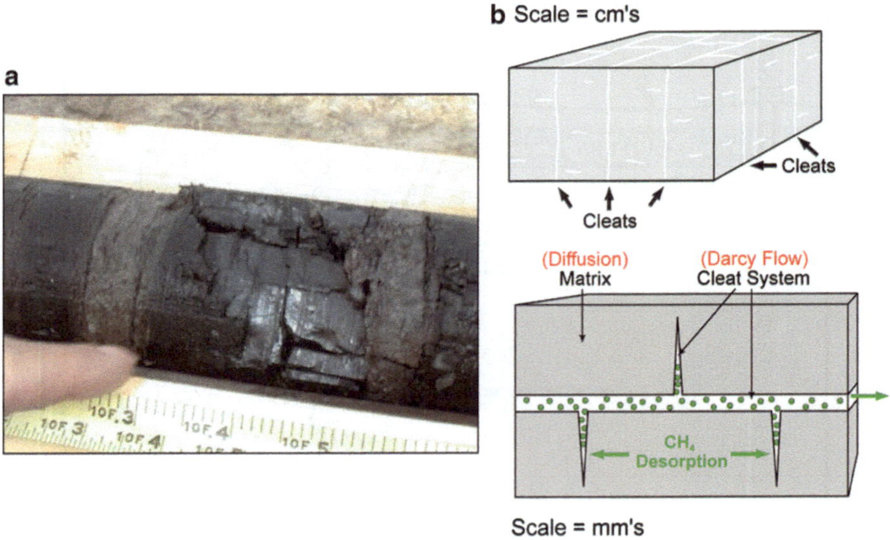

Fig. 3.9 (**a**) Montana (United States) cretaceous coal core showing cleats (*Source*: http://mt.water.usgs.gov/pub/cbm3.jpg, modified 30-Mar-2004 17:47). (**b**) Schematic coal structure displaying matrix and cleats system (Modified from Schurger et al. 2006)

organic matter and tends to reduce the adsorption capacity of coal. Because, in geological conditions, temperature and pressure increase in parallel as a function of depth, the two factors are competing. In practice at shallow depth the adsorption capacity is mainly controlled by pressure and increases rapidly with depth, but at greater depth temperature becomes the dominant factor implying a decrease in the adsorption capacity (Juch et al. 2004). A tentative CBM resource assessment by the IEA, is provided in Fig. 2.14.

3.3.5.2 Shale Gas

During the primary migration process, corresponding to the expulsion of the hydrocarbons from the source rock, a part of the generated hydrocarbons reaches the carrier system and is available for secondary migration. Another part remains trapped within the source rock, according to a phenomenon tentatively explained by retention/adsorption and by the concept of saturation needed for expulsion (Darcy's law extended to polyphasic systems). These remaining hydrocarbons are subjected to further cracking, allowing additional expulsion but never exhausting the internal system. In this respect the highly mature source rocks still contain gas. The gas occurring in shale gas can be stored as free gas within the pore spaces and the fractures, as adsorbed gas on organic matter and clays or as dissolved gas in formation water. The factors controlling the quantity of gas are complex and numerous. Among them, thanks to their high specific surface, the clay content and nature play

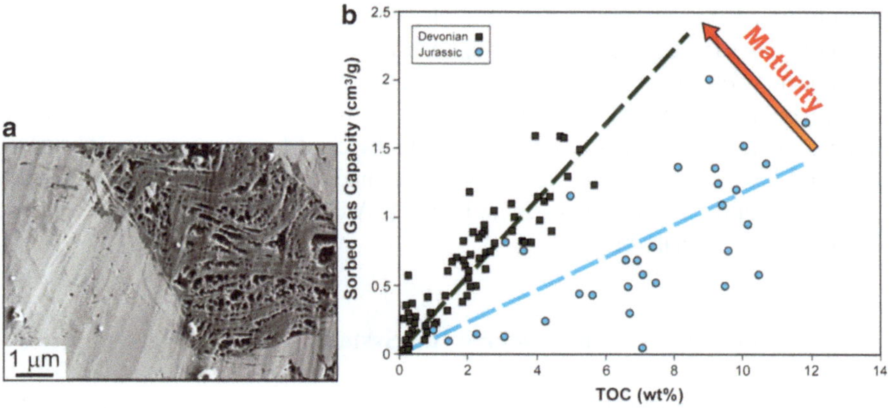

Fig. 3.10 Role of organic matter in the retention of gas in shale gas (**a**) nanoporosity associated with organic material in Mississippian Barnett shale, Texas, United States (Loucks et al. 2009), (**b**) relationship between organic content (TOC%), maturity and sorbed gas capacity in Devonian and Jurassic shale gas from British Columbia, Canada (Modified from Bustin et al. 2009)

a role. However, the organic matter itself appears to be instrumental in the storage capacity which increases with the organic content and its maturity in relation to the nanoporosity created within the kerogen as a result of the thermal generation of hydrocarbons (Fig. 3.10). Temperature and pressure act the same way as in the case of coalbed methane (Loucks et al. 2009; Bustin et al. 2009).

This gas can be produced using technologies such as horizontal drilling and engineered fracturing. In this respect the clay content that can be, to a limited instance, a favoring factor for the adsorption of gas, becomes detrimental for the petrophysic properties amenable to fracturing which is more efficient for not swelling and brittle material such as biogenic silica. It should be noted that the biogenic silica is also often positively correlated with the organic content, inasmuch as both can be inherited from the same source, namely the organic and inorganic bioproduction of phytoplankton. A tentative shale gas resource assessment by the IEA is provided in Fig. 3.14.

3.3.5.3 Tight Oil (Oil from Mature Source Rock)

Tight oil is also referred as “shaly oil,” as in Fig. 3.13, “oil in shale,” “oil producing shale,” “light tight oil,” “oil from source rock,” or confusingly, but commonly, as “shale oil”.

In the event that a source rock is isolated from a regional carrier system, hindering a proper expulsion and secondary migration, the generated oil will stay trapped within the source rock itself or within more brittle low permeability adjacent rocks (Fig. 3.2). The progressive pressure increase might eventually lead to moderate lateral migration along internal or external sedimentary beds with poor petrophysical quality, as described locally in the Lias of the Paris basin (Espitaliè et al. 1988). In any instance the result is an overpressured system that can be produced by taking advantage of the overpressure within the source rock environment and using technologies such as directional drilling and fracturing of the source rock itself or of adjacent beds. The most relevant example of such a situation is located in parts of the Devonian Bakken source rock formation in the Williston Basin in the United States, where a low-porosity dolomite is sandwiched by two organic-rich members and has been producing for 30 years (Sonnenberg and Pramudito 2009). A tentative tight oil resource assessment by the IEA is provided in Fig. 3.13, where it is referred to as "shaly oil."

3.3.5.4 Tight Gas in Basin-Centered Gas Systems

Tight gas occurs in sandstone or limestone reservoir rocks with unfavorable petrophysical properties. It should be noted that tight gas can occur in conventional gas accumulations as well as in continuous gas accumulations, the latter being usually referred to as "basin-centered gas accumulation." In continuous-type reservoirs, they are typically characterized by regional reservoirs that are gas saturated, abnormally (high or low) pressured, commonly lacking a downdip water contact (in a dipping, not flat-lying, hydrocarbon reservoir that contains gas, oil, and water, the gas is updip, the gas–oil contact is downdip from the gas, and the oil–water contact is still farther downdip), and exhibiting restricted permeability. A generally accepted industry definition is that they correspond to reservoirs that do not produce economic volumes of natural gas without assistance from massive stimulation treatments or special recovery processes and technologies, such as horizontal wells and fracturing (Holditch 2006). Low permeability is primarily due to the fine-grained nature of the sediments, compaction, or infilling of pore spaces by carbonate or silicate cements. There is still much that is not currently understood about the origin and development of these accumulations. In petroleum system terms they have been proposed to be considered as a system involved in a dynamical evolution, where gas loss due to secondary migration was compensated for by contemporary gas generation. In other words, what was previously a carrier rock corresponds today to a reservoir altered by diagenetic processes. As a consequence the declining secondary migration that has been affected by cementation, dramatically lowering the permeability, reduces the gas loss with time, so that significant gas trapping can be obtained in the present-day situation (Fig. 3.2). Overall, the accumulations can be described in terms of an "impedance trap" (in the sense of flow resistance) where the presence of gas is a transient phenomenon in terms of geologic time. The free flow of gas from the source rock to surface escape is impeded by the tight formation, to the point where a commercial volume of gas can be trapped in monoclinal dip on the basin flanks or even in the basin syncline. Tight gas was first developed in the western US San Juan basin; large exploitable reserves are assessed and produced in the United States: the Greater Green River basin of southwest Wyoming, and Canada: west central Alberta and British Columbia. Enormous resources are expected all over the world where geological conditions are met, including geodynamic settings such as

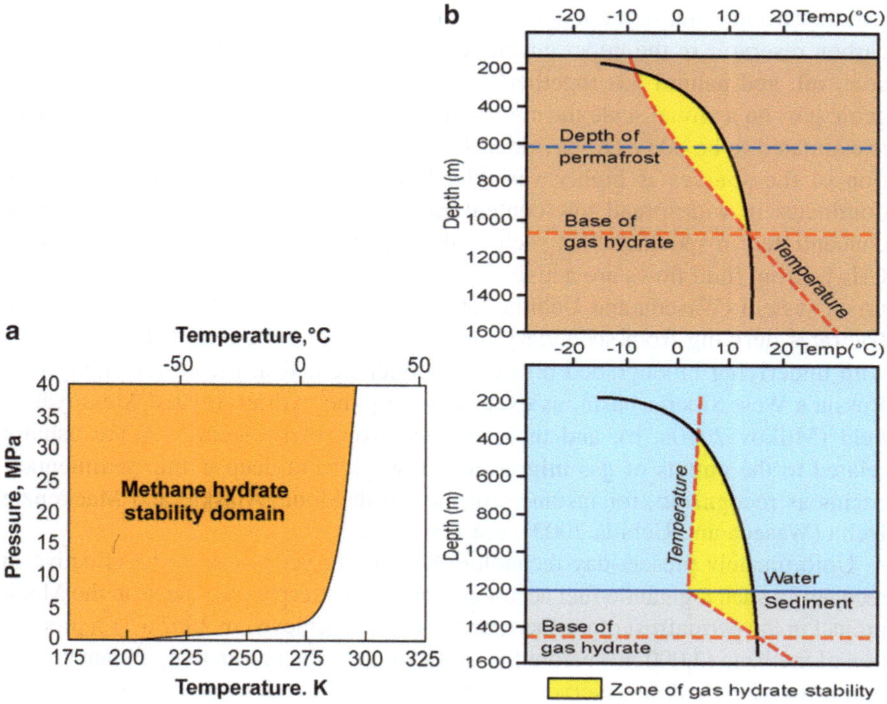

Fig. 3.11 (**a**) Phases diagram for methane hydrates, (**b**) Gas hydrate stability zone in permafrost and deepwater situations (depth assuming hydrostatic pressure) (Modified from Collett 2008; Kvenvolden 1988)

axial parts of rift basins, foredeep of foreland basins, synclinal of orogenic belts; where they are associated with highly mature source rock. A tentative tight gas resource assessment by the IEA is provided in Fig. 3.14.

3.3.6 Methane Hydrate[11]

Whatever its origin (primary biogenic gas, secondary biogenic gas, thermal gas) the methane following a secondary migration pathway or dismigration and reaching shallow depth can escape to the sea water or the atmosphere (gas seepages) or can meet adequate thermodynamic conditions leading to the formation of gas hydrates (Fig. 3.11).

These conditions include low temperature in the permafrost of polar regions and high pressure in the upper layer of marine sediment in water depth greater than about 500 m (Fig. 3.11).

[11] *Methane hydrate* is a solid in which a large amount of methane is trapped within a crystal structure of water (up to 180 m^3 of methane for 1 m^3 of hydrate), forming a solid similar to ice.

The carbon sequestered in these "thermodynamic traps" represents the largest carbon reservoir in the upper envelopes of our planet and twice the resource of coal, oil, and natural gas together! Currently, and in comparison with thermogenic gas, on a global scale the contribution of the biogenic gas to these hydrates is estimated to be 80 % (Kvenvolden 1988, 1995). Locally the relative contribution of the sources is highly variable. Usually primary biogenic gas largely dominates in widespread low-content gas hydrate deposits, but can be highly concentrated in specific areas such as the Japan Nankai trough where biogenic CH_4-bearing fluid flows are active throughout thrust features in the accretionary prism system (Waseda and Uchida 2003; Namikawa et al. 2004; Collett 2008). Hydrates deriving from secondary biogas are likely to be regionally associated with underlying biodegraded oil fields as documented in the eastern margin of Russia's West Siberia basin, as exemplified by the hydrate-related Messoyakha field (Milkov 2010a, b), and those derived from thermogenic gas are usually related to the outlets of gas migration pathways from deep within sedimentary basins as recognized, for instance, in the North Slope, Alaska, and Mackenzie Delta (Waseda and Uchida 2003); see Fig. 3.12.

Unfortunately present-day technologies have not yet proven to be effective in recovering such a gigantic fuel accumulation. In this respect we mention the Malik project in the permafrost of northern Canada which achieved in 2002 a 51 h production of methane (3,000 m^3) before the degradation of the productivity parameters of the surrounding hydrate reservoir induced sand mobilization and prevented further operation. After a modification of the pumping system, the Malik experiment in 2008 resulted in an improved production of 13,000 m^3 for a six-day test; no degradation of the reservoir was observed, but the quantity of produced methane did not meet a profitable level. However, in spite of the difficulties, R&D is still undergoing and pilot drilling projects are in progress in Mt Elbert (North Slope, Alaska), Ignik Sikumi (Alaska) and, on a larger scale, in the Nankai Trough offshore southeast Japan, where a huge resource of gas hydrates has been assessed (Moridis et al. 2009) and positively tested in March 2013. When gas hydrate becomes technically and economically producible, environmental aspects would need to be addressed. Methane is a hydrocarbon gas, still producing CO_2, however, the consequences of the extraction, such as possible ground subsidence, have to be assessed and remediated. In addition, because of the uncertainty on the use of hydrates as a source of energy and according to the very limited pilot experiments, field feedback does not allow proper study of the environmental issues.

3.3.7 Aquifer Gas

As previously quoted, in the subsurface the source of the hydrocarbon gases is multiple: thermogenic or biogenic, including methanogenesis associated with the early diagenesis of sedimentary organic matter and biodegradation of hydrocarbons.

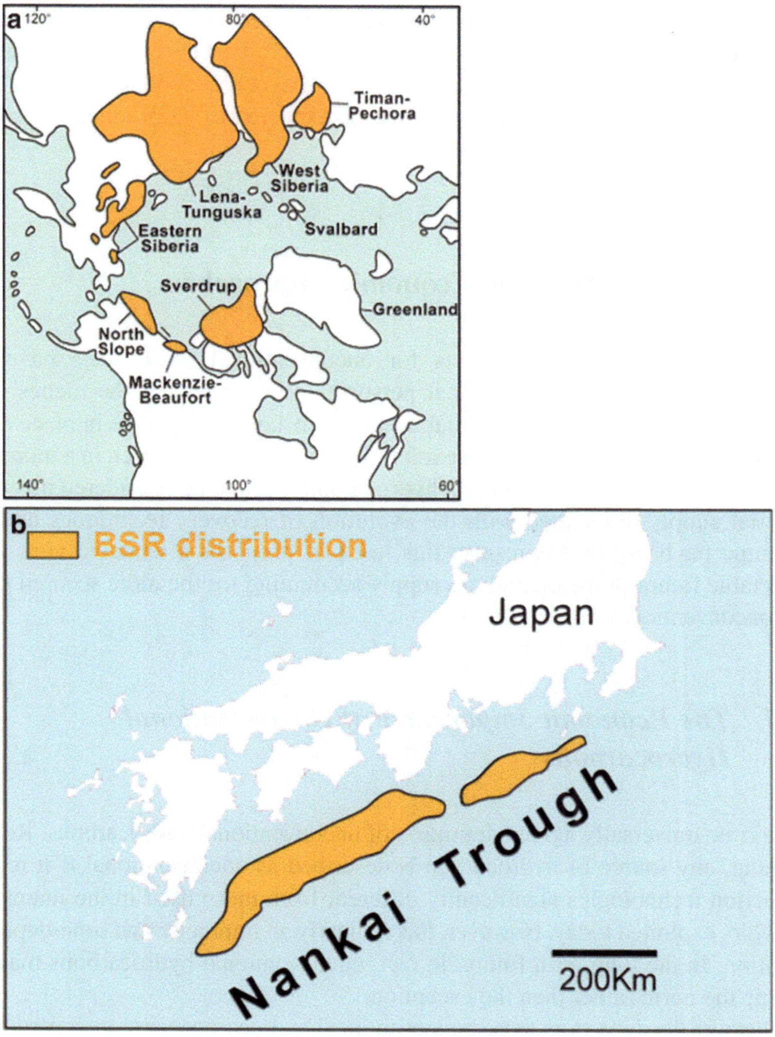

Fig. 3.12 (**a**) Permafrost-associated gas hydrate (After Collett and Dallimore 2000) and (**b**) distribution of deepwater gas hydrate in the Nankai Trough, Japan as inferred from indirect indicators involving the interpretation of seismic reflection data method: occurrence of a bottom simulating reflector: *BSR* (After Namikawa et al. 2004)

The porous system of the sediments is generally filled with water and these gases can migrate as a separate phase as soon as they reach saturation level. However, a part of them (mainly methane) stays in solution in the aquifers. The solubility of the methane in water is limited but is a function of temperature, pressure, and salinity. For instance at 5,000 m the solubility of methane in the groundwater ranges from 2 to 12 m^3/m^3 (Bonham 1978). Even considering this relatively low solubility and according to the estimate proposed by the BGR (Bundesanstalt für Geowissenschaften und Rohstoffe, Germany) the world's aquifer gas resources would reach a huge amount ranging from 24 to 1,500 tcm (Gerling 2005). However, it is clear that most of it is far away from available technology and beyond economic interest. Only a very marginal part of this resource might be expected to be produced in specific geological and economical conditions.

3.4 The Production and Economic Approaches

In previous sections the prospects for unconventional fossil fuels have been addressed according to a geological perspective by reviewing the niches of the "petroleum system" that are potential hosts of substantial resources in place and do not correspond to the conventional exploration guidelines. However, in a more pragmatic approach the nonconventional hydrocarbons need to be considered in terms of potential supply associated with the evolution of recovery techniques and with economic payback. In this respect this last part proposes a tentative view of the predictable future of the oil and gas supply accounting for the more secured part of the nonconventional hydrocarbons.

3.4.1 The Economic Importance of Unconventional Hydrocarbons

There is no universally agreed definition of unconventional hydrocarbons. Roughly speaking, any source of hydrocarbon is described as unconventional if it requires production technologies significantly different from those used in the mainstream reservoirs exploited today. However, this is clearly an imprecise and time-dependent definition. In the long-term future, in fact, unconventional hydrocarbons may well become the norm rather than the exception.

Whatever the definition, these unconventional hydrocarbons present a particularly interesting economic potential.

Different reporting systems and standards in use around the world give rise to large variations in definitions and estimates of oil and gas resources and reserves. In the following figures, we present an estimation of the ultimate recoverable resources.

The ultimate recoverable resources refer to the total volume of a resource that is both technically and economically recoverable. They include proven, probable, and possible reserves in discovered fields, as well as hydrocarbons that have yet to be found.

Most works on quantifying resources of natural oil and gas have focused on conventional sources. Although the amount of recoverable unconventional resources worldwide is thought to be very large, they are currently poorly quantified and mapped. This is true even in the United States where, despite significant effort, large uncertainties remain.

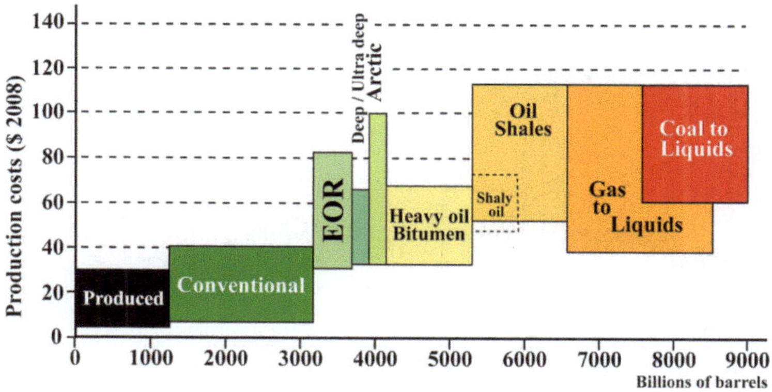

Fig. 3.13 Long-term oil supply cost curve (*Source*: IEA 2009, slightly modified). Oil includes conventional production and nonconventional sources. *Note*: "Shaly oil" refers to "tigh oil"/"shale oil" (oil from mature source rock), as opposed to oil shale corresponding to immature source rock

Assessments of conventional resources typically include volumetric calculations and the measuring and modeling of pressure changes within the reservoir. This method is harder to apply to unconventional gas resources, as assessments are complicated by the heterogeneity of the rock formations, their extremely low permeability, and uncertainty as to the volume of the reservoir that can be connected to a production well. Accurate unconventional resource quantification requires geological modeling and study of the production behavior of several wells or assessment by analogy to other, known, resources. As unconventional gas begins to play an increasingly important role in worldwide supply, more accurate assessments of recoverable resources become more important.

3.4.2 Oil Supply

In the future the fossil oil supply will rely on a large diversification of the sources (Fig. 3.13).

3.4.2.1 Conventional Resources

More than half the conventional resources are in the Middle East and North Africa, and amount to about 2.1 trillion barrels (based on the latest USGS assessments). The cost of exploiting these resources is expected to be much lower on average than the cost of all other costs of supply. The cost of exploiting conventional resources typically ranges from less than $10 to $40 per barrel. In term of proven reserves the conventional resources represent 1.4 trillion barrels (Fig. 3.13). By comparison with the 1.2 trillion barrels already produced, the era of cheap oil is on the ending edge.

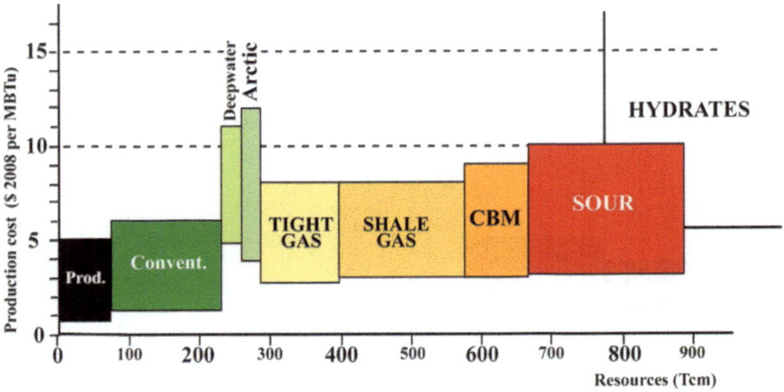

Fig. 3.14 Long-term gas supply cost curve (*Source*: IEA 2009, slightly modified). Gas will come from conventional production and increasing nonconventional sources

Additional resources include:

- Enhanced oil recovery (EOR) is estimated at between 400 and 500 billion barrels assuming the introduction of new EOR technologies is accelerated, lowering unit cost.
- Deepwater and ultra and ultradeepwater resources could deliver over 160 billion barrels and Arctic resources could amount to 90 billion barrels

3.4.2.2 Unconventional Resources

- Extra-heavy oil and oil sand resources total more than a trillion barrels and could be produced at a cost \$40–\$80 per barrel.
- Because of the lack of major commercial projects, the prospects for improving production of oil retorted from immature oil shale remains very uncertain. The IEA does not expect oil shale to make any significant contribution to world oil before 2030.
- In recent years, the Williston Basin (Canada/United States) has produced natural oil from mature shale (referred as “tight oil” in Sect. 3.3.5.3 and Fig. 3.2, and shaly oil in Fig. 3.13). This production, very similar to that of shale gas, could grow significantly in other basins and has a reasonable cost.
- Coal-to-liquid and (CTL) and gas-to-liquid (GTL) have large potential but will be in competition for their feedstock with other potential applications (Fig. 3.13).

3.4.3 Gas Supply

Over the long term fossil gas will be produced from a large variety of sources (Fig. 3.14).

Although unconventional gas currently accounts for only 4 % of the world total of proven gas reserves, it made up nearly 12 % of global production in 2008. The United States accounted for three quarters of global unconventional output, having expanded production nearly fourfold since 1990 to reach just less than 300 billion cubic meters (bcm), more than half of total US gas production. Canada was the next biggest producer with nearly 60 bcm, around one third of its total gas output. Exploitation of unconventional resources is gathering momentum elsewhere as the experience gained is transmitted to other regions. Tight gas, CBM, and shale gas resources have followed different routes from initial discovery to commercial exploitation, but the common factor has been the successful development and deployment of technologies that enable unconventional resources to be produced at costs similar to those of conventional gas.

The total long-term potential gas resource base from these sources (conventional and unconventional) is estimated at approximately 850 trillion cubic meters (tcm). Of this total, some 66 tcm have already been produced (or unfortunately flared and vented) at costs of up to \$5/MBtu. The most easily accessible part of the remaining conventional resources amounts to about 155 tcm, with typical production costs between \$0.50/MBtu and \$6/MBtu. Resources in the Arctic Circle could amount to 50 tcm and deepwater resources of 80 tcm could be produced at costs ranging from \$5/MBtu to \$11/MBtu.

Unconventional resources totalling 380 tcm (including 110 tcm tight gas, 180 tcm shale gas, and 90 tcm CBM) could be produced at costs between \$2.70/MBtu and \$9/MBtu. Sour gas resources, with high concentrations of hydrogen sulphide (H_2S) or carbon dioxide (CO_2), total some 220 tcm and could be produced at costs between \$3.10/MBtu and \$10/MBtu.

Gas is currently produced from all of these resource categories. Factors determining the order and intensity of future resource exploitation include regional availability and accessibility, emerging technologies (which could alter their relative costs of production), and market dynamics. In some of North America's more mature basins, exploitation of more than one type of resource is now occurring simultaneously through wells specifically designed for this purpose.

3.5 Conclusion

The petroleum system paradigm is sufficiently robust to describe the rationale of the occurrence of conventional petroleum and gas in sedimentary basins, and by extension to accommodate the geological settings of hydrocarbons referred to as "unconventional". The latter include coal, oil shale/"kerogen oil", heavy oils and bitumen, primary and secondary biogenic gas, gas hydrate, shale gas, coalbed methane, oil from source rock (tight oil/oil producing shale/shale oil), tight gas, aquifer gas, and so on.

Within this framework, expertise in geology, chemistry, physics, and mechanics, but also in microbiology and biochemistry are still mandatory to identify, quantify, and model the mechanisms explaining the occurrences of these unconventional resources and to devise technologies to produce them using environmentally friendly methods.

The challenge at stake is the size of the resources to unlock. These "new" hydrocarbons can greatly contribute to help in managing the energy transitional period.

However, environmental issues attached to the production and use of these resources should be faced and managed in a proper manner. Social acceptance is dependent on best practice deployment.

Political, environmental, regulatory, and fiscal factors will strongly influence the extent to which the potential of unconventional hydrocarbon is exploited. Production of unconventional resources leaves a greater environmental footprint than conventional ones and generates additional production of CO_2. In this respect CO_2 management programs (including capture and sequestration, CCS) need to be supported with tenacity. Emission-reduction incentives would have a major impact on the cost curve especially for unconventional oil. Technology developments will also continue to change the size and the timing of these productions.

Acknowledgments We thank I. Moretti, R. Eschard, and J. B. Saulnier for providing constructive remarks that greatly improved the initial manuscript.

References

Battani A, Prinzhofer A, Deville E, Ballentine CJ (2011) Trinidad mud volcanoes: the origin of the gas. In: Woods Lesli J (ed) Shale tectonics, vol 93, AAPG memoir. AAPG, Tulsa, pp 225–238. ISBN 9780-89181-375-0

Bernard BB, Brooks JM, Sackett WM (1977) A geochemical model for characterization of hydrocarbon gas sources in marine sediments. In: Offshore technology conference. May 1977, OTC 2934. OTC (Offshore Technical Conference), Houston, pp 435–438

Bonham LC (1978) Solubility of methane in water at elevated temperatures and pressures. AAPG Bull 62(12):2478–2481

Bustin RM, Bustin A, Ross D, Chalmers G, Murthy V, Laxmi C, Cui X (2009) Shale gas opportunities and challenges. Search and discovery articles #40382. http://w.searchanddiscovery.net/documents/2009/40382bustin/ndx_bustin.pdf

Clayton C (1991) Carbon isotope fractionation during natural gas generation from kerogen. Mar Pet Geol 8:232–240

Collett TS (2008) Geology of marine gas hydrates and their global distribution. In: Offshore technology conference paper # 19241, 5–8 May 2008 . OTC (Offshore Technical Conference), Houston

Collett TS, Dallimore SR (2000) Permafrost-associated gas hydrate. In: Max MD (ed) Natural gas hydrate in oceanic and permafrost environments. Kluwer, Boston, pp 43–60

Dyni JR (2005) Geology and resources of some world oil-shale deposits. USGS, scientific investigations report 2005–5294

Eschard R, Huc AY (2008) Habitat of biodegraded heavy oils: industrial implications. Oil Gas Sci Technol 63(5):587–607

Espitaliè J, Maxwell JR, Chenet PY, Marquis F (1988) Aspects of the hydrocarbon migration in the Mesozoic of the Paris basin as deduced from an organic geochemical survey. Org Geochem 13(1–3):467–481

Gerling JP (2005) Non-conventional hydrocarbons: what, where, how much. In: van Geuns LC, Groen LA (eds) Fossil fuels reserves and alternatives – a scientific approach. Extended abstracts of the lectures presented during the symposium fossil fuels: reserves and alternatives – a scientific approach, organized by the Earth and Climate Council of the Royal Netherlands Academy

of Arts and Sciences and Clingendael International Energy Programme, Amsterdam, 9 Dec 2004, pp 19–29
Holditch SA (2006) Gas tight sands. Society of Petroleum Engineers publication, paper n° SPE 103356, Distinguised author series, 86–93 June 2006, Society of Petroleum Engineers, Richardson
Huc AY (2011) Heavy crude oils, from geology to upgrading, an overview. Ed Technip, Paris, 442p
IEA (2009) World Energy Outlook 2009 International Energy Agency report. IEA, Paris, 578 p
International Energy Agency – IEA (2010) Key world energy statistics. http://www.iea.org/stats
Juch D, Gaschnitz R, Thiellemann T (2004) The influence of geological history on coal mine gas distribution in the Ruhr district; a challenge for future research and recovery. Geol Belg 7(3–4):191–199
Katz B (1995) Biogenic gas – its formation and economic significance. In: Proceedings Indonesian Petroleum Association 24th annual convention, vol 1. IPA, Jakarta, pp 461–474
Kvenvolden KA (1988) Methane hydrate-a major reservoir of carbon in the shallow geosphere? Chem Geol 71:41–51
Kvenvolden KA (1995) A review of the geochemistry of methane in natural gas hydrate. Org Geochem 23:997–1008
Lin CM, Gu LXG, Li Y, Zhao YY, Jiang WS (2004) Geology and formation mechanism of late quaternary shallow biogenic gas reservoirs in the Hangzhou Bay area, eastern China. AAPG Bull 88(5):613–625
Loucks RG, Reed RM, Ruppel SC, Jarvie DM (2009) Morphology, genesis and distribution of nanometer-scale pores in siliceous mudstones of the Mississippian Barnett shale. J Sediment Res 79:848–861
Milkov AV (2010a) Worldwide occurrences and significance of secondary microbial methane during petroleum biodegradation. In: AAPG annual convention and exhibition, New Orleans, 11–14 Apr 2010
Milkov AV (2010b) Methanogenic biodegradation of petroleum in the West Siberian Basin (Russia): significance for formation of giant Cenomanian gas pools. AAPG Bull 94(10):1485–1541
Moridis J, Collett TS, Boswell R, Kurihara M, Reagan MT, Sloan ED, Koh C (2009) Toward production from gas hydrates: current status, assessment of resources, and simulation-based evaluation of technology and potential. SPE J 12(5):745–771. doi:10.2118/114163-PA
Namikawa T, Nakamizu M, Ochiai K, Tsuji Y (2004) An appraisal project for offshore methane hydrate in Japan. In: AAPG international conference, Barcelona, 21–24 Sept 2003. Search and Discovery Article #10064 (2004). JNOC (Japan National Oil Company, Chiba). http://www.searchanddiscovery.com/documents/2004/namikawa/
Parkes RJ, Cragg BA, Wellbury P (2000) Recent studies on bacterial populations and processes in subsea floor sediments: a review. Hydrogeol J 8:11–28
Prinzhofer A, Battani A (2002) 8th Latin-American congress on organic geochemistry. In: Rangel A, Gonçalves FTT, Parra P, Hernández CS (eds) Tracing petroleum systems with noble gases: a new challenge for oil and gas geochemistry, Cartagena de Indias, 20–24 Oct 2002, pp 2002–2004
Prinzhofer A, Battani A (2003) Gas isotopes tracing: an important tool for hydrocarbons exploration. Oil Gas Sci Technol Rev IFP 58(2):299–311
Rice DD, Claypool GE (1981) Generation, accumulation, and resource potential of biogenic gas. Am Assoc Pet Geol Bull 65:5–25
Schurger SG, Newell KD, Blencoe JG, Carr TR (2006) Integrated subsurface carbon sequestration and enhanced coalbed natural gas recovery using cement Kiln emissions, Wilson County, Kansas. Poster presented at AAPG annual convention, Houston (Kansas Geological Survey KGS Open-file 2006-13)
Shurr GW, Ridgey JL (2002) Unconventional shallow biogenic gas systems. AAPG Bull 86(11):1939–1969
Sonnenberg SA, Pramudito A (2009) Petroleum geology of the giant Elm Coulee field, Williston Basin. AAPG Bull 93(9):1127–1153
Waseda A, Uchida T (2003) Gas origin and migration in marine and terrestrial gas hydrate-bearing sediments. In: Proceedings of ICGG7, 2003. Copernicus org., Göttingen, pp 42–43. © Copernicus GmbH 2003

Chapter 4
Urban Physical Infrastructure Adaptation to Climate Change

Nikolai Bobylev

Abstract The chapter discusses urban physical infrastructure (UPI) which is a complex set of systems and networks that provide vital services to a city. UPI is vulnerable to climate change (CC) impacts, and on the other hand can help cities to adapt to CC. The analysis focuses on CC adaptation options of physical components of infrastructure: sewer conduits, bridges, flood barriers, electricity poles, and the like. UPI characteristics and development trends (interconnectivity, interdependence, convergence, sustainability, efficiency, vulnerability, resilience, critical elements, evolution, and flexibility) have been considered in the CC adaptation context. The chapter concludes that extreme weather events represent a major threat to UPI worldwide in the short and long term. Key recommendations for UPI CC adaptation were identified as: (1) early consideration of UPI adaptation in spatial development plans; (2) allowing UPI flexibility and considering the infrastructure development trends during planning and design processes; (3) mainstreaming CC adaptation into relevant legislation, giving particular attention to revision of reliability coefficient values in building codes and norms; and (4) developing managerial measures in a wider context of UPI

N. Bobylev (✉)
School of Geography and the Environment,
St. Petersburg State University, St. Petersburg, Russia

School of Civil Engineering, University of Birmingham, UK

St. Petersburg Research Centre for Ecological Safety of the Russian Academy of Sciences, Russia

School of Innovation Science, St.Petersburg State Polytechnical University,
P. O. Box 45, 195267 St. Petersburg, Russia
e-mail: nikolaibobylev@gmail.com; n.bobylev@bham.ac.uk

J.B. Saulnier and M.D. Varella (eds.), *Global Change, Energy Issues and Regulation Policies*, Integrated Science & Technology Program 2, DOI 10.1007/978-94-007-6661-7_4,

operational safety and reliability. Analysis of projections for world expenditure on infrastructure development and the global growth of built-up areas revealed a huge gap between needs and actual provision of UPI, which should be bridged for successful UPI adaptation to CC.

Keywords Adaptation • City planning • Climate change • Critical infrastructure • Urban physical infrastructure

List of Acronyms

AR4	IPCC Assesment Report N° 4 (2007)
AVHRR	Advanced Very High Resolution Radiometer
CC	Climate change
G-cans	The G-Cans project (Shutoken Gaikaku Housui Ro or the Metropolitan Area Outer Underground Discharge Channel) is a massive underground waterway and water storage area built by the Japanese government to protect Tokyo from flooding during the monsoon seasons.
GDP	Gross domestic product
IPCC	Intergovernmental Panel on Climate Change
OECD	Organisation for Economic Cooperation and Development
SLR	Sea-level rise
SMART	Stormwater Management and Road Tunnel
UNISDR	United Nations International Strategy for Disaster Reduction
UK	United Kingdom
UNFCCC	United Nations Framework Convention on Climate Change
UPI	Urban physical infrastructure
WG II	IPCC AR4 Working Group (Impacts Adaptation and Vulnerability)
SPOT	Satellite Pour l'Observation de la Terre (Satellite for Earth Observation)

4.1 Introduction

The chapter discusses urban physical infrastructure (UPI), a complex set of systems and networks that provide vital services to a city. UPI is vulnerable to climate change (CC) impacts, and on the other hand can help cities to adapt to CC. The chapter aims to give readers an overview of UPI, giving particular attention to discussing its characteristics and development trends. A review of observed and projected impacts of CC on UPI follows.

Key research findings are presented as analysis of infrastructure development trends in the context of CC and recommendations for UPI adaptation. In conclusion an analysis of factual data on global infrastructure perspectives is given.

4.2 Urban Physical Infrastructure

4.2.1 Definition and Description

The notion of infrastructure can be explained as a collection of physical objects and operational arrangements that allow these objects to function as a system or network to provide services. Urban infrastructure is a broad notion and can be associated with provisioning such diverse services as healthcare, flood protection, and telecommunication among others. Infrastructure includes physical and institutional components. For example, flood protection infrastructure would include water barriers, drainage conduits, retainer walls, embankments, and other structures as *physical infrastructure*; and monitoring, emergency response, public information, and contingent plans as *institutional infrastructure* components. Institutional flood protection infrastructure can also be presented by laws, norms, and regulations that are implemented by designated departments at different tiers of government. Such institutional infrastructure can be presented at a federal or national level (emergency response, environmental regulation), regional (contingent plans), city (water supply/sewerage and transport planning), or municipal (shelter, drainage). The concept of *infrastructure* is led by associated function; that is, the same physical objects can pertain to different infrastructures. For instance, pumping stations can be a part of the sewerage, flood management, or electricity infrastructure.

Urban infrastructure is a vital component of a city; it provides a variety of human, environmental, and economic services. In this chapter we focus on urban physical infrastructure (UPI), which can include bridges, roads, conduits, pipelines, hydraulic structures, electricity, and telecommunication lines and equipment, as well as other engineered structures that provide services to the city.

Different types of infrastructure can be described by several common characteristics, which are explained in the next section.

4.2.2 Characteristics

To study UPI in the changing environment and its adaptation to climate change it is useful to look at how UPI can be characterized in relation to urban development and the interaction between different infrastructure types. We explain several UPI characteristics below and analyze CC impact on them in later sections.

4.2.2.1 Interconnectivity

In urban areas of high population and built stock densities different infrastructure systems are physically close to each other and have the potential to interact and affect each other. This interaction can be positive, as well as negative. Some examples of potential positive interaction (Kirshen et al. 2008) may concern economical use of reclaimed wastewater for industrial cooling; the use of waste heat from power plants for district heating; installation of dual quality water supply systems; and reducing storm runoff by using temporary water storage tanks and recharging sewer systems during low flow periods. The negative side of interconnectivity is associated with structural failures of infrastructure elements that can physically affect operations of other infrastructure (e.g., water pipe leaks can damage electric power cable collectors).

4.2.2.2 Interdependence

Interdependence means that the service provided by one infrastructure is used by another infrastructure, rather than by its end users. For example, an electricity line provides power for telecommunication equipment as well as to households, who are end users of electricity but not the electricity for a telecommunication line. The above example illustrates functional interdependence. Physical interdependence means physical connections of structural elements of different infrastructure: as an example, storm water sewers can be connected or adjacent to motor/rail transport tunnels. This physical interdependence increases vulnerability of both infrastructures to, for example, floods, accidents with chemical leakages, or geotechnical failures.

4.2.2.3 Convergence

Convergence means that several originally independent infrastructures or services converge into one physical infrastructure. Examples: (1) "triple play" offers, where the operator simultaneously provides via one cable telephone, internet, and access to broadcast services (OECD 2006); (2) urban underground collectors, that include sewer, electricity lines, and water pipes; (3) a storm water management and road tunnel in Kuala Lumpur, Malaysia that normally functions as a double-deck motorway, however, during flash floods the tunnel is closed to traffic and functions as a storm water collector (SMART Project 2006).

4.2.2.4 Sustainability

Sustainable UPI means UPI that aspires to impose minimal adverse impacts on the urban environment, including the social sphere, landscape aesthetics, and human health in particular.

4.2.2.5 Efficiency

Efficient UPI would deliver required services with minimal use of resources, including energy and occupation of physical space.

4.2.2.6 Vulnerability

Vulnerability is the condition determined by physical, social, economic, and environmental factors or processes, which increase the susceptibility of UPI to the impact of hazards (modified after UNISDR 2004). Considering climate change, it is pertinent to refer to the definition of vulnerability given by the Intergovernmental Panel on Climate Change (IPCC): vulnerability is the degree to which a system is susceptible to, and unable to cope with, adverse effects of CC, including climate variability and extremes. Vulnerability is a function of the character, magnitude, and rate of CC and the variation to which a system is exposed, its sensitivity, and its adaptive capacity (IPCC 2007).

4.2.2.7 Resilience

Resilience is the ability of infrastructure to withstand unusual impacts, that is, impacts that were not calculated (or calculated of a smaller magnitude) during the design process. Resilience of UPI can be increased by implementing technical and operations management measures.

4.2.2.8 Critical Elements

Many of the urban infrastructure elements can be considered as critical, meaning that a city, as a system, depends on uninterrupted provision of their services. For instance, artificial coastal defenses play a critical role in the cities of Tokyo, Shanghai, Hamburg, Rotterdam, New Orleans, and London (Nicholls et al. 2007). A significant flood defense barrier project is under development in the city of Venice.

Most recently construction of a flood protection scheme in Saint Petersburg (Russia) has been completed. This sizable project, with a dam length of more than 25 km, started back in 1979, was delayed for a variety of reasons, then resumed in 2002, and finally has been fully completed and put into operation in 2011. The scheme has undergone a successful test on December 26–27, 2011, when the flood gates were closed to separate the river Neva from the upstream wave coming from the Gulf of Finland, preventing an estimated 281-cm flood (the water level of 160 cm above the norm is considered to be a flood event in Saint Petersburg).

The G-Cans project (G-Cans 2006) in Tokyo is a remarkable example of critical infrastructure. The G-Cans is an underground infrastructure for prevention of flooding during rain and typhoon seasons. Extreme weather events that cause flooding are becoming more frequent in many regions; and we may see more development of projects similar to G-Cans of storm water infrastructure.

4.2.2.9 Evolution

UPI evolution means technological progress and increase in diversity and complexity of infrastructure as a whole. Interdependence and convergence characteristics are going to increase during continuing UPI evolution (DTI Foresight 2006).

4.3 Climate Change Observed Impacts and Projections

4.3.1 Global Climate Change Impacts

The IPCC names the following effects caused by climate change: greater frequency of heat waves; increased intensity of storms, floods, and droughts; rising sea levels; a more rapid spread of disease; and loss of biodiversity (IPCC 2007). In the context of impacts on UPI, extreme weather events and sea-level rise (SLR) pose the major threats.

Anthropogenic greenhouse gas and aerosol emissions are already affecting average climate conditions and climate extremes (Hegerl and Zwiers 2007). For example, 11 of the 12 warmest years globally occurred between 1995 and 2006 (Trenberth and Jones 2007).

Recent data show that SLR should be considered as gaining several meters in the next 100 years. The World Bank suggests that for precautionary planning purposes, SLR in the range of 1–3 m should be regarded as realistic (Dasgupta et al. 2007). As an example, Fig. 4.1 illustrates the relative vulnerability of coastal deltas as shown by the indicative population potentially displaced by current sea-level trends to 2050 (IPCC 2007).

Nicholls estimates (Nicholls 2004) that by 2080 up to 561 million people (under IPCC A2 scenario, involving a 0.28 m SLR) could be living in coastal flood plains (area below the 1 in 1,000-year flood level). Mc Granahan estimated (Mc Granahan et al. 2007) that over 600 million people (360 million of whom are urban dwellers) are

Fig. 4.1 Relative vulnerability of coastal deltas as shown by the indicative population potentially displaced by current sea-level trends to 2050 (Extreme≥1 million; High=1 million to 50,000; Medium=50,000–5,000; following Ericson et al. 2006; Nicholls et al. 2007, p. 327, IPCC AR4-WG II)

living in the less than 10 m elevated above the sea-level coastal zone. Comparison of these two findings suggests that a number of populations in coastal areas affected during extreme weather events such as flooding could be much higher than 600 million.

Currently almost two thirds of the world's cities with more than five million inhabitants fall in the 10 m elevated zone, at least partly (Mc Granahan et al. 2007). The directly affected SLR urban land areas constitute 4.68 % of the total world urban area in the case of the 5-m SLR scenario (Dasgupta et al. 2007). However, it is likely that over the years this figure will increase due to strong population growth in coastal zones, which is driven by urban sprawl, growing demand for waterfront properties, and coastal resorts development (IPCC 2007).

Coastal areas have quite high density of physical infrastructure. Gross domestic product (GDP) can be used as a proxy that reflects density of infrastructure: coastal areas generate about 6 % of the world's GDP on about 1 % of the area.

Satterthwaite distinguishes three ways in which climate change has the potential to increase flooding risks in cities: SLR and storm surges, heavier rainfall or rainfall that is more prolonged than in the past, and changes that increase river flows, including increased glacial melt (Satterthwaite 2008).

The rate of global warming in the next few decades is projected to be substantially faster than in the last few decades, due to the accumulation of greenhouse gases emitted in the past and the inertia of the climate system (Meehl and Stocker 2007).

4.3.2 *Observed Impacts on Urban Physical Infrastructure*

Observed impacts of CC on UPI can be considered in two categories: (1) extreme weather events that resulted in physical damage, and (2) prolonged unusual weather conditions that affected operations and/or resulted in physical damage.

The majority of damage to UPI has been done so far by impacts that fall into the first category. Presently it is impossible to establish a direct link between a concrete extreme weather event and the anthropogenic impact on the environment. Therefore we refrain here from further discussing examples in the first category of impacts.

Climate change impacts on UPI that fall into the second category can be directly associated with human activity with more certainty. This at least as concerns local climate and urban heat waves. Three examples of prolonged unusual weather conditions that affected UPI are given below.

4.3.2.1 Example 1

Impacts of the 2003 heat wave in Europe: In France, the cold storage systems of 25–30 % of all food-related establishments were found to be inadequate (Létard et al. 2004). Electricity demand increased with the high heat levels, but electricity production was undermined by the facts that the temperature of rivers rose, reducing the cooling efficiency of power plants (also nuclear) and that flows of rivers were diminished; six power plants were shut down completely (Létard et al. 2004).

The crisis illustrated how infrastructure can be unable to deal with complex, relatively sudden environmental challenges (Lagadec 2004).

4.3.2.2 Example 2

Impacts of an urban heat wave in London, United Kingdom: During prolonged heat waves, the difference in temperature between central London and the green and rural areas surrounding the city can be as much as 9 °C (Prasad et al. 2009). The reasons for a heat wave are (1) absorption of solar energy by buildings during the day and their radiation at night, and (2) creation of heat in the city (minor factor). An urban heat wave deteriorates air quality and increases the amount of electricity used for cooling in the summer months.

4.3.2.3 Example 3

Impacts on Russian Arctic towns: Structural failures in transportation and industrial infrastructure are becoming more common as a result of permafrost melting in northern Russia, the effects being more serious in the discontinuous permafrost zone (Arctic Climate Impact Assessment 2004)

4.3.3 Projected Impacts on Urban Physical Infrastructure

Industrial sectors, including infrastructure, are generally thought to be less vulnerable to the impacts of CC than other sectors, such as agriculture (IPCC 2007).

4.3.3.1 Case Studies

Published case studies of projected impacts of climate change on UPI include the New York Metropolitan Area (Bloomfield et al. 1999; Rosenzweig et al. 2000), the Boston Metropolitan Area (Kirshen et al. 2008), Los Angeles (Koteen et al. 2001), and a number of cities in the United Kingdom (Holman et al. 2005).

A case study of the Boston Metropolitan Area (Kirshen 2002; Kirshen et al. 2008) concluded that compared to conditions of just population growth, CC impacts are significant on many infrastructure systems. The systems analyzed included energy use, SLR, river flooding, surface vehicle transportation, water supply, and public health (heat–stress mortality). Localized case studies were carried out for water quality, tall buildings, and bridge scour. The authors of this study used dynamic modeling of the period 2000–2100 to analyze the performance of UPI in the Boston area (Kirshen et al. 2008). The authors concluded that many components of UPI are critical infrastructures; the impacts on each infrastructure system give rise to secondary impacts on other systems. These secondary impacts tend to be mutually reinforcing (negative impacts on one system create negative impacts on other systems): impacts measured for a single system in isolation will tend to be underestimated (Kirshen et al. 2008).

4.3.3.2 Floods

Impacts of floods on UPI structural elements are very dependent upon their physical strength, that is, whether they were designed to withstand increased surface water flow. Groundwater or near ground flows result in erosion (on the surface) and suffusion (in the ground) processes that lead to washing away small particles of soil and the rest of the soil body loses its strength which can result in uneven settlements and cracks in structures. These processes affect roads, bridge scour, sewers, and conduits. Concrete UPI elements are most susceptible; metal pipelines can better withstand erosion and suffusion processes.

4.3.3.3 Water Supply

Increased temperatures and changes in precipitation can contribute to increases in water demand, for drinking, for cooling systems, and for garden watering (Kirshen 2002).

4.3.3.4 Sewerage

In spite of the estimations that the effect of climate change on sanitation is likely to be less than on the water supply (Wilbanks et al. 2007), cities' sewerage systems are a great concern as CC advances. Sewage treatment works are exposed to damage during floods; SLR will affect the functioning of sea outfalls (Wilbanks et al. 2007).

Sewerage conduits operate best in the range of intake for which they were designed. Droughts and reduced intake lead to decreased waste and rainwater flows in servers, which lead to increased sedimentation in the servers and reduces their capacity. A high concentration of sediment can also lead to premature wear and tear of pumps. The process of reducing sewerage capacity due to less water intake has been observed in "shrinking" towns (e.g., in Germany), where the population moves out of the town for socioeconomic reasons. Less use of sewerage has led to the necessity to arrange more frequent sewer cleaning, to keep its normal operation.

Sewerage systems can be combined or separated. A separated sewerage system divides household sewers and rainwater ones. Rainwater sewers can be damaged by too high water velocities during flood. Household sewerage is not supposed to transport rainwater, thus it can remain unaffected. Cities with a combined sewerage system face higher flood vulnerability due to high functional damage if sewerage becomes temporarily unoperational. The worst case CC scenario for sewerage is a change towards more extreme seasonal precipitation, whereas long periods of drought would be followed by heavy rainfall. In this case, clogged by sediment, sewers will not be able to accommodate an overcapacity rainwater intake.

4.3.3.5 Structural Damages

Extreme weather events represent the major threat to UPI in the short term. SLR of some meters represents the major threat to UPI in coastal areas in the long term. Major structural failures of infrastructural elements exposed to rising surface and groundwater levels due to SLR are unlikely, except for inundation. However, UPI that would not be inundated, such as conduits, collectors, and sewers, can get minor damage. This damage will be caused by exposure of previously "dry" parts of underground structures to groundwater, and increased hydraulic pressure to lower parts of the structures; both of these phenomena will result in leakages. Leakages need to be eliminated in a timely manner to avoid serious damage. Thus, SLR will require increased spending on UPI underground structure maintenance.

4.3.3.6 Wind

High wind velocities are associated with a variety of extreme weather events that are likely to occur as climate change advances. A case study of a typical tall building in Boston found that if designed wind velocities increased by 30 % over those presently specified in the relevant building codes, large wind-induced sways potentially could cause human discomfort and costly architectural damage (Sanayei et al. 2003). Sways could also cause cracking and spilling of fire protection materials from the surface of steel structural elements leading to higher vulnerability to fires.

4.3.3.7 Rare Weather Events

There is a specific group of rare extreme weather events, the frequency of which might increase due to climate change. Details of projected frequencies, or indeed solid connection with CC has not been sufficiently explored yet, however, we think that at least two examples of these events are worth mentioning: wildfires and freezing rain.

4.3.3.8 Freezing Rain

Freezing rain is a weather phenomenon when a warm cyclone brings rain to cold and frosty lands. Freezing rain is rain that falls in liquid form but freezes upon impact to form a coating of glaze upon the ground and on exposed objects (Glossary of Meteorology 2000). The temperature of the high clouds in this case is higher than on the surface, which is unusual. If the surface temperature is below the freezing point, rainwater freezes immediately as it reaches objects on the ground. This phenomenon is particularly damaging to electric power lines. Metal wires and poles are most seriously affected: ice accretion occurs; icicles are formed and can cover

Fig. 4.2 Ice accretion on electric power line (*Source*: Ministry of Civil Defence, Emergencies and Disaster Relief of Russian Federation. Copyright: http://www.mchs.gov.ru/. Copyright «МЧС России», 2010. Все права на материалы, находящиеся на сайте, охраняются в соответствии с законодательством РФ. При полном или частичном использовании материалов ссылка на mchs.gov.ru обязательна [Copyright "Ministry of Civil Defence, Emergencies and Disaster Relief of Russian Federation", 2010. The rights to all the materials on this web site are protected in accordance with the Russian legislation. If using the materials, the reference to the mchs.gov.ru is required])

air-exposed metal parts completely (Fig. 4.2). Due to the accumulation of ice, wires and poles themselves become very heavy, and can break down eventually. This accumulation of ice sometimes completely destroys electric power lines for long distances, leaving cities and regions without an electricity supply. Railways, city trolleybus, and tramway lines can suffer as well (Fig. 4.3).

There have been recent reports of freezing rain phenomena occurring in Canada (Laflamme and Périard 1998; Roberts and Stewart 2008) and Russia. It has been reported that due to freezing rain on December 26, 2010 and subsequent days about 400,000 households were cut off from the electricity supply in the Moscow region of Russia. Freezing rain that occurred in the Ontario, Quebec, and New Brunswick provinces of Canada on January 5, 1998 resulted in the destruction of 130 power transmission towers and the fall of about 30,000 utility poles affecting the electricity supply to over four million people (Environment Canada 2010).

4.3.3.9 Wildfires

Wildfire is a natural phenomenon and can be prompted by human negligence. For example, in Siberia wildfires cover an area of at least one million hectares annually (United Nations Economic Commission for Europe 2002). Climate change creates favorable conditions for wildfires (Zhanga et al. 2003; Sukhinin et al. 2004; IPCC 2007).

Fig. 4.3 Ice accretion on electric contact wire of a railway line (*Source*: Ministry of Civil Defence, Emergencies and Disaster Relief of Russian Federation. http://www.mchs.gov.ru/. Copyright: http://www.mchs.gov.ru/. Copyright «МЧС России», 2010. Все права на материалы, находящиеся на сайте, охраняются в соответствии с законодательством РФ. При полном или частичном использовании материалов ссылка на mchs.gov.ru обязательна [Copyright "Ministry of Civil Defence, Emergencies and Disaster Relief of Russian Federation", 2010. The rights to all the materials on this web site are protected in accordance with the Russian legislation. If using the materials, the reference to the mchs.gov.ru is required])

Most recently, intensive and unusual wildfires occurred in the central areas of the Russian Federation and covered a total area of about six million hectares in the period of July 22–August 30, 2010 (Sukachev Institute of Forest 2010). Exceptionally hot and dry weather instigated wildfires in regions that have been traditionally considered as having a low wildfire risk (Bobylev 2010b). According to the Russian Federal Service for Hydrometeorology and Environmental Monitoring, in 2010 Russia has seen the longest unprecedented heat wave for at least 1,000 years. The city of Moscow reached the highest-ever recorded temperature of 38.2 °C on July 29, and there were 19 absolute day temperature maximums recorded during summer 2010. Wildfires directly damaged 199 human settlements (Ministry of Civil Defence 2012).

Moscow suffered from smog coming from wildfires. Continuous smog conditions resulted in increased demand for air-conditioning and electricity supply. Public facilities with large ventilation systems, including the Metro (underground public rail system), were unable to purify the air and remove smoke, which created an uncomfortable and potentially dangerous situation for human health. The Metro ventilation infrastructure serves large numbers of people confined in a limited space, and is designed with an assumption that intake air is clean. This case of a prolonged smog condition has raised some important questions: (1) is smog accumulating in underground volumes and what are the particularities (depth, difference in pressure, smog chemical composition)? (2) Should ventilation systems serving underground have filters to remove smog, and particles of what size/which components ought to be removed?

Table 4.1 Climate change impacts on UPI

Climate-related impacts	Impacts on UPI (projected and possible, not an exhaustive list)
Freshwater flooding SLR and associated flooding Groundwater level and mode changing due to changes in surface water and precipitation	Inundation of structures; road erosion; bridge scour erosion; electric equipment damage; hydraulic structure damage due to increased water flow (surface parts) and erosion (foundations); structural damage of concrete parts due to changing stress–strain conditions (foundations)
Extreme temperatures (air)	Overuse of ventilation systems; damage to facades and roofs; damage to suspended bridges (possibly in combination with extreme wind)
Extreme temperatures (water)	Damage of intake pumps
Extreme winds	Variety of structural damages
Rare weather events (e.g., ice rain, forest fires)	Damage of surface electric power lines, ventilation systems
Saltwater intrusion into freshwater areas	Disabling water supply intakes; corrosion of structures
Change in flora and fauna (invasive species)	Reducing capacity of water supply intakes
Changing average temperatures	Structural damage of pipelines in permafrost regions; changing of energy providing UPI operation modes (average and peak demands)
Change in precipitation	Extensive change in surface and groundwater modes, may cause droughts or flooding (see impacts for those)
Droughts	Soil erosion, damage to structures' foundations, uneven settlements of structures

The Moscow 2010 wildfire smog case raises clear concerns about the preparedness of UPI ventilation systems. It is worth noting that actual fires took place hundreds of kilometers away from the city.

4.3.3.10 Summary Climate Change Impacts

Table 4.1 summarizes observed and projected impacts on UPI reported in the case studies and IPCC assessments; these impacts are explained and analyzed in the following paragraphs.

4.3.4 Favorable Impacts on Urban Physical Infrastructure

Climate change will bring not only adverse impacts on UPI, but beneficial ones as well. Among them might be more frugal energy consumption in regions with a cold climate where warming is projected. Less energy consumption, fewer operating costs, and less wear and tear can be among the beneficial impacts. SLR may provide new opportunities for innovative hydropower production solutions. However, the key issue in this discussion about benefits and disadvantages is the local specifics of CC. For instance, stronger winds associated with more extreme climate can provide additional opportunities for renewable wind energy production, however, sporadic strong winds can damage or destroy windmill equipment.

4.4 Urban Physical Infrastructure Adaptation

4.4.1 Adaptation Typology, Research, and Governance

4.4.1.1 Adaptation Typology

Numerous approaches to conceptualizing, defining, and classifying adaptation have been developed in recent decades. These approaches are summarized in Table 4.2 and discussed below. Disaster risk reduction and climate change adaptation communities are the two most notable research groups that have been developing adaptation studies. Extensive discussions highlight different approaches of these research communities to defining and conceptualizing vulnerability, risk, and adaptation (Renaud and Perez 2010). A widely used definition of adaptation (IPCC 2007; UNISDR 2009) explains the adjustment in natural or human systems in response to actual or expected climatic stimuli or their effects, which moderates harm or exploits beneficial opportunities.

Table 4.2 Summary of adaptation types in the context of climate change and examples for UPI

Adaptation typology suggestions with references	Adaptation types	Explanations
Adaptation (IPCC 2007)	Anticipatory	Takes place before significant impacts of climate change are observed
	Autonomous (spontaneous)	Does not constitute a conscious response to climatic stimuli, but rather is triggered by ecological changes in natural systems and by market or welfare changes in human systems
	Planned	Uses information about present and future CC to review the suitability of current practices
Adaptation (IPCC 2007; UNISDR 2009)	Reactive	Occurs as a response to CC effects, such as natural disasters
	Proactive	Measures that are taken before actual effects of CC are observed, magnitude of effects is based on predictions
Adaptation assessment approaches (Burton et al. 2005)	Hazards-based	Considers the incremental impacts of climate change
	Vulnerability-based	Assesses future climate change in the context of current climate risks

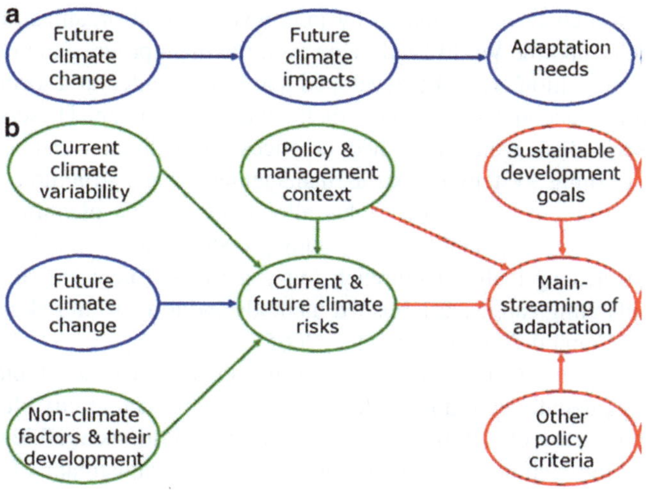

Fig. 4.4 Evolution of approaches for determining adaptation needs: (**a**) linear hazards-based approach; (**b**) complex integrative approach (Füssel 2007, Figure 3, p. 272; with kind permission of Springer Science)

UPI can be considered vulnerable to the CC system (as described in the previous section), hence it has to be targeted in adaptation activities (*adaptation* means actions targeted at the vulnerable system (McCarthy et al. 2001)). Adaptation also implies expanding the range of variability with which the system can cope (Wilbanks et al. 2007). Planning adaptation to CC mean actions undertaken to reduce the risks and capitalize on the opportunities associated with global climate change (Füssel 2007).

4.4.1.2 Research Needs

Key adaptation research needs (Yohe and Schlesinger 2002) include reviewing options, considering local conditions and uncertainty, and calculating the costs of adaptation. These needs can be further detailed as methodological studies, monitoring and indicator studies, empirical research, field and experimental research, predictive modeling, scenario development, economic costing, integrated assessment, quantification of the impacts of extreme weather events, modification of existing coping strategies, testing and evaluation of adaptation measures, and stakeholder participation (Carter 2007). A suggested graphical interpretation (Füssel 2007) of the contemporary CC research agenda is shown by Fig. 4.4.

4.4.1.3 Adaptation Governance

Institutions concerned with climate change adaptation include intergovernmental organizations, national governments, and local authorities. Currently the CC mitigation and adaptation debate is dominated by intergovernmental organizations, such as UNFCC, IPCC, and the World Bank. However, the main policy and practical steps towards CC adaptation are to be developed and implemented at a local (e.g., city) level, taking into account specific geography and local climate. These steps can be broadly described as well-established practices from disaster risk management (e.g., early-warning systems), coastal management (e.g., structural protection), resource management (e.g., water rights allocation), spatial planning (e.g., flood zone protection), urban planning (e.g., building codes), public health (e.g., disease surveillance), and agricultural outreach (e.g., seasonal forecasts) (Füssel 2007). Adaptation planning can use established methods and tools from risk management (Willows and Connell 2003; Burton et al. 2005).

UPI adaptation to CC needs to be locality specific, however, developing common standards (e.g., updating building codes) is equally important as city planning practice. Thus in the case of UPI the important role of institutions at all levels is accumulating specific technical studies and using them to mainstream CC adaptation into relevant legislation (e.g., codes and norms).

4.4.2 Urban Physical Infrastructure Development Trends

Characteristics of UPI that were defined and explained in the previous sections are worth considering in the adaptation context. Table 4.3 lists UPI characteristics and analyzes them taking into account climate change and adaptation types. Here we analyze what UPI development trends (based on the characteristics) may be pursued under different external factors (climate change, need to adapt to climate change).

As Table 4.3 specifies, there are clear conflicts in numerous UPI development trends with and without CC. It is suggested that this analysis be taken into account while planning UPI, and planning adaptation of urban areas to CC. Concrete recommendations for UPI adaptation to CC would depend on the location (main factors to be considered are geography and local climate) and industrial sector (i.e., specific UPI type). For instance, urban underground infrastructure adaptation to CC needs specific analysis (Bobylev 2009).

4.4.3 Recommendations for Urban Physical Infrastructure Adaptation

Here we try to avoid giving specific adaptation solutions, inasmuch as any recommendation of this type would require locality-specific extensive study. On the other hand, some general ideas are given regarding UPI CC adaptation options. Recommendations for UPI adaptation to climate change were systematized according to the UPI lifecycle stage (Table 4.4) and break down to technical and managerial measures (Table 4.5).

Table 4.3 UPI development trends under climate change adaptation

UPI characteristic	Hypothetical global development scenarios/options to assess UPI		
	No climate change (based on current trends)	Climate change without adaptation: "Do Nothing"	Adaptation to climate change
Interconnectivity	Increase	Decrease due to risks decoupling	Increase to lesser extent to allow technological progress and efficiency, but taking into account risks and vulnerabilities
Interdependence	Increase	Decrease greatly to withstand CC adverse impacts and continue to provide reliable services	Decrease to allow reliable performance and risk diversification
Convergence	Increase	Increase/decrease depending on particular sectors, governed by service delivery under CC effects	Increase to allow technological progress and energy efficiency
Sustainability	Increase due to technological progress and resource pressures	Greatly decrease due to a need to overcome adverse CC impacts and spend more resources on reliable performance	Increase/decrease depending on success and innovation in adaptation strategies
Efficiency	Increase due to technological progress and resource pressures	Decrease due to CC adverse impacts	Decrease due to need to spend resources on adaptation, however, can increase due to technological progress and mitigation aspirations
Vulnerability	Increase due to complexity, interconnectivity, and interdependence of networks/systems	Increase greatly due to multiple adverse synergetic factors, including CC	Increase slightly if adaptation measures successful
Resilience	Increase due to technological progress	No change	Increase slightly if adaptation measures successful
Critical elements	No change	Increase, because more elements become of critical importance due to CC stress	Depends on the adaptation measures
Evolution (technological progress, complexity)	Ongoing	Ongoing driven by emergency response	Ongoing driven by anticipatory adaptation

Table 4.4 Recommendations for UPI adaptation to CC: measures depending on life-cycle stage

UPI life-cycle stages	Issues to consider regarding adaptation to CC
City/regional planning	Spatial vulnerability to extreme weather events and SLR
Infrastructure systems/networks planning	Flexibility: there is a substantial uncertainty as to what concrete local effects CC would have, thus some room should be left for UPI alteration
	UPI development trends under CC adaptation (see Table 4.3)
Design solutions	Reliability coefficient enhancing
Construction	Safety standards in construction process
Operation	Disaster risk management
	Anticipatory assessment of vulnerability
Upgrading/modernization	UPI development trends (see Table 4.3)
	Technical measures to increase reliability
Utilization/demolition	Disaster risk management

Table 4.5 Recommendations for UPI adaptation to CC: technical and managerial measures

Adaptation measures type	Type of UPI	Explanations and examples
Technical (or structural)	Existing structures	Enhancing load-bearing capacity by structural elements, for example, stronger joints in waterproof gates
	New development	Construction of specific civil defense (disaster prevention) structures, for example, flood barriers
Managerial (Nonstructural)	Existing structures	Development and review of infrastructure operation practices, taking into account UPI development trends under CC adaptation (see Table 4.3)
		Disaster risk management plans
	New development	Review of design and construction norms and standards, including review of reliability coefficients
		City planning, disaster management preparedness, sustainable policies in a broad sense, for example, "green" floodplain management

Tables 4.4 and 4.5 contain several important cross-cutting issues, which are discussed below.

4.4.3.1 Spatial Planning

Spatial planning is an important component of planned adaptation (for a definition see Table 4.2, adaptation types). Adaptation by means of spatial planning brings many benefits in terms that potentially adverse climate change impacts are avoided in the first place, thus specific technical and disaster prevention measures would not be needed. Some case studies showed that it would also be the most "green" and cost-effective option (Kirshen et al. 2008). However, CC adaptation by spatial planning strategy can be implemented only for long-term adaptation, and requires strong policies in built environment regulation and land property rights, which could require change in, for example, national legislation. Examples of such extensive policy initiatives could be prohibition of development on flood plains or creating buffer (any development exclusion) zones around vulnerable critical infrastructures (e.g., rainwater discharge open canals). Spatial planning strategy could be difficult to realize because it requires significant policy shifts and legislative alterations, which would require bold commitments from institutional actors from various sectors of governance.

4.4.3.2 Flexibility and UPI Development Trends

There is substantial uncertainty as to what CC effects would be in specific locations, thus UPI flexibility is an important issue. By flexibility we mean the possibility of altering UPI design and operation modes in realistic time and cost frames to adjust to CC impacts. Considering UPI development trends specified in Table 4.3 could be helpful in planning and designing adaptable UPI.

4.4.3.3 Reliability Coefficients

Review of reliability coefficients specified in construction codes and norms is an important task of enhancing anticipatory adaptation. This activity is closely linked to adaptation research needs. Research on modeling and prediction of CC would inform a sectoral regulatory base (e.g., building codes) on CC-related uncertainties and projected variability of environmental conditions (e.g., SLR, temperatures) which would allow adjusting the reliability coefficients accordingly.

4.4.3.4 Managerial Measures

Managerial (nonstructural) measures are equally important in UPI climate change adaptation as physical upgrading of infrastructures. By managerial measures we mean actions that do not require any physical changes (e.g., strengthening a dam by geotechnical ground improvement), but rather relate to policies, operation guidelines, and community engagement. One example of such measures can be river flood management (Bobylev 2010a; Cörvers 2009). Services to human welfare can be provided by ecosystems as well as by artificial infrastructure (Bobylev 2010a). River flood regulation can be achieved using physical infrastructure. This infrastructure includes a variety of artificial structures: dams, dikes, canals, locks, pumping stations, and flood barriers. Alternatively, natural river retention areas can be expanded and used to accommodate floodwaters.

The Netherlands has an advanced physical infrastructure system for water resource management, however, the recent increase in flood occurrence and projected effects of CC prompted the Dutch government to develop a number of policies that would involve ecosystem restoration and nature development in an attempt to use ecosystem services to substitute failures or inefficiency of physical infrastructure (Bobylev 2010a). A Room for the River concept, introduced by the Dutch government in 2000 (Dutch Ministry of Transport Public Works and Water Management 2006) can illustrate approaches to managerial measures development to adapt UPI to climate change. A Room for the River concept combines technical (or structural) and managerial measures: it calls for more space for the rivers by excavating the river forelands, widening riverbeds, removing obstacles, creating retention areas, returning previously reclaimed land to the river system, creating side-channels, and repositioning dikes farther inland. Explicitly managerial measures include working with local communities to move their activities back from the floodplain (e.g., agriculture), as well as policies limiting construction and reconstruction in the floodplains.

Managerial measures also include disaster risk management, preparedness (identifying weak components of UPI), analysis of UPI vulnerabilities and interdependences, and development of emergency response plans.

4.4.4 Urban Physical Infrastructure as a Tool for Urban Adaptation

Being vulnerable to climate change, UPI in its turn can help cities to adapt to CC. Actually, UPI includes many civil defense and emergency response facilities.

Certain types of infrastructure can provide opportunities to mitigate specific hazards; for example, underground facilities can provide a variety of opportunities for CC mitigation and adaptation (Bobylev 2009; Sterling et al. 2012). The G-Cans project (see description in the section "Critical Elements" and G-Cans 2006) can be one example; another is flood control using underground rivers in the western part of Tokyo (Prasad et al. 2009). This UPI will be constructed based on a comprehensive flood control measure plan that includes measures for each river basin. In addition, the installation of adjustment reservoirs (water storage tanks) along rivers is planned (Prasad et al. 2009).

One can find opportunities for CC mitigation and adaptation by using a variety of compact urban structures, including tree screens, buildings, and artificial landscapes (Wende et al. 2010).

4.5 Outlook for Urban Physical Infrastructure Development

Previous sections of the chapter detailed how important UPI is, and the challenges that UPI faces to successfully adapt to climate change. This section discusses projections for UPI development, and explores how these projections correlate with the need for CC adaptation.

Table 4.6 Estimated average annual world infrastructure expenditure (additions and renewal) for selected sectors, 2000–2030, as a percentage of world GDP

Type of infrastructure	World GDP (approximate %) Years 2000–2010	World GDP (approximate %) Years 2020–2030	Difference in average annual infrastructure expenditure (as % of GDP)
Road	0.38	0.29	–23
Rail	0.09	0.06	–33
Electricity	0.22	0.24	+9
Water (only OECD countries, Russia, China, India, and Brazil)	1.01	1.03	+2

Data source: OECD (2006)

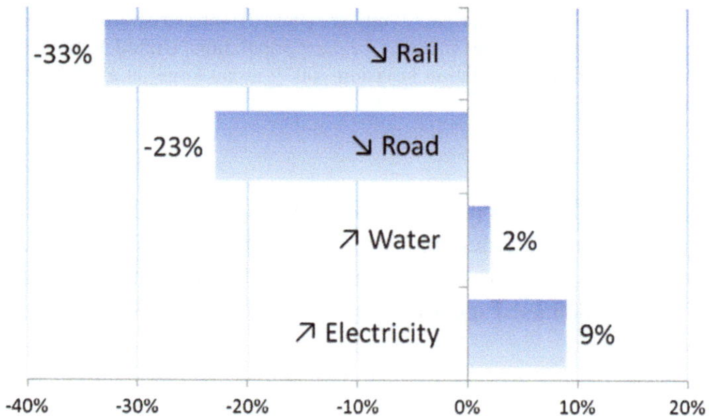

Fig. 4.5 Graphical interpretation of the differences in average annual infrastructure expenditures as percentage of GDP (Table 4.6) (*Data source*: OECD 2006)

Estimates about the UPI future face uncertainties associated with local effects of CC and technological progress. Indeed, past experiences suggest that technological change is highly unpredictable and can have far-reaching impacts on infrastructure (e.g., the impact of mobile telephony on fixed line infrastructure) (OECD 2006). Table 4.6 and Fig. 4.5 present data estimates on average annual world infrastructure expenditure which shows modest growth or even decline in some sectors. This contradicts global urban area and population projections presented in Fig. 4.6.

The estimated global increase in built-up areas in industrialized countries (from 300,000 km^2 in year 2000 to 700,000 by 2030), and in developing ones (from 250,000 km^2 in year 2000 to 820,000 by 2030) (Angel et al. 2005) show that urbanization in terms of global physical city area expansion (276 % by year 2030) will happen much quicker than cities' population growth (66 % by 2030) (United Nations Department of Economic and Social Affairs Population Division 2007). Both growth in built-up areas and population growth imply significant expansion

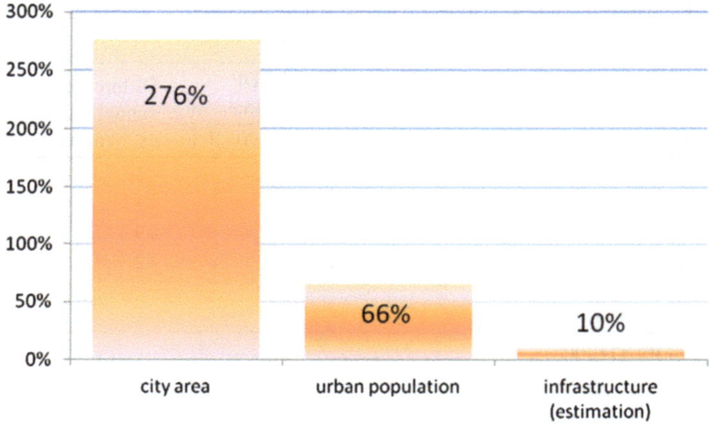

Fig. 4.6 Estimated (1) urban population, (2) city area, and (3) infrastructure global growth by year 2030, as percentage of 2000 year level (*Data sources*: population (United Nations Department of Economic and Social Affairs Population Division 2007); area (Angel et al. 2005); infrastructure (OECD 2006))

of UPI, which is not predicted according to the OECD. Thus, there is a clear discrepancy between future demand for UPI and its actual provision (as presented in Table 4.6 and Figs. 4.5 and 4.6). Furthermore, UPI adaptation to climate change is likely to require increased spending on existing UPI and its new development.

4.6 Conclusion

UPI is a complex set of systems and networks upon which cities depend in order to function, and UPI definitely requires serious attention in the face of climate change.

UPI characteristics, interconnectivity, interdependence, convergence, sustainability, efficiency, vulnerability, resilience, critical elements, evolution, and flexibility are helpful in the analysis of current and future UPI development trends, as well as in the analysis of how these trends may or should change to allow successful UPI adaptation to climate change.

There has already been strong evidence of climate change and its impacts on UPI, and more diverse impacts are projected. Among the adverse impacts are freshwater floods, SLR, wildfires, groundwater disturbance, extreme air and water temperatures, winds, ice rains, saltwater intrusions into freshwater areas, invasive species, and droughts. Extreme weather events represent a major threat to UPI worldwide in the short and long term, and SLR threatens UPI on up to 5 % of the world urban area in the long term. Climate change may have a positive impact on UPI associated with energy consumption, operational costs, and innovative hydropower solutions.

Key adaptation research needs were identified as tackling uncertainty in locally specific CC impacts, upon which the adaptation strategy in this location would depend.

Key recommendations for UPI climate change adaptation were identified as:

1. Early consideration of UPI adaptation in spatial development plans.
2. Allowing UPI flexibility and considering the infrastructure development trends during planning and design processes.
3. Mainstreaming climate change adaptation into relevant legislation; particular attention should be given to revision of reliability coefficient values in building codes and norms.
4. Developing managerial measures in a wider context of UPI operational safety and reliability

We summarized UPI climate change adaptation measures in accordance with UPI elements' life-cycle stage and distinguished between technical and managerial measures.

We concur with Kirshen et al. (2008) in our conclusion that land use planning may hold the greatest potential for adaptation policy; however, we give equal importance to consideration of the infrastructure development trends in planning, design, and operation processes.

UPI can be an important tool that would help to adapt urban areas to CC, particularly in terms of providing critical operational and emergency response facilities.

Analysis of the projections for world expenditure on infrastructure development and the global growth of built-up areas revealed a huge gap between needs and actual provision of UPI, which should be bridged for successful UPI adaptation to climate change.

Acknowledgments The author would like to acknowledge the contribution of (1) the Alexander von Humboldt Foundation and (2) the European Union Marie Curie Fellowship grant PIIF-GA-2010-273861, which both provided financial support for conducting a substantial part of this research.

References

Angel S, Sheppard SC, Civco DL (2005) The dynamics of global urban expansion. World Bank, Washington, DC

Arctic Climate Impact Assessment (2004) Impacts of a warming arctic: arctic climate impact assessment. Cambridge University Press, Cambridge, 1042 pp

Bloomfield J, Smith M, Thompson N (1999) Hot nights in the city: global warming, sea-level rise and the New York metropolitan region. Environmental Defence Fund, Washington, DC

Bobylev N (2009) Urban underground infrastructure and climate change: opportunities and threats. In: 5th urban research symposium, Marseille, 28–30 June 2009, 17p. http://www.urs2009.net/docs/papers/Bobylev.pdf

Bobylev N (2010a) Ecosystem services in the context of human development. In: Liotta PH et al (eds) Achieving environmental security: ecosystem services and human welfare. Ios Press, Amsterdam, pp 183–205. ISBN 978-1-60750-578-5

Bobylev N (2010b) Environmental change and migration: governmental compensation policies to natural disasters victims and urbanization process. A case study of wildfires in Russian

Federation in 2010. In: Proceedings of the ESF-ZiF-Bielefeld 2010 conference on environmental change and migration: from vulnerabilities to capabilities, Bad Salzuflen, 5–9 Dec 2010, 17p. http://www.esf.org/activities/esf-conferences/details/2010/confdetail328

Burton I, Malone E, Huq S, Lim B, Spanger-Siegfried E (2005) Adaptation policy frameworks for climate change: developing strategies, policies and measures. Cambridge University Press, Cambridge

Carter T (2007) Adaptation: local climate change impacts, adaptation and vulnerability. In: The future climatic window: local impacts of climate change. Leibnitz, Austria

Cörvers R (2009) Vulnerable to flooding? Nature development and 'Room for the River': a governance perspective. In: Fra Paleo U (ed) Building safer communities. Risk governance, spatial planning, and responses to natural hazards. IOS Press, Amsterdam

Dasgupta S, Laplante B, Meisner C, Wheeler D, Yan J (2007) The impact of sea level rise on developing countries: a comparative analysis. World Bank Policy Research Working Paper 4136, February 2007. World Bank, Washington, DC

DTI Forersight (2006) Intelligent infrastructure futures. Office of Science and Technology, United Kingdom, 23 Mar 2009. www.foresight.gov.uk

Dutch Ministry of Transport, Public Works and Water Management (2006) Spatial planning key decision room for the river. http://www.ruimtevoorderivier.nl

Environment Canada (2010) Statistics. Media releases. http://www.ec.gc.ca/default.asp?lang=En

Ericson JP, Vorosmarty CJ, Dingman SL, Ward LG, Meybeck M (2006) Effective sea-level rise and deltas: causes of change and human dimension implications. Glob Planet Chang 50:63–82

Federal Service for Hydrometeorology and Environmental Monitoring (2010) Meteorological. http://meteoinfo.ru/

Füssel H-M (2007) Adaptation planning for climate change: concepts, assessment approaches, and key lessons. Sustain Sci 2:265–275. doi:10.1007/s11625-007-0032-y, Springer

G-Cans Project (2006) Tokyo. Retrieved 1 June 2006. http://www.g-cans.jp/

Glossary of Meteorology (2000) Cambridge, MA. http://amsglossary.allenpress.com/glossary

Hegerl GC, Zwiers FW (2007) Chapter 9: Understanding and attributing climate change. In: IPCC (ed) Climate change 2007: the physical science basis. Cambridge University Press, Cambridge

Holman IP, Nichols R, Berry P, Harrison P, Audsley E, Shackley S, Rounsevell M (2005) A regional, multisectoral and integrated assessment of the impacts of climate and socio-economic change in the UK, part II, results. Clim Chang 71:43–73

IPCC (2007) In: Parry ML, Canziani OF, Palutikof JP, van der Linden PJ, Hanson CE (eds) Climate change 2007: impacts, adaptation and vulnerability. Contribution of working group II to the fourth assessment report of the intergovernmental panel on climate change. Cambridge University Press, Cambridge

Kirshen PH (2002) Potential impacts of global warming in eastern Massachusetts. J Water Resour Plan Manag 128:216–226

Kirshen P, Ruth M, Anderson W (2008) Interdependencies of urban climate change impacts and adaptation strategies: a case study of Metropolitan Boston USA. Clim Chang 86:105–122. doi:10.1007/s10584-007-9252-5

Koteen L, Bloomfield J, Eichler T, Tonne C, Young R, Poulshock H, Sosler A (2001) Hot prospects: the potential impacts of global warming on Los Angeles and the Southland. Environmental Defense Fund, Washington, DC

Laflamme J, Périard G (1998) The climate of freezing rain over the province of Québec in Canada: a preliminary analysis. Atmos Res 46(1–2):99–111

Lagadec P (2004) Understanding the French 2003 heat wave experience. J Conting Crisis Manag 12:160–169

Létard V, Flandre H, Lepeltier S (2004) La France et les Français face à la canicule: les leçons d'une crise (France and French people facing a heat wave : the lessons of a crisis). Report No. 195 (2003–2004) to the Sénat, Government of France, 391 pp

Mc Granahan G, Balk D, Anderson B (2007) The rising tide: assessing the risks of climate change and human settlements in low-elevation coastal zones. Environ Urban 19(1):17–37

McCarthy JJ, Canziani OF, Leary NA, Dokken DJ, White KS (eds) (2001) Climate change 2001: impacts, adaptation and vulnerability. Cambridge University Press, Cambridge

Meehl GA, Stocker TF (2007) Chapter 10: Global climate projections. In: IPCC (ed) Climate change 2007: the physical science basis. Cambridge University Press, Cambridge

Ministry of Civil Defence (2012) Emergencies, and disaster relief of Russian Federation. http://www.mchs.gov.ru/. Accessed 3 Jan 2012

Nicholls RJ (2004) Coastal flooding and wetland loss in the 21st century: changes under the SRES climate and socio-economic scenarios. Glob Environ Chang 14:69–86

Nicholls RJ, Wong PP, Burkett VR, Codignotto JO, Hay JE, McLean RF, Ragoonaden S, Woodroffe CD (2007) Coastal systems and low-lying areas. In: Parry ML, Canziani OF, Palutikof JP, van der Linden PJ, Hanson CE (eds) Climate change 2007: impacts, adaptation and vulnerability. Contribution of working group II to the fourth assessment report of the intergovernmental panel on climate change. Cambridge University Press, Cambridge, pp 315–356

OECD (2006) Infrastructure to 2030: telecom, land transport, water and electricity. OECD, Paris. ISBN 92-64-02398-4

Prasad N, Ranghieri F, Trohanis F, Kessler E, Sinha R (2009) Climate resilient cities 2009. The international bank for reconstruction and development and the World Bank

Renaud F, Perez R (2010) Climate change vulnerability and adaptation assessments. Sustain Sci 5:155–157. doi:10.1007/s11625-010-0114-0

Roberts E, Stewart R (2008) On the occurrence of freezing rain and ice pellets over the eastern Canadian Arctic. Atmos Res 89(1–2):93–109

Rosenzweig C, Solecki W, Paine C, Gornitz V, Hartig E, Jacob K, Major D, Kinney P, Hill D, Zimmerman R (2000) Climate change and a global city: an assessment of the Metropolitan East Coast Region. The U.S. national assessment of the potential consequences of climate variability and change. U.S. Global Change Research Program, USA

Sanayei M, Edgers L, Alonge J, Kirshen P (2003) Effects of increased wind loads on a tall building. Civ Eng Pract 18(2)

Satterthwaite D (2008) Climate change and urbanization: Effects and implications for urban governance. United Nations expert group meeting on population distribution, urbanization, internal migration and development, Switzerland

SMART Project (2006) Stormwater management and road tunnel. Retrieved 1 June 2006. http://www.smarttunnel.com.my/

Sterling R, Admiraal H, Bobylev N, Parker H, Godard JP, Vähäaho I, Rogers CDF, Shi X, Hanamura T (2012) Sustainability issues for underground space in urban areas. Proc ICE – Urban Des Plann 165(4):241–254 (14). doi:10.1680/udap.10.00020

Sukachev Institute of Forest of Siberian Branch of the Russian Academy of Sciences (2010) Maps catalogue. http://forest.akadem.ru/engindex.html

Sukhinin A, French N, Kasischke E, Hewson J, Soja A, Csiszar I, Hyer E, Loboda T, Conrad S, Romasko V, Pavlichenko E, Miskiv S, Slinkina O (2004) AVHRR-based mapping of fires in Russia: new products for fire management and carbon cycle studies. Remote Sens Environ 93(4):546–564

Trenberth KE, Jones PD (2007) Chapter 3: Observations: surface and atmospheric climate change. In: IPCC (ed) Climate change 2007: the physical science basis. Cambridge University Press, Cambridge

UNISDR (2004) Living with risk – a global review of disaster reduction initiatives. UN/ISDR, Geneva

UNISDR (2009) UNISDR terminology on disaster risk reduction. http://www.unisdr.org/eng/terminology/UNISDR-Terminology-English.pdf

United Nations (2007) Department of economic and social affairs. Population division. World urbanization prospects: the 2007 revision. http://www.un.org/esa/desa/

United Nations Economic Commission for Europe (2002) Forestry statistics. ECE/TIM/BULL/2002/4 LV (4). http://w3.unece.org/pxweb/

Wende W, Huelsmann W, Marty M, Penn-Bressel G, Bobylev N (2010) Climate protection and compact urban structures in spatial planning and local construction plans in Germany. Land Use Policy 27:864–868. doi:10.1016/j.landusepol.2009.11.005, Elsevier

Wilbanks TJ, Romero Lankao P, Bao M, Berkhout F, Cairncross S, Ceron J-P, Kapshe M, Muir-Wood R, Zapata-Marti R (2007) Industry, settlement and society. In: Parry ML, Canziani OF, Palutikof JP, van der Linden PJ, Hanson CE (eds) Climate change 2007: impacts, adaptation and vulnerability. Contribution of working group II to the fourth assessment report of the intergovernmental panel on climate change. Cambridge University Press, Cambridge, pp 357–390

Willows R, Connell R (2003) Climate adaptation: risk, uncertainty and decision-making. UKCIP technical report. United Kingdom Climate Impacts Programme, Oxford

Yohe G, Schlesinger M (2002) The economic geography of the impacts of climate change. J Econ Geogr 2(3):311–341

Zhanga Y, Woostera M, Tutubalinab O, Perrya G (2003) Monthly burned area and forest fire carbon emission estimates for the Russian Federation from SPOT VGT. Remote Sens Environ 87(1):1–15

Chapter 5
Sustainable Nuclear Energy Helps Europe to Meet Its Energy Challenges

Hamid Aït Abderrahim

Abstract The European Sustainable Nuclear Energy Technology Platform (SNETP) now gathers more than 100 organizations (research organizations, utilities, vendors, technology providers, technical safety organizations, universities, consulting companies, and nongovernmental organizations). Its first Strategic Research Agenda (SRA) was edited by a specific Task Group drawing on contributions from more than 160 scientists and engineers from more than 60 member organizations of SNETP and taking into account the feedback obtained from an open public consultation: the SRA provides the foundation for the establishment of joint research priorities that will enable European stakeholders, with the support of the European Commission, to transform a shared vision into reality, thus contributing to European energy policy and in particular, via the European Sustainable Nuclear Industrial Initiative (ESNII), to the objectives of the European Strategic Energy Technology Plan (SET Plan).

This chapter summarizes the contents of the agenda and presents the prospects for the need for hot labs and their application to the different generations of reactors. The implications of the Fukushima accident for SNETP is discussed and the imperative necessity of increased research, education, and training, to reinforce nuclear energy sustainability is also emphasized.

Keywords Nuclear energy • Fission • Reactor safety • Gen IV • ADS

List of Acronyms

ADS	Accelerator-driven systems
ASTRID	Advanced Sodium Technological Reactor for Industrial Demonstration
CCS	Carbon capture and storage

H. Aït Abderrahim (✉)
SCK·CEN, Boeretang 200, 2400 Mol, Belgium
e-mail: haitabde@sckcen.be

J.B. Saulnier and M.D. Varella (eds.), *Global Change, Energy Issues and Regulation Policies*, Integrated Science & Technology Program 2, DOI 10.1007/978-94-007-6661-7_5,

CEA	Commissariat à l'énergie atomique et aux énergies alternatives (Alternative Energies and Atomic Energy Commission).
dpa	Displacement per atom (irradiation exposure unit)
ENEF	European Nuclear Energy Forum
ENEN	European Nuclear Education Network
EPR	European pressurised water reactor
ESNII	European Sustainable Nuclear Industrial Initiative
ETKM	Education, training, and knowledge management
ETPP	Experimental Technological Pilot Plant
EU	European Union
FNR	Fast neutron reactor
FP6	Framework Program 6
Gen III	Third generation reactors
Gen IV	Fourth generation reactors
GFR	Gas fast reactor
GIF	Generation IV International Forum
JRC EU's	Joint Research Centre
LFR	Lead fast reactor
LWR	Light water reactor
MOX	Mixed oxide fuel (uranium and plutonium) involving spent fuel treatment
MFSR	Molten salt fast reactor
MYRRHA	Multi-purpose hybrid research reactor for high-tech applications
P&T	Partitioning and transmutation
PSA	Probabilistic safety assessment
PWR	Pressurized water reactors
SCK·CEN	StudieCentrum voor Kernenergie – Centre d'Etude de l'énergie Nucléaire – Belgian Nuclear Research Centre
SCWR	Supercritical water reactor
SET Plan	Strategic Energy Technology Plan
SFR	Sodium fast reactor
SNETP	Sustainable Nuclear Energy Technology Platform
SRA	Strategic Research Agenda
VHTR	Very High Temperature reactor
WETO	World Energy Technology and Climate Policy Outlook

5.1 Introduction

The Sustainable Nuclear Energy Technology Platform (SNETP 2007) was officially launched in September 2007, and at this event, the vision of the technology platform was presented. It highlighted the role nuclear energy plays in Europe's energy mix as the main provider of low carbon electricity (providing 31 % of EU's electricity and representing a nonemission of almost 900 million tonnes of CO_2 per year), and identified future research, development, and demonstration (RD&D) tracks that the nuclear fission sector must follow in order to address three objectives:

1. Maintain the safety and competitiveness of today's technologies.
2. Develop a new generation of more sustainable reactor technologies, the so-called Generation IV (Gen IV) fast neutron reactors with closed fuel cycles.
3. Develop new applications of nuclear power such as the industrial-scale production of hydrogen, desalination, or other heat applications in industrial processes.

The SNETP aims to support fully through RD&D programs the role of nuclear energy in Europe's energy mix, its contributions to the security and competitiveness of the energy supply, as well as to the reduction of greenhouse gas emissions. To achieve this objective, SNETP has elaborated a Strategic Research Agenda (SRA 2009), that identifies and prioritizes research topics which are presented here.

We first emphasize the role of nuclear fission in Europe's low carbon energy policy, then present the 2020 objectives (maintain the competitiveness of nuclear energy with long term waste management solutions), and the 2050 vision (Gen IV fast neutron reactors with closed fuel cycle for increased sustainability). The case of high-temperature heat processes and developing other applications of nuclear energy are discussed, and we identify the needs in terms of research infrastructures and competences, to ensure the success of all these new developments. Before concluding, we also examine the implications of the Fukushima accident for SNETP.

5.2 The Role of Nuclear Fission in Europe's Low Carbon Energy Policy

In January 2007 the European Commission published a seminal communication, EPE (2007), that underlined for the first time the benefits of nuclear energy: low carbon emissions, competitiveness, and stable prices. In the context of an anticipated increase in the use of nuclear energy in the world, the Commission also recognised that "There are therefore economic benefits in maintaining and developing the technological lead of the EU in this field." This communication was endorsed by the Council in March 2007, and also committed the European Union to meet ambitious objectives by 2020 of a 20 % reduction in greenhouse gas emissions (compared to 1990), 20 % renewable energies in the energy mix, and 20 % reduction in energy consumption through better energy savings and management.

In order to achieve these goals and realize the longer-term vision of a low carbon society by 2050, the Commission identified RD&D prospects of key low carbon energy technologies in a follow-up communication, SET Plan (2007), published in November 2007. "Europe needs to act now, together, to deliver sustainable, secure and competitive energy. The inter-related challenges of climate change, security of energy

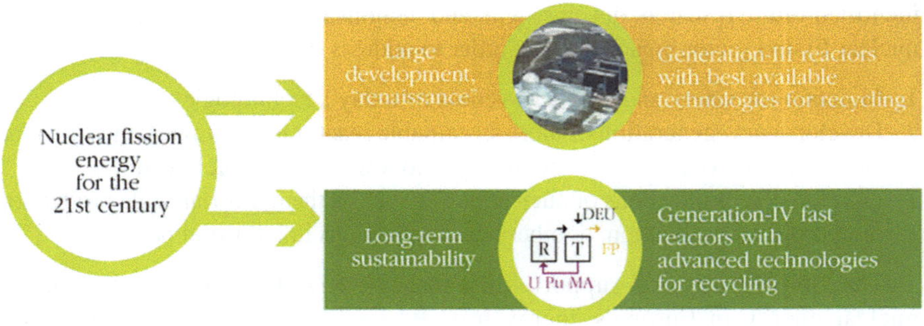

Fig. 5.1 Two main roads for nuclear of future: improvement of the Gen III (Sect. 5.3) and development of the new Gen IV technologies (Sect. 5.4)

supply and competitiveness are multifaceted and require a coordinated response … [We] need a dedicated policy to accelerate the development and deployment of cost-effective low carbon technologies."

Nuclear fission is cited together with other low carbon technologies, such as renewables and carbon capture and storage (CCS) technology, as one of the contributors to meet the 2020 challenges. By maintaining "competitiveness in fission technologies, together with long- term waste management solutions," fission energy will continue to lead low carbon energy technology in Europe. Projections published in the WETO (World Energy Technology and Climate Policy Outlook) report (WETO 2003) indicate that by 2030, nuclear energy will continue to produce more than half of the electricity produced by nonfossil fuel-based technologies.

Beyond the 2020 objectives, the SET Plan also identifies fission energy as a contributor to the 2050 objectives of a low carbon energy mix, relying on a new generation of reactors and associated fuel cycles. This objective is to be achieved by acting now to "Complete the preparations for the demonstration of a new generation (Gen IV) of fission reactors for increased sustainability."

From 2040 onwards, it is envisaged that this new generation of fast neutron reactors will be operating in parallel to the advanced Generation III (Gen III) light water reactors (LWRs) now being built in Europe and China, thereby maintaining the current one third share of nuclear electricity in Europe (Fig. 5.1).

5.3 The 2020 Objectives: Maintain the Competitiveness of Nuclear Energy with Long-Term Waste Management Solutions

Maintaining competitiveness should necessarily assure safe, secure, and economical operation of existing and future light water reactors but at the same time, the problems of waste minimization and resource optimization should also be considered as high priorities.

5.3.1 *How to Assure Safe, Secure, and Economical Operation of Existing and Future Light Water Reactors (LWRs)?*

Given the present share of low carbon electricity produced by nuclear reactors, it is essential that the European energy policy support the long-term operation of current plants, among which, about 75 % are pressurized water reactors (PWR). To achieve this objective, priority actions must be undertaken:

- Enhance knowledge to understand, prevent, and mitigate the effects of ageing.
- Harmonize long-term operation justification methodologies at the European level.

In addition to the operation of existing plants, it is essential to facilitate the construction of new Gen III light water reactors, among which are the European pressurized water reactors (EPR). Design certification should be harmonized so that requirements necessary for licensing should be the same throughout Europe aiming at European harmonized plant design and justification methodology.

Gen III reactors will contribute significantly to Europe's low carbon electricity production. Future units shall benefit from experience feedback from the first ones and from integration of RD&D results addressing:

- Improvement of system, structure, and component design
- Upgraded human–system interface, simplified reactor systems
- Advanced fuel and power performance

The impact of external issues, such as industrial obsolescence, impact of the environment on power generation, or evolution of regulatory requirements are also taken into account.

5.3.2 *How to Develop Advanced Fuel Cycles for Waste Minimization and Resource Optimization*

Nuclear waste is often perceived by the general public as a problem without a solution. However, the technical feasibility and safety of geological disposal sites are now undeniable, and within a decade the first geological repositories for conditioned high-level nuclear waste are expected to be in operation in the European Union.

However, to increase the sustainability of nuclear energy, more efforts should be dedicated to the development of advanced fuel cycles. This will further improve the competitiveness of nuclear energy, for instance, through use of more efficient cores and fuels for an optimal exploitation of the energy content of uranium fuel (improve uranium and plutonium usage in LWRs).

To minimize the high-level long-lived waste, research on partitioning and transmutation (P&T) must be continued, with the view to separate ("partition") from the spent fuel the *trans*-uranic elements (plutonium and minor actinides) that are responsible for the highest heat loads and radiotoxicity inventory in the long term. The next step is to burn or "transmute" these minor actinides, something that can only be envisaged in fast neutron spectrum systems. We need for this to continue the research on partitioning

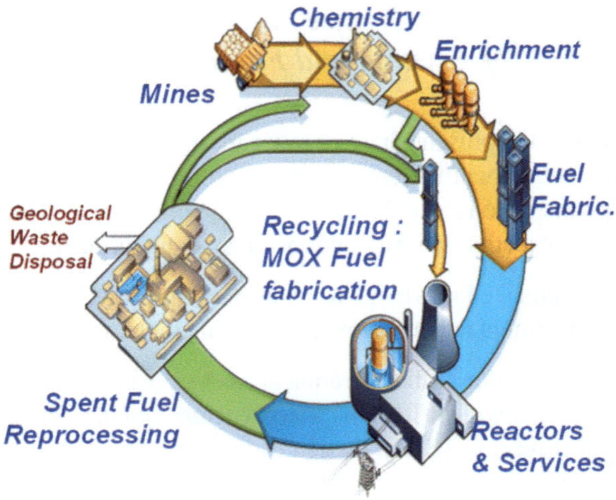

Fig. 5.2 The nuclear fuel cycle

technologies and fast neutron systems (reactors and accelerator-driven systems, ADS) well adapted to transmutation. The objective of this research is to assess the industrial feasibility of the minor actinide reprocessing and burning option (Fig. 5.2).

Generation IV International Forum (GIF 2001)
The GIF, founded in 2001, now has 13 members: Argentina, Brazil, Canada, China, France, Japan, Russia, South Korea, South Africa, Switzerland, United Kingdom, United States, and Euratom.

The goals adopted by the GIF provided the basis for identifying and selecting six nuclear energy systems for further development. The six selected systems employ a variety of reactor, energy conversion, and fuel cycle technologies. Their design may involve thermal or fast neutron spectra, closed or open fuel cycles, and a wide range of reactor sizes from very small to very large. They plan to improve economy and safety, but also the minimization of both fuel consumption and waste production, and finally to enhance the resistance to proliferation. Depending on their respective degrees of technical maturity, the Generation IV systems are expected to become available for commercial introduction in the period between 2020 and 2030 or beyond (CEA Clefs 2007; ADS 2011).

Fast neutron[1] reactors (FNR).
The Gas Fast Reactor (GFR) cooled with helium, has an outlet temperature of 850 °C and can deliver electricity, hydrogen, or process heat with high efficiency.

(continued)

[1] *Fast neutrons* are those neutrons generated by nuclear reaction, moving at a very high velocity (~20,000 km/s) corresponding to a kinetic energy of ~2 MeV.

(continued)

The sodium fast reactor (SFR) is sodium cooled and is designed for management of high-level wastes and, in particular, management of plutonium and other actinides.

The lead fast reactor (LFR) cooled with lead allows an optimal use of uranium and burning plutonium and minor actinides.

The molten salt fast reactor (MSFR) uses molten salt fluorides both as fluid fuel and coolant (favorable thermal-hydraulic properties, high boiling temperature, optical transparency, online separation of poisoning fission products)

Thermal neutrons[2] reactors.
The very-high temperature reactor (VHTR) is cooled with helium and operates at a high temperature (outlet up to 1,000 °C). It is a high-efficiency system, and can supply electricity and heat to a broad spectrum of high-temperature and energy-intensive processes.

The supercritical water reactor (SCWR) cooled with supercritical water has a thermal efficiency about one third higher than current light water reactors (outlet up to 550 °C).

5.4 The 2050 Vision: Gen IV Fast Neutron Reactors with Closed Fuel Cycle for Increased Sustainability

To address the issue of sustainability of nuclear energy, in particular, the use of natural resources, fast neutron reactors (FNRs) must be developed, inasmuch as they can typically multiply by over a factor of 50 the energy production from a given amount of uranium fuel compared to current reactors. FNRs, just as today's fleet, will be primarily dedicated to the generation of low carbon base-load electricity. Demand for electricity is likely to increase significantly in the future, as current fossil fuel uses are being substituted by processes using electricity. For example, the transport sector is likely to rely increasingly on electricity, whether in the form of fully electric or hybrid vehicles, either using battery power or synthetic hydrocarbon fuels. Here, nuclear power can also contribute, via generation of either electricity or process heat for the production of hydrogen or other fuels.

FNRs have been operated in the past (especially in Europe), but today's safety, operational, and competitiveness standards require the design of a new generation of fast reactors. Important R&D is currently being coordinated at the international level through initiatives such as GIF. Europe, through SNETP, has defined its own strategy and priorities for FNRs: The sodium-cooled fast reactor (SFR) as a proven concept, as

[2] *Thermal neutrons* are also called slow or thermalized neutrons (in equilibrium with the atoms of the matter) and move at a low velocity (#2–3 km/s) corresponding to a kinetic energy of ~1 eV.

well as the lead-cooled fast reactor (LFR) and the gas-cooled fast reactor (GFR) as alternative technologies. The French Commissariat à l'Energie Atomique (CEA) has chosen the development of the SFR and GFR technologies. Other countries including Italy, Belgium, Sweden, and Romania are focusing their R&D efforts on the LFR.

R&D topics for all three FNR designs include:

- Primary system design simplification
- Improved materials
- Innovative heat exchangers and power conversion systems
- Advanced instrumentation, in-service inspection systems
- Enhanced safety

Those for fuel cycle issues pertain to:

- Partitioning and transmutation
- Innovative fuels (including minor actinide-bearing) and core performance

Beyond the R&D, demonstration projects are planned in the frame of the SET Plan ESNII (European Sustainable Nuclear Industrial Initiative, ESNII 2010) for sustainable fission.

These demonstration projects include the SFR prototype ASTRID (Advanced Sodium Technological Reactor for Industrial Demonstration) whose construction is planned to be finished in France in 2020 and the construction of a demonstrator of an alternative technology—either LFR or GFR—to be decided around 2012. The MYRRHA project proposed in Belgium by SCK·CEN can play the role of an Experimental Technological Pilot Plant (ETPP) for the LFR technology. In addition, supporting research infrastructures, irradiation facilities, experimental loops, and fuel fabrication facilities, will need to be constructed.

Regarding transmutation purposes, the ADS technology must be compared to FNR technology from the point of view of feasibility. It is the objective of the MYRRHA (Multi-purpose hYbrid Research Reactor for High-tech Applications) project in Belgium to be an experimental demonstrator of ADS technology. From the economical point of view, the ADS industrial solution should be assessed in terms of its contribution to closing the fuel cycle.

ADS Technology

The concept of an accelerator-driven system combines a particle accelerator (protons) with a subcritical core reactor that produces fission without achieving criticality. Instead of a sustaining chain reaction, a subcritical reactor uses additional neutrons from an outside source that can be a particle accelerator producing neutrons by spallation: a high energy proton (~1 GeV[3]) impinges upon a

(continued)

[3] 1 eV is 1 electron volt. Its value is defined as the kinetic energy of an electron accelerated from rest through a potential difference of one volt. So 1 eV = 1.6 10^{-19} J, 1 keV = 10^3 eV, 1 Mev = 10^6 eV, 1 Gev = 10^9 eV.

(continued)

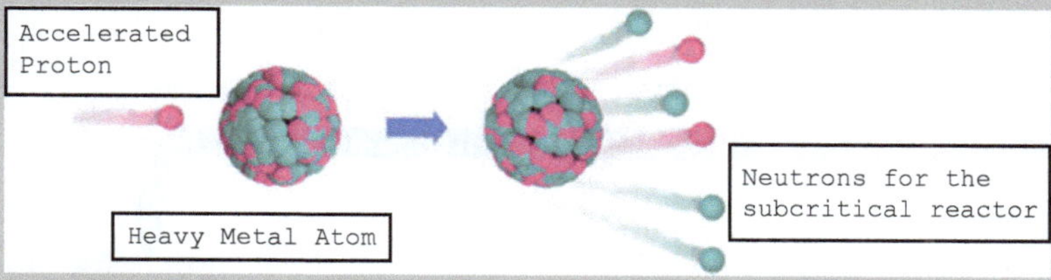

Fig. 5.3 Schematic of the spallation process

heavy metal atom and initiates the spallation process[4] during which neutrons and protons are set free (Fig. 5.3).

The spallation neutron flux can be used to cause a fission process (chain: neutron and energy emission/fission/neutron and energy emission, etc.) of thorium or uranium atoms mixed with minor actinides to be burned, but the chain reaction can be controlled by controlling the beam intensity of incident protons and there is no longer the risk of a runaway or criticality accident. Such a device with a reactor coupled to an accelerator is called an *accelerator-driven system* (ADS), or a hybrid system. It can be used, of course, as a generator to produce energy.

But the neutron flux can also be used for nuclear waste burning (leading to the reduction of the radioactive decay period of the high-level waste): it can thus help to destroy, or "burn," plutonium or waste even more troublesome than the actinides that are currently generated in the power generation reactors.

These properties make hybrid reactors particularly attractive and therefore SNETP in its SRA recommends foreseeing a demonstration program at a reasonable power scale (~100 MWth) that would allow realistic projection towards an industrial scale. Ambitious research programs have been undertaken to validate the principles. A first demonstrator, before a prototype on an industrial scale, could emerge by 2020: the MYRRHA project, supported by the European Community and developed in the laboratory of SCK·CEN at Mol, Belgium

MYRRHA: Multi-purpose hYbrid Research Reactor for High-tech Applications

MYRRHA is a Belgium project developed in Mol (SCK•CEN, StudieCentrum voor Kernenergie – Centre d'Etude de l'énergie Nucléaire), including a flexible fast spectrum research reactor (50–100 MWth) conceived as an accelerator-driven system (ADS) able to operate in subcritical and critical modes. It contains a proton accelerator of 600 MeV, a spallation target, and a multiplying core with MOX fuel, cooled by liquid lead–bismuth (Pb–Bi; Fig. 5.4).

(continued)

[4] *Spallation* is a process in which fragments of material (spall) are ejected from a body due to impact or stress. In the present case, a high-energy proton coming from an accelerator impinges on a heavy metal atom (the spallation target: lead, bismuth, etc.), producing in particular a flux of neutrons.

(continued)

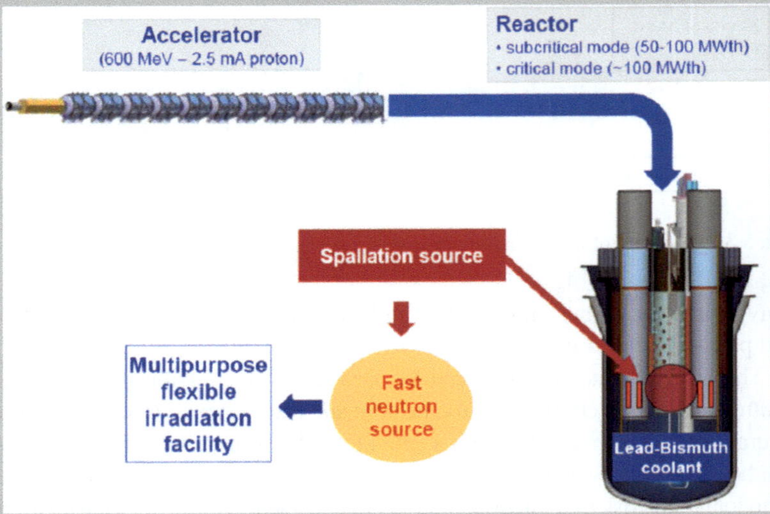

Fig. 5.4 The principle of MYRRHA (Aït Abderrahim 2011)

5.5 High-Temperature Heat Processes, Developing Other Applications of Nuclear Energy

Increasingly, fossil fuel-based industrial processes will be substituted by processes that use low carbon energy supplies. These processes typically require large and continuous amounts of energy in the form of heat, electricity, and hydrogen, all of which can be supplied by a nuclear reactor. Examples of such processes include the large-scale production of hydrogen for synthesizing fertilizers, for refining heavy crude oil, for optimizing the production of synthetic hydrocarbon fuels from coal or biomass, or for other industrial processes.

High-temperature gas-cooled reactors have long been identified as the most appropriate supplier of nuclear heat, and a first prototype of such a reactor coupled to the process heat application could be built around 2020. Other types of advanced reactors may also be suitable. The main R&D challenges lie with the technology of the coupling of the reactor to the industrial processes, and with the licensing issues:

- Technology developments: heat exchangers, heat transport systems, adaptation of industrial processes to specific aspects of nuclear heat supply
- Material and fuel improvement for very high temperature and qualification
- Tools and methodologies for licensing of nuclear reactor coupled to industrial process
- Management of waste (especially graphite)

5.6 A Need for Research Infrastructures and Competences

In order to carry successfully out the above R&D programs and demonstration projects, the nuclear sector must address the need to reinforce and further develop its competence pool, manage existing knowledge, and organize a network of research infrastructure.

5.6.1 Basic Research Needs for Cross-Cutting Topics

- *Material research*
  Material research is one of the most important topics for energy research, in particular for nuclear fission, where ageing, performance, and safety issues all need to be addressed. New materials as well as fabrication and welding processes need to be developed to achieve higher performance levels and longer lifetimes, as well as to withstand more extreme conditions such as higher temperatures (beyond 500 °C) and higher irradiation exposure (up to 100 dpa). Challenges remain in the area of multiscale modeling of material behavior under irradiation, which together with irradiation experiments will be the key techniques in the development of new materials.
- *Prenormative research*
  For the development of European codes and standards to be used for the future construction of Gen IV reactors, prenormative research must also be performed. R&D performed under quality assurance will contribute to this objective.
- *Modeling, simulation, and methods*
  The development of more advanced physical models and computational approaches benefiting from the increase of computational power allows for very detailed simulation of reactor behavior over a range of scenarios from normal to accident conditions and provide best estimate safety evaluations.

  A further area of application of best estimate methods with statistical analysis is the mechanical analysis of components. To exploit fully the potential of these tools, new basic data and specific separate and integral effect validation experiments using advanced measurement techniques will be required (Fig. 5.5).

5.6.2 Fuel Research

Basic research is needed to develop and improve modeling tools for innovative fuels (including minor actinide-bearing fuels) for Gen IV reactors. This research aims at establishing fuel properties and behavior under representative nominal operating, incidental, and accidental conditions, as well as addressing fabrication processes. Experimental programs aiming at qualifying the fuel must also be carried out.

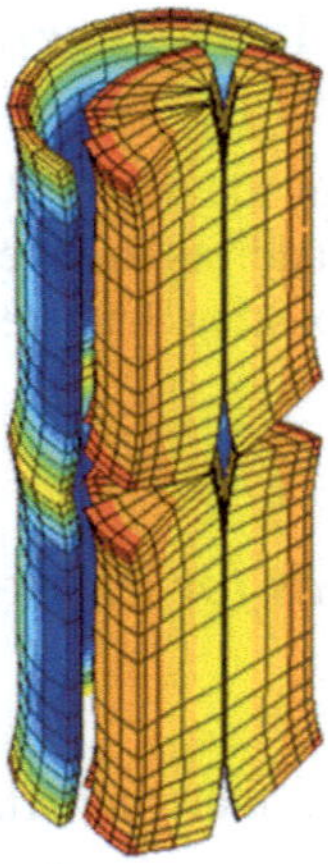

Fig. 5.5 Advanced modeling and simulation methods. This can be achieved by coupling neutronics, thermal-hydraulics, and fuel performance codes, at various physical and time scales. Particular efforts shall be directed at the development of CFD (computational fluid dynamics) methods for reactor design and safety analysis, and at the development of uncertainty and sensitivity analyses

5.6.3 Nuclear Safety

Nuclear reactor research in Europe has always had a strong focus on safety, and this will continue so as to ensure that European reactors continue to operate at the highest level of safety. In addition to further research to increase knowledge in the basic nuclear sciences, research on human and organizational factors and plant-relevant issues such as instrumentation and control (I&C) and electrical equipment, or external hazards, will be addressed. Research must also be carried out specifically to:

- Support long-term operation of nuclear power plants.
- Contribute to the design of intrinsically safe Gen IV FNRs.

5.6.4 Building a European Research Area of Nuclear R&D Infrastructures

Fission research has always relied on experimental programs for validating models, qualifying materials, and, more generally, for developing knowledge. Because the cost of maintaining research infrastructures is high, and following a more integrated approach to carrying out research programs, a network of complementary facilities must be established in support of the Strategic Research Agenda. Some facilities will need to be upgraded to support the R&D programs. New facilities will also need to be constructed to replace old ones.

Let us mention, among the new facilities:

- *Very large scale nuclear research infrastructures*
They provide irradiation capabilities that are essential for material and fuel development, and safety experiments. Three major facilities are being planned in Europe, the Jules Horowitz Material Testing Reactor (whose construction started in Cadarache, France, in March 2007), the MYRRHA fast spectrum irradiation facility (planned in Mol with a 40 % support from the Belgian government with a provisional date for start of construction in 2015), and the Pallas reactor that will replace the JRC (EU's Joint Research Centre) High Flux Reactor of Petten as Europe's leading provider of radioisotopes (RI) for medical applications and a back-up material test reactor. In addition to these facilities, the fast spectrum Gen IV demonstrators will also provide supplementary experimental irradiation and minor actinide transmutation capabilities.
- *Fuel cycle facilities*
Gen IV demonstration reactors and associated irradiation facilities also call for the construction of pilot manufacturing facilities for their driver (MOX fuel) and experimental fuels (minor actinides bearing fuels).

5.6.5 Education and Training

Education and training of young researchers and engineers is necessary to maintain existing knowledge and to carry out the research and development programs described above. SNE-TP has set up a specific Working Group dedicated to education, training, and knowledge management (ETKM 2010) issues with essential support in this area being provided by the European Nuclear Education Network (ENEN) Association, through its activities in FP6 (ENEN-II) and FP7 (ENEN-III) programs. This trained workforce will in part also provide qualified staff to Europe's nuclear industrial sector to accompany the development of the sector in the next decades, although this need will primarily be addressed by specific industry-led initiatives discussed in the European Nuclear Energy Forum (ENEF). More detailed information on ETKM activities can be found on the platform's website.

5.7 The Implications of the Fukushima Accident for SNETP

The accident that occurred at the Fukushima Dai-Ichi nuclear power plant on March 11, 2011, has raised public concern on nuclear energy and drawn new attention to the safety of nuclear power plants, in particular in the case of extremely severe external hazards.

A number of initiatives have been undertaken in many countries and at an international level in order to take into account the very first lessons learned from this accident for the improvement of nuclear reactor design and for the organization to manage an accidental situation. SNETP decided to empower a Task Group to investigate how the first lessons learned from the Fukushima accident could affect safety-related R&D orientations and priorities. The Task Group has concentrated on the developments, updating and validating of methods and tools for areas that are not considered well-enough understood. The report (IFA 2011) issued by the Task Group gives high-level orientations on the main challenges revealed by the accident, on the identification of relevant research areas, and finally provides a vision on post-Fukushima nuclear energy.

5.7.1 What Are the Main Challenges Revealed by the Fukushima Accident?

The Fukushima accident was triggered by the combination of two main initiating events:

- An exceptional magnitude earthquake which caused the sudden total loss of almost all the off-site power supply
- The associated tsunami caused flooding of the site under a wave of about twice the size considered previously in the evaluation of risk. It led to both the loss of all emergency power supply systems and of the cold sink.

The immediate challenge for the emergency response team was to recover cooling capabilities, in a situation where the off-site power supply has required about 11 days to be effective.

More generally, the challenges identified from the first lessons learned from the accident are the following.

- To extend even more in depth the safety approach to any type of initiating event, especially severe natural hazards and any combination of them. It shall be done to current reactors, Gen III reactors, and the development of GEN IV reactors.
- To include more systematically, at the design stage, the beyond-design basis accidents to ensure robustness of the defense in depth and to avoid cliff-edge effects. The approach shall include situations where all units on the same site are affected by a beyond-design event.
- To develop wider and more robust lines of defense with respect to the design basis aggressions and beyond-design basis events to define additional measures to consider in the design.

A specific emphasis has to be put on emergency management which has been very challenged during the accident due to:

- The concomitance of many events, the severe environmental conditions, and the mutual interaction between the affected units on site
- The complexity and the difficulty of the decision-making process which has altered the effectiveness and the promptness of the actions and has generated both confusion and delay.
- The practical impossibility to recover a suitable and stable electrical supply source during several days

The improvement of the emergency preparedness and response shall include the consideration of several items:

- The availability of more sophisticated tools to provide to the operators more reliable and quick indications/measurements on the reactor status to help in the implementation of an appropriate recovery strategy
- The availability of redundant intervention means in the vicinity of the site
- Better international cooperation/expertise which could provide help on the plant status diagnostic for the situation evolution and on the mitigation strategy

A careful investigation of the Fukushima accident outcomes will generate a new scale of priorities with a specific focus on extreme external events and their combinations, on common mode failure and human behavior, and with the assessment of their impact on the robustness of the defense in depth.

5.7.2 Identification of Relevant Research Areas

The Fukushima event especially reveals the importance of enhancing the analysis of human and organizational factors under high stress and harmful conditions in order to identify operational ways to improve emergency preparedness and the response to a severe nuclear accident.

Following a review of the available information on the Fukushima accident, nine main areas of research have been identified focused on siting, design, and operation of nuclear power plants:

- Systematic assessment of vulnerability in the defense in depth
- Advanced method for the assessment of external hazards
- Probabilistic safety assessment (PSA) application to external hazards
- Advanced method for the analysis of severe accidents
- Enhanced methods for accident management
- Improved modeling of fuel degradation in the spent fuel pool
- Radiological impact of serious reactor accident
- Advanced safety systems
- Advanced materials for nuclear reactors

Special attention shall be paid as to how the research outcomes will be implemented and so transferred into normal industrial practice.

5.7.3 A Vision of Post-Fukushima Nuclear Energy

Despite the Fukushima accident, nuclear energy remains an important component for today and for the future. But it is the prime responsibility of the nuclear energy stakeholders—and SNETP is an appropriate forum—to take benefit from all the lessons learned from the accident.

Research and development are essential tools for a better understanding of the phenomena and thus to enhance the prevention and the mitigation of severe accidents. No really new phenomena were revealed from the Fukushima accident and the basic orientations of the Strategic Research Agenda are still valid. However, the specific research areas, as identified above, shall be considered with the appropriate priority in the update of the SRA to be developed by the end of 2012. In particular, the issues related to extreme severe and rare accidents shall be considered in a more global approach to safety in order to better understand the design margins and the behavior of nuclear reactors under beyond-design scenarios.

With the perspective of the worldwide development of nuclear energy, the implementation of Generation III reactors should be accelerated and, with a longer perspective, the development of Generation IV reactors remains an important goal keeping a high safety level as an uppermost priority.

5.8 Conclusion

Let us start with a very simple exercise: the current global energy consumption is about 500 EJ[5] annually, which is equivalent to an instantaneous yearly average consumption of 16 TW. Projected population and economic growth will more than double this global energy consumption by the mid-twenty-first century, asking probably for a prospective new resource of more than 15 TW. This would require, for instance, the construction of a new 1 GW nuclear fission plant somewhere in the world every day for the next 40 years (BESW 2005). This is, of course, unrealistic, but it is just helpful to give an idea of our energy challenge! Yet, when faced with the problem of CO_2 mitigation, combined with the diminishing resources of fossil fuels, and with a slow development of renewable energy now, nuclear power cannot be swept aside with the back of a hand. To enlighten the reader, we have first emphasized the role of nuclear fission in Europe's low carbon energy policy, and its place in the future Europe's energy mix.

Nuclear electricity is already providing 31 % of the EU's electricity, representing a nonemission of 900 million tons CO_2 per year. Projections indicate that by 2030, nuclear energy will continue to produce more than half of the electricity produced by nonfossil fuel-based technologies. Beyond the 2020 objectives, the SET Plan also identifies fission energy as a contributor to the 2050 objectives of a low carbon energy mix, relying on a new generation of reactors and associated fuel cycles.

The 2020 objectives are oriented towards maintaining competitiveness that should necessarily assure safe, secure, and economic operation of existing and future light water reactors (Gen III). But at the same time, the problems of waste minimization and resource optimization will also be considered as high priorities.

---

[5] 1 EJ = 1 ExaJoule = 10^{18} J.

The 2050 vision considers the Gen IV fast neutron reactors with closed fuel cycle for increased sustainability. Fast neutron reactors must be developed, because they can typically multiply by over a factor of 50 the energy production from a given amount of uranium fuel compared to current reactors. This is particularly compatible with the transport demand for electricity which is likely to increase significantly in this future (fully electric or hybrid vehicles, either using battery power or synthetic hydrocarbon fuels) and nuclear power can also contribute, via the generation of electricity or of the accompanying heat, to the production of hydrogen or other fuels.

Despite the Fukushima accident, the nuclear energy remains an important component for today and for the future European energy mix and also a very significant contribution to fulfill the worldwide energy needs. But it is the prime responsibility of the nuclear energy stakeholders—and SNETP is an appropriate forum—to take benefit from all the lessons learned from the Fukushima accident.

In a more general sense, and it has been reinforced by the recent events in Fukushima, there is a permanent need for increased research, education, and training in nuclear R&D, with a strong cooperation between European and world experts.

Acknowledgments The author would like to thank the European nuclear energy research community that was represented by the colleagues listed in the SRA report of SNETP and those who authored the IFA SNETP report whose works served to synthesize this chapter.

References

ADS (2011) Réacteurs hybrides Les ADS: des réacteurs pilotés par un accélérateur. Retrieved 27 Dec 2011 from www.laradioactivite.com/fr/site/pages/reacteurshybrides.htm

Aït Abderrahim H (2011) Progress of ADS projects in the world. The MYRRHA project in Belgium. In: Proceedings of GLOBAL 2011, Makuhari, 11–16 Dec 2011

BESW (2005) Basic energy science workshop. Basic research needs for solar energy utilisation. Retrieved 27 Dec 2011 from science.energy.gov/~/media/bes/pdf/reports/files/seu_rpt.pdf

CEA Clefs (2007) Les enjeux d'une production durable d'énergie (Issues of a sustainable production of energy). Clefs CEA N° 55 Eté 2007

ENEF European Nuclear Energy Forum. http://ec.europa.eu/energy/nuclear/forum/enef_news_en.htm

ENEN European Nuclear Education Network. http://www.enen-assoc.org/

EPE (2007) An energy policy for Europe. Communication from the commission to the European council and the European parliament 2007. Retrieved 27 Dec 2011 from http://ec.europa.eu/energy/energy_policy/doc/01_energy_policy_for_europe_en.pdf

ESNII (2010) The European sustainable nuclear industrial initiative, a contribution to the EU low carbon energy policy: demonstration program for fast neutron reactors, concept paper SNETP. Retrieved 27 Dec 2011 from http://www.snetp.eu/www/snetp/images/stories/Docs-ESNI/ESNII-Folder-A4-oct.pdf

ETKM (2010) Education, Training and Knowledge Management. Nuclear education and training – key elements of a sustainable European strategy. http://www.snetp.eu/www/snetp/index.php?option=com_content&view=article&id=21&Itemid=19

GIF (2001) Generation IV International Forum. Retrieved 27 Dec 2011 from http://www.gen-4.org/

IFA (2011) Implications of the Fukushima accident for SNETP, November 2011. Retrieved 27 Dec 2011 from http://www.snetp.eu/www/snetp/images/stories/Docs-Newsflash/Implication_of_Fukushima_SNETP.pdf

SET Plan (2007) A European strategic technology plan (SET-Plan). Communication from the Commission to the Council, the European Parliament, The European Economic and Social committee and the Committee of the Regions. Retrieved 27 Dec 2011 from http://eur-lex.europa.eu/LexUriServ/LexUriServ.do?uri=COM:2007:0723:FIN:EN:PDF

SNETP (2007) The sustainable nuclear energy technology platform. Retrieved 20 Dec 2011 from www.snetp.eu

SRA (2009) Strategic research agenda SNETP. Retrieved 27 Dec 2011 from http://www.snetp.eu/www/snetp/images/stories/Docs-AboutSNETP/sra2009.pdf

WETO (2003) World energy technology and climate policy outlook 2030. European Commission Directorate-General for Research Energy. Retrieved 27 Dec 2011 from ec.europa.eu/research/energy/pdf/weto_final_report.pdf

Chapter 6
CO_2 Capture Transport and Storage, a Promising Technology for Limiting Climate Change

Christian Fouillac

Abstract On a world-scale basis, fossil fuels are likely to remain the main sources for electricity generation in the twenty-first century, and many industrial processes that are also large CO_2 emitters will still be active for many decades. Therefore, carbon capture and storage (CCS) is generally considered a necessary option for reducing CO_2 emissions to the atmosphere. This chapter introduces the concept, which consists in the separation of CO_2 from energy-related and industrial gas streams, and its transport to a geological storage location where it is permanently and safely stored. The characteristics of the main capture processes: postcombustion, oxycombustion, and precombustion are summarized in terms of energy consumption and costs. Some other possible technological options are briefly described. The methods utilized for CO_2 transport are also presented with some cost estimates. The main formation for geological storage—depleted oil and gas fields, deep saline aquifers, and nonexploitable coal seams—are briefly described, with the mechanisms involved in storage operations. Storage capacity evaluations, methodologies for risk assessment and management, are also briefly summarized. The chapter discusses the 14 large-scale integrated CCS projects which are in operation or under construction today, with a rough total storage capacity of 33 million tons a year. This could indicate that provided public awareness, social acceptance, and economic drivers evolve favorably, CCS could play a very significant role in the transition to a future low emission energy use.

Keywords CO_2 capture • Geological storage • Implementation costs • Risk assessment

C. Fouillac (✉)
Fouillac Consultant, 28 Rue du Change, 45190 Beaugency, France
e-mail: cfouillac.conseil@orange.fr

J.B. Saulnier and M.D. Varella (eds.), *Global Change, Energy Issues and Regulation Policies*, Integrated Science & Technology Program 2, DOI 10.1007/978-94-007-6661-7_6,

List of Acronyms

ASU	Air separation unit
CCS	Carbon capture and storage
CDM	Clean development mechanism
COP	Conference of the Parties
CSLF	Carbon Sequestration Leadership Forum
DOGF	Depleted oil and gas fields
ECBM	Enhanced coalbed methane
EGR	Enhanced gas recovery
EOR	Enhanced oil recovery
ETS	Emission trading scheme
EU	European Union
GHG	Greenhouse gas
IEA	International Energy Agency
IGCC	Integrated gasification combined cycle
IPCC	Intergovernmental Panel on Climate Change
LNG	Liquefied natural gas
MIT	Massachusetts Institute of Technology
MMV	Measure monitoring and verification
MPa	MegaPascal (1 MPa = 10^6 Pa)
Mt	Megaton (1 Mt = 10^6 ton)
MtCO2/y	Megaton of CO_2 per year
MW	MegaWatt (1 MW = 10^6 W)
MWth	Thermal megaWatt
NER300	EU program involving both CCS and renewable energy sources (RES) technologies
NGO	Nongovernmental organization
OECD	Organisation for Economic Co-operation and Development
Pa	Pascal (Pressure unit:1 atmosphere # 10^5 Pa)
SA	Deep saline aquifer
UNFCCC	United Nations Framework Convention on Climate Change
ZEP	Zero emission platform

6.1 Introduction

Due to world population evolution and economic growth in developing countries, the world has experienced a rapid increase of its global energy consumption. Energy forecasts predict that this trend will go on in the short to medium term, although various growth rates are considered according to different economic, technological, and societal scenarios. For several decades human societies have been searching for economical and clean new energy sources, and have promoted energy efficiency.

Table 6.1 Profile by process or industrial activity of worldwide large stationary CO_2 sources with emissions of more than 0.1 million tons of CO_2 per year

Process	Emissions ($MtCO_2/y$)
Power	10,539
Cement production	932
Refineries	798
Iron and steel industry	646
Petrochemical industry	379
Oil and gas processing	50
Other sources	33
Global	**13,377**

Other sources (buildings, transport, etc.) contribute to an emission of about 15,000 $MtCO_2/y$, but in a more dilute way. Data from IPCC (Metz et al. 2005)

In spite of these efforts, it is generally acknowledged that fossil fuels will remain as the primary energy source used throughout the twenty-first century. Fossil fuel resources are still abundant, especially for unconventional oil, natural gas, and coal (see Chap. 3), and can often be exploited at affordable prices. Therefore the elevation of atmospheric CO_2 concentrations resulting from human-caused emissions seems inevitable. Hence global temperature rise and ocean acidification could attain levels for which severe and irreversible effects could threaten our planet ecology (see Chaps. 1 and 2). Therefore, carbon capture and storage (CCS) may be generally considered as a promising technological option for reducing CO_2 emissions to the atmosphere. The major application for CCS is in reducing CO_2 emissions from power generation from fossil fuels, principally coal and gas. However, CCS can also be applied to CO_2-intensive industries such as cement, iron and steel, petrochemicals, oil and gas processing, and others (cf. Table 6.1) After capture, the CO_2 is transported to a suitable geological formation where it is injected, with the aim of isolating it from the atmosphere for good.

The use of CO_2 for commercial enhanced oil recovery (EOR) began in United States in the early 1970s but it is only in 1989 that the Carbon Capture and Sequestration Technologies Programs were launched at MIT (Massachusetts Institute of Technology). The International Energy Agency (IEA) Greenhouse Gas (GHG) R&D program which gathers government and company funding to study CCS technology was launched in 1989. A few years after, in 1991, the Norwegian government instituted a tax on offshore CO_2 emissions, and the European Commission financed the first research program on geological storage, "The underground disposal of carbon dioxide," from 1993 to 1995, which adhered to the promises of the concept. The real industrial birth of this technology dates from1996 when the Sleipner project (natural gas field in the North Sea, about 250 km west of Stavanger, Norway) began to store 1 Mt of CO_2 a year in a deep offshore aquifer after its separation from the production of a natural gas field production. In 2000 the injection of roughly 1.1 Mt of CO_2 per year began in the Weyburn oil field for EOR, the CO_2 being produced in a synfuel plant in North Dakota (United States), and then transported by gas pipe to the Weyburn field in Saskatchewan (Canada).

After this infancy period, several remarkable events took place in the 2000s. The year 2003 saw the first US Department of Energy budget for CCS research, and the inaugural meeting of the Carbon Sequestration Leadership Forum (CSLF), a large ministerial-level international initiative focused on the development of CCS. In 2005, the IPCC (Intergovernmental Panel on Climate Change) issued its extremely important Special Report on Carbon Dioxide Capture and Storage (Merz et al. 2005), the European Union (EU) commenced operations under the EU Emission Trading Scheme (ETS) for CO_2, and CCS was integrated into the Chinese national development plan. During the whole decade, the European Commission supported multimillion euros worth of public/private research programs under the succession of its fifth, sixth, and seventh framework programs. Several public and private initiatives were also taken in CCS research in many countries during the decade including, among others, the creation of the Global Carbon Capture Institute by the Australian government in 2008. Although its future role is still disputed by certain scientists, economists, and nongovernmental organizations (NGOs), CCS doubtless became a major element in the toolbox that can be used for the limitation of CO_2 emissions and is receiving more and more attention by governments and industries. For example the IEA in its "Blue map scenario" of 2009 considers that CCS could account for as much as 20 % of the total reduction of CO_2 emission in 2050.

As we show below, the question raised by the deployment of this new technology lies in the scientific, technological, and economic domains, but legal and regulatory aspects along with social acceptance will also play major roles in the road to deployment.

6.2 CO_2 Capture Processes

6.2.1 General Considerations

There are three main methods for capturing CO_2 in power plants: postcombustion capture, precombustion capture, and oxycombustion. For CO_2 emitted by industrial uses the situation mostly corresponds to postcombustion capture, but with CO_2 content varying from low content to almost pure CO_2 gas in the ammonia production processes or for hydrogen production from steam reforming of natural gas. The principles of the most common capture processes are briefly described in the following sections.

6.2.2 Postcombustion Capture

6.2.2.1 Conventional Processes

Postcombustion capture refers to the separation of CO_2 from low-pressure flue gas, after the combustion process is complete. The concentration of CO_2 in the flue gas varies between 4 % for natural gas combustion to 12–15 % with coal or heavy fractions of liquid hydrocarbon. A favored technique is to separate CO_2 from nitrogen in the flue gas with chemical solvents. The most common chemical solvents used for postcombustion CO_2 capture are alkanolamines. The CO_2 reacts with the solvent in an absorption vessel, then the CO_2-rich solvent is heated with steam in a stripping column to release the high-purity CO_2, and the CO_2-free, regenerated amine is recycled to the scrubber. The CO_2 released in the stripper is compressed for transport and storage.

This technology has been used for almost 60 years in the energy, refining, and chemical industries, and for sour natural gas purification.

In coal-fired power stations, a desulfurization stage, prior to CO_2 capture is often necessary to prevent the impurities in the flue gas from contaminating the CO_2 solvent. Postcombustion capture utilizes already well-proven technologies, and can be used on new or retrofitted power plants. The very large gas streams that must be handled in power plants, requiring large-scale equipment with high capital costs, and the amount of additional energy needed to operate the process constitute today the two major difficulties preventing a widespread use of the technology.

In the search to lower the energy penalty of the amine scrubbing process, alternative solvents that require lower energy for regeneration and at the same time present better absorption–desorption and corrosive properties are being developed. Currently chilled ammonia seems to be the most promising, but other options are being studied theoretically and in the laboratory or at small pilot scales.

6.2.2.2 Cryogenic Processes

CO_2 can be separated from other gases by cooling and condensation. Although cryogenic separation is now used commercially for purification of CO_2 from streams having high CO_2 concentrations (typically >90 %), it is not used for more dilute CO_2 streams because of high-energy requirements. In addition, components such as water must be removed before the gas stream is cooled to avoid freezing and blocking flow lines.

6.2.3 Oxycombustion

6.2.3.1 Conventional

Using pure oxygen instead of air for combustion, in a boiler or in a gas turbine, increases the concentration of CO_2 in the flue gas, and makes the separation step easier. To avoid the excessively high flame temperature associated with combustion in pure O_2, a part of the CO_2-rich flue gas is recycled to the combustion chamber. Oxycombustion for power generation has thus far been demonstrated on a medium scale (up to about 30 MWth) in the Schwarze Pumpe demonstration plant in Germany for brown coal combustion, and in the Lacq pilot plant in France on natural gas combustion. A large-scale integrated project is scheduled in Germany.

Presently, oxygen is produced by cryogenic air separation, in large air separation units (ASU), and is already used industrially on a large scale. A disadvantage of oxycombustion is that producing cryogenic oxygen is expensive in capital cost and energy intensive. To limit these disadvantages, other means of producing pure oxygen are actively being studied.

Gas separation membranes such as porous inorganics, nonporous metals (e.g., palladium), polymers, and zeolites can be used to separate oxygen from the air–gas mixture.[1] Many membranes cannot achieve the high degrees of separation needed in a single pass, so multiple stages and/or stream recycling are necessary. This leads to increased complexity, energy consumption, and costs. This concept has been subject to long-term tests in a commercial test facility. Development of a membrane capable of separating O_2 and N_2 in air could play an important indirect role in CO_2 capture. Lower O_2 production costs would be important in technologies involving coal gasification and in oxyfuel combustion. Much development and scale-up is required before membranes could be used on a large scale for capture of CO_2 in power stations.

6.2.3.2 Chemical Looping Combustion

One radical but attractive technology is chemical looping combustion, in which direct contact between the fuel and combustion air is avoided by using a metal oxide to transfer oxygen to the fuel in a two-stage process. In the first reactor, the fuel is oxidized by reacting with a solid metal oxide, producing a mixture of carbon dioxide (CO_2) and water vapor (H_2O). The reduced solid is then transported to a second reactor where it is reoxidized using air. Efficiencies comparable to those of other natural gas power generation options with CO_2 capture have been estimated. The major issue is development of materials able to withstand long-term chemical and thermal cycling.

6.2.4 Precombustion Capture

In precombustion capture the fuel is first partially reacted at high pressure with oxygen, air, or steam, to produce hydrogen (H_2) and carbon monoxide (CO). The CO is reacted with steam in a catalytic shift reactor and produces additional H_2 and CO_2. The CO_2 is then separated and H_2 is used as fuel in a combined cycle plant for electricity generation. Precombustion capture implies a great change to power station design, and large integrated plants have still to be demonstrated. However, most elements of the technology are already well proven in other industrial processes. It is hoped that pure, or nitrogen-diluted, H_2 can be burned in an existing gas turbine

[1]Air is a mixture of oxygen (O_2), Nitrogen (N_2), Argon (Ar).

with only little modification; this technology has to be demonstrated on a large scale, but turbine testing has already been carried out by manufacturers.

The conditions for CO_2 separation in precombustion capture processes are quite different from those in postcombustion capture. For example, the feed to the CO_2 capture unit in an IGCC (integrated gasification combined cycle)[2] process, located upstream of the gas turbine, would have a CO_2 concentration of about 35–40 % and a total pressure of 20 bar or more. Under these precombustion conditions, physical solvents that result in lower-regeneration energy consumption could be advantageous.

6.2.5 *Capture Costs, Energy Penalty, Impacts of Flue Gas Composition*

The capture stage is the most important in determining the overall cost of CCS. Cost reductions of solvent absorption systems, new separation systems, new ways of deploying existing separations, and new plant configurations to make capture easier and less costly can deliver incremental cost decreases. However, novel approaches, such as rethinking the power generation process, are needed if substantial reductions in the cost of capture are to be achieved.

The presence of fuel contaminants and specific combustion products impose additional constraints on the choice and operation of CO_2 capture and storage. With coal-fired systems, particulates can erode turbine blades in IGCC plants, contaminate solvents, foul heat exchangers in absorption processes, and foul membranes or sorbents in the new capture processes. Sulfur and nitrogen compounds must also be reduced to low levels before CO_2 capture because these impurities tend to react with amines to form heat-stable salts, and may interact with membrane materials or sorbents to reduce the separation or capture efficiency. In contrast, natural gas and its combustion products are much more benign and tend to create fewer problems for all potential CO_2 capture options. Current work on "ultraclean coal" products aims to address impurity and particulate issues so that coal–water mixtures can be used directly in reciprocating and turbine power generation systems.

In oxycombustion processes, incondensable residual argon can change the conditions of transport operations.

Depending on the process (pre-/postcombustion, oxycombustion), or of the fuel (natural gas, coal, etc.), the capture cost may be about 55/80 US$/ton CO_2. Realistic improvement could lead to a decrease of between 10 and 20 % (Finkelrath 2011).

---

[2] A gasification process can produce syngas from high-sulfur coal, heavy petroleum residues, and biomass. The plant is called *integrated* because its syngas is produced in a gasification unit in the plant which has been optimized for the plant's combined cycle (gas + steam). The syngas produced is used as fuel in a gas turbine that produces electrical power. To improve the overall process efficiency heat is recovered from both the gasification process and also the gas turbine exhaust in "waste heat boilers" producing steam. This steam is then used in steam turbines to produce additional electrical power.

6.3 Transport of CO_2

6.3.1 Transportation Needs

The potential CO_2 storage sites will not always match the emission points throughout the world. This can be observed within a given country but also among neighboring states. For example, in Europe some member states show significant levels of CO_2 emissions, and rather limited storage potential within their national boundaries. Hence the construction of domestic or transborder CO_2 transport infrastructures, onshore or in the maritime environment may become necessary (Morbee et al. 2010). On a state scale, Japan, South Korea, and to a lesser extent, India, may also have problems finding sufficient storage capacity in the neighborhood of their main emission locations. Therefore CO_2 transport will be an important component of the whole chain.

6.3.2 Transport by Pipelines

Pipelines have been used for several decades to transport CO_2 obtained from natural geological or industrial sources to oil fields for enhanced oil recovery (EOR) operations. More than 25 Mt/year of pressurized CO_2 are transmitted through pipelines in North America at distances of 5,600 km. The 330 km pipeline, which transports 1.8 Mt/year of CO_2 from a coal gasification plant in North Dakota (United States), to an EOR project in Weyburn, Saskatchewan (Canada), is the first demonstration of large-scale integrated CO_2 capture, transmission, and storage associated with EOR purposes. The composition of the gas carried by the pipeline is typically CO_2 96 %, with some impurities (H_2S 0.9 %, CH_4 0.7 %, other hydrocarbons 2.3 %, CO 0.1 %, N_2 less than 300 ppm, O_2 less than 50 ppm and H_2O less than 20 ppm). The delivery pressure at Weyburn is 15.2 MPa, and there are no intermediate compressor stations.

Since 2009, the natural gas produced in the Snøvhit gas field in the Barents Sea offshore Norway, is transported via a 145-km pipeline to a land treatment facility. After CO_2 separation from natural gas, a second 145-km pipeline transports the captured carbon dioxide back to the Snøhvit field where it is injected into the deep saline aquifer (see Sect. 6.4.1) of a porous sandstone formation. When operating at full capacity, 700,000 tons of CO_2 will be transported each year. During a 2-year demonstration project in Lacq, South of France, the transport by a 27-km pipeline of 120,000 tons of CO_2, after oxycombustion and capture, is being carried out for ultimate CO_2 storage in a depleted gas reservoir.

Some technological questions are raised for pipeline transport operations, due to chemical and physical properties of the transported gas. As shown above, CO_2 captured from power plant and industrial sources is not pure, and the amount and type of impurities depend on the combustion process and the capture technology used.

The effects of impurities on industrial transport of CO_2 is first to modify the critical pressure of the transported phase with respect to those of pure CO_2. This can imply a change in the operating regime of the pipeline. In order to maintain a dense single-phase fluid, supercritical or liquid, the transport may have to be operated at a higher pressure than would be required for pure CO_2.

The impurities present in the CO_2 stream could also affect the density and viscosity of the fluid. Some combinations of CO_2 with hydrogen, argon, or nitrogen can cause higher pressure and temperature drops for a given pipeline length than for pure CO_2. Sudden temperature drops can potentially cause steel embrittlement or hydrate formation, which could both damage the pipeline. These risks are controlled by shortening the distance between compressor stations along the pipeline, which in turn, increases the overall pipeline cost. Note that, in any case, compressor stations are not viable for subsea pipelines.

The solvent properties of supercritical CO_2 which are possibly detrimental to the elastomers commonly used for sealing purposes in oil and gas pipelines can prove to be inadequate for CO_2 transportation. Except for the high-purity CO_2 gas phase and/or transportation over short distances (e.g., the Lacq Project), all these considerations render very improbable the idea of natural gas pipeline reuse. Hence it is extremely likely to consider the implementation of designed-for-purpose CO_2 pipelines. Eventually, CO_2 pipeline grids, similar to those used for natural gas transmission, will be built as CCS becomes widely deployed.

6.3.3 Transport by Tankers

Tankers on the scale of 1,500 cubic meters (m^3) of CO_2 have been operating in the North Sea area for more than 10 years. Hence, the infrastructure design has been studied in detail. It comprises large-scale vessels to transport CO_2 from capture sites located nearby, and appropriate port facilities. Such schemes may occur in the future, particularly if high geological storage potential exists offshore at reasonable distances. The CO_2 would be transported in marine vessels such as those currently deployed for liquefied natural gas (LNG), although under less energy-consuming conditions. Compared to pipelines, ships offer increased flexibility, but as they function intermittently, a temporary storage facility must be included in the global infrastructure. However, they may be cheaper than off-shore transportation pipelines particularly for longer distances.

For example, the North Sea Basin has great potential for the permanent storage of CO_2 in subseabed geological formations. So in November 2005, the United Kingdom and Norway agreed to establish the North Sea Basin Task Force. In 2008, The Netherlands and Germany joined, raising its membership to four countries.

It is estimated (ZEP 2009) that the transport of 20 Mt/y of CO_2 over a distance of 500 km would cost, by ship, about US$ 15/ton of CO_2, whereas transporting the same amount of CO_2 the same distance would cost about US$ 8/ton (offshore pipe) or US$ 6/ton (onshore).

6.4 Geological Storage

6.4.1 General Consideration and Storage Potential

Most of the world's carbon is held in geological formations: locked in minerals, in hydrocarbons, or dissolved in water. Naturally occurring CO_2 is frequently found with petroleum accumulations, having been trapped either separately or together with hydrocarbons for millions of years.

Owing to their geological properties, several types of geological formations can be used to store CO_2. Deep saline-water (SA) contained in the porous sedimentary formations, depleted oil and gas fields (DOGF), and unmineable coalbeds have the greatest potential capacity for CO_2 storage. CO_2 can be injected and stored as a supercritical fluid in deep saline formations and depleted oil and gas fields. Supercritical conditions for CO_2 occur at 31.1 °C and 7.38 megaPascal (MPa). These conditions are encountered at approximately 800 m below surface level, for a geothermal gradient close to 30 °C/km. Supercritical CO_2 has properties of both a gas and a liquid, with a density up to about 700 kg/m^3; it is nearly 600 times more dense than at surface conditions, but still remains more buoyant than the brine filling the geological formations. Storing CO_2 under a supercritical state is very favorable. In fact, 1 ton of CO_2 occupies 509 m^3 at surface conditions of 0 °C and 1 bar, whereas the same amount of CO_2 occupies only 1.39 m^3 at 1,000 m subsurface conditions of 35 °C and 102 bar.

CO_2 can also be injected into coalbeds that are either too deep or too thin to be economically exploited, where it is stored by adsorption onto the coal surface.

Location and type of field (available knowledge and reusable infrastructure), reservoir capacity, and quality are the main determinants for costs:

- Onshore storage is cheaper than offshore.
- Depleted oil and gas fields (DOGF) are cheaper than deep saline aquifers (SA).
- Larger reservoirs are cheaper than smaller ones.
- High injectivity is cheaper than poor injectivity.

Costs vary significantly from US$1–9/ton CO_2 stored for onshore DOGF, to US$8–26/ton for offshore SA (ZEP 2011).

6.4.2 Depleted Oil and Gas Reservoir

Conventional oil and gas reservoirs are a subgroup of porous sedimentary formations, with several interesting properties. The reservoir is a permeable rock formation with an impermeable cap rock (seal). The reservoir has a bottom layer filled with brine topped by oil and gas layers of variable proportions, with sometimes, no oil phase. This reservoir zone, contained within a structural or stratigraphic closure (e.g., an anticline or dome), is therefore physically able to trap and store a concentrated amount of oil and/or gas.

Utilization of many of the thousands of depleted oil and gas fields for CO_2 storage is possible when the fields reach the end of economic hydrocarbon production. Such reservoirs have contained oil and gas for millions of years and their integrity with respect to CO_2 storage has a high degree of certainty. Another advantage of storing CO_2 in depleted oil and gas fields could lie in lower costs than for storage in poorly known deep saline aquifers. Indeed, mature oil and gas fields have already been explored, their geology is reasonably well known, and some of the oil and gas production equipment and infrastructure could be used for CO_2 injection.

In addition, CO_2 injection could be combined with enhanced hydrocarbon recovery. The situation is more favorable in oil fields because many of them still contain large volumes of unproduced oil after primary and water flooding enhanced oil recovery production has ceased. For gas fields, most of the initial gas in place can be produced. Enhanced gas recovery (EGR) is not physically very favorable inasmuch as mixing between the injected CO_2 and the remaining natural gas is likely to occur. Consequently, gas fields are usually only considered for storage. Large-size depleted gas fields possess significant storage capacity due to their high recovery factor, frequently as high as 80 %. Oil reservoirs with recovery factors ranging from 25 to 65 % possess lower storage capacity. But if CO_2 injection could allow additional oil production, it may partially offset the cost of CO_2 storage. In summary, storage in depleted oil reservoirs will involve an element of EOR, whereas CO_2 injection into depleted gas reservoirs may not result in additional gas production.

However, note that oil and gas reservoirs are penetrated by many wells of variable age, quality, and integrity, which may constitute leakage paths for the stored CO_2 (see Sect. 6.4.6).

As shown above, the storage capacity of depleted oil and gas fields is small relative to the potential capacity of deep saline formations and to CO_2 emissions. However, they do present an early opportunity for CO_2 storage, particularly where associated with EOR. In addition, deep saline formations around, beneath, or above depleted oil and gas fields have frequently been mapped and studied and could be used for CO_2 storage at reduced costs.

6.4.3 Deep Saline Aquifers

Deep saline formations found in sedimentary basins around the world provide, by far, the largest potential volumes for geological storage of CO_2 and are therefore described in a more detailed way than other storage concepts. Sandstone and limestone reservoirs containing saline fluid could be suitable for CO_2 storage, provided that they are sufficiently porous and permeable. In addition the depth of the formation and its fluid salinity must be high enough to prevent any technical and economic exploitation of the water supply. Good storage sites are located in geologically stable areas, with low seismic hazards that could increase the risks for the CO_2 potentially to migrate from the storage zone to the surface.

The reservoir must be large enough to store the CO_2 emissions planned for the lifetime of, at least, one power plant or industrial facility. As CO_2 is injected into the porous reservoir rock, it displaces the in situ pre-existing pore fluid. If the permeability of the rock is low, or if there are some structural barriers to fluid flow, injection will cause a progressive increase in the fluid pressure in the vicinity of the injection well. This will limit the CO_2 injection rate, and may ultimately limit the amount of CO_2 that can be safely stored. Therefore, highly structurally compartmentalized reservoirs are less favorable for CO_2 storage than large unfaulted, permeable reservoirs.

The reservoir must be overlain by a "cap rock," or seal, that is impermeable to CO_2 migration. The CO_2 injected into the reservoir is less dense than the reservoir brine. It will rise to the top of the reservoir, and be trapped underneath the cap rock that prevents its further vertical migration. The cap rock, usually made of shale, mudstone, or evaporite layers, provides the main trapping mechanism for CO_2. Ideally, the cap rock should be unfaulted and continuous over long distances, to prevent migration of the CO_2 out of the reservoir. A detailed analysis of each storage site would be needed to characterize its cap rock integrity.

Provided that a good cap rock exists, storage of CO_2 can be carried out in both confined and unconfined aquifers. Closely analogous to gas storage schemes in hydrocarbon fields mentioned in Sect. 6.4.2, storage in confined aquifers results from trapping of the buoyant CO_2 by structural or stratigraphic features of the sedimentary sequence. Storage in unconfined aquifers involves the injection of CO_2 into large regional aquifers with no specific large structural or stratigraphic traps. After injection, the CO_2 migrates upwards until the impermeable cap rock stops further vertical movement. The CO_2 then migrates laterally, in the uppermost parts of the aquifer, filling successively smaller domes and undulations of the reservoir underneath the cap rock. This mechanism is analogous to repeated minor size structural trapping. The volumes of CO_2 that can be stored in these small perturbations could be quite large if the CO_2 is distributed across a large area. Consequently, in unconfined aquifers, the CO_2 will become gradually distributed in low concentrations over a large area.

In both confined and unconfined aquifers, the combination of three other processes will contribute to the containment of the injected CO_2 in the reservoir rock. Their relative importance will be site specific, but together, they can produce safe permanent subsurface storage.

Capillary trapping, which is sometimes referred to as residual-phase trapping, mainly traps CO_2 after injection stops, when immiscible droplets of CO_2 and brine start to cofill the pores. Capillary trapping could be particularly important for sequestration in unconfined dipping aquifers (Hesse et al. 2008; Ide et al. 2007).

The dissolution of CO_2 into the pore brine leads to solubility trapping. The amount of gas that can dissolve into the water depends on pressure, temperature, and chemistry of the brine (Spycher et al. 2003; Lagneau et al. 2005; Koschel et al. 2006; Oldenburg 2007). Under the conditions displayed by most storage in aquifers (30 to ~150 °C, few hundred bars total pressure), CO_2 solubility increases with increasing pressure (i.e., depth) but decreases with increasing temperature and salinity (i.e., depth equally). The rate of dissolution is accelerated by the amount of mixing of CO_2 and formation water. As the formation water becomes saturated with dissolved CO_2 its density increases and it tends to sink in the reservoir, initiating slow convection movements which will, in turn, favor fluid mixing and dissolution of the CO_2. In any case, dissolution could be slow, on the order of a few thousand years for some injection scenarios (Ennis-King and Paterson 2003; Audigane et al. 2007). The principal benefit of solubility trapping is that once the CO_2 is dissolved, it is much less subject to the buoyancy and to migration.

Once it is dissolved, CO_2 could enter into mineral trapping. It can form carbonate minerals through reaction with calcium, magnesium, or reduced iron, initially present in the brine, or originating from dissolution of silicate minerals of the reservoir rock (Oelkers et al. 2008). Mineral trapping is very favorable for storage security because it could immobilize CO_2 for an extremely long time period (Gunter et al. 1997). Dissolution and mineral trapping are very favorable for long-term security. To which extent and at what rate these processes will occur, is very site specific. Detailed modeling of coupled physical and geochemical reactions is necessary to predict, tentatively, the respective contribution of each process for a given site. If they are not accelerated artificially, thousands of years seems necessary to produce significant dissolution of a CO_2 plume. A timescale, one or two orders of magnitude longer, is probably necessary to achieve important mineral trapping that will be ultimately limited by cation availability in the reservoir.

6.4.4 *Unmineable Coal Seams*

Coalbeds below economic exploitation conditions, either too deep or too thin, could be used to store CO_2. When injected into unmineable coalbeds, CO_2 is adsorbed onto the coal seam surfaces and remains stored as long as the coal is not mined. To allow adsorption, prior to CO_2 injection it is necessary to extract the water originally present in variable quantities in the coalbeds. Methane, naturally occurring with coal in variable quantity, will be displaced when CO_2 is injected and can result in enhanced coalbed methane (ECBM) production. Because of its radiative power, 21 times stronger than that of CO_2, methane should be captured and used. Otherwise, methane release into the atmosphere will produce a process with a negative global balance. This limits CO_2 storage in coal to places where a natural gas pipe network already exists nearby, or where gas turbines can be economically installed. From the physical and geological viewpoints, the process is limited to a relatively narrow depth range: shallow beds less than 600 m can economically be mined, and coal beds at depths greater than 1,000–1,200 m usually display insufficient permeability for viable injection. The horizontal variability of coal permeability may require many wells for CO_2 injection, with subsequent cost increases. Coal may also swell with adsorption of CO_2, which induces further decrease of initial permeability. It is difficult to envisage fracturing operations to increase coal permeability because in doing so, there is the risk of damaging the integrity of the cap rock layer, which increases the potential for CO_2 migration out of the storage zone. Another barrier to widespread use of CO_2 storage in coal is that the optimum storage depth interval may coincide with the zone of protected groundwater.

Taking all these considerations into account the estimated global CO_2 storage capacity in coalbeds is rather limited. However, in the situations where favorable injection properties and good methane recovery rates can be found, the process remains attractive. Consequently, storage in unmineable coalbeds has been, and is being, investigated in several pilot projects worldwide.

6.4.5 Unconventional Geological Storage Options

Several unconventional geological CO_2 storage options are being investigated. Storage in a basaltic formation constitutes the most frequently studied one, with many theoretical and experimental works and a small-scale project in the demonstration stage. Other concepts using deep sea sediments, engineered salt caverns, and lignite seams are also considered. Some other ideas try to combine CO_2 storage with some beneficial process such as the exploitation of oil shale with subsequent CO_2 storage in the shale formations, or the exchange of CO_2 with methane from hydrates in the permafrost formation. CO_2 storage in coal seams followed by in situ methanogenesis has also been envisaged in coal seams. Today, the most advanced scheme of energy production combined with CO_2 storage may probably be found in CO_2-based heat extraction from geothermal reservoirs.

However, all these options are in early stages of development, and appear to have limited capacity, although they may constitute niche opportunities for emission sources located far from the higher-capacity conventional storage options.

At the crossing between CO_2 storage and CO_2 reuse, ex situ carbonatation reactions have been studied among the early technologies for reduction of CO_2 emissions. The idea is to mimic the natural carbonatation of silicate rock which occurs in nature in aqueous environments but proceeds extremely slowly. Once formed, the carbonate mineral will be stable for millions of years in a perfectly safe form. Considering the available volume of basalt and ultramafic[3] rocks, which present the highest carbonatation capacity, the world potential is immense. However, establishing an industrial process is necessary to speed the reaction. This is extremely energy intensive because the rocks must be quarried and finely ground before reacting with water and $CO_{2\,at}$ high temperature and pressure. In spite of numerous research efforts it seems extremely difficult to obtain an acceptable capture cost. For that reason other concepts plan to react CO_2 in aqueous media with solid wastes that are often ready to react without the need to quarry and grind them. This route should be less costly but will have a much more limited potential than reaction with silicate rocks. Today only a pilot scale of these processes has been carried out and this concept is regarded as a safe but long-term option.

[3]Mafic is used for silicate minerals, magmas, and rocks that are relatively high in the heavier elements. The term is derived from the "ma" of magnesium and the "fic" of ferric oxide. Ultramafic rocks contain higher amounts of Fe and Mg.

6.4.6 Risk Assessments, Monitoring, and Mitigation

The occurrence of natural CO_2 accumulations in geological formations where CO_2 has remained trapped for millions of years demonstrates that safe permanent storage is a rather common geological process. However, the rate at which CO_2 has filled these natural reservoirs was probably very different from what it will be in industrial storage projects. Hence, both for purposes of security and overall environmental efficiency, geologic storage operations must be conducted in carefully chosen sites and operated to avoid leaks. Thanks to the experience gained through natural gas storage of natural gas and CO_2 EOR operations, leaks can usually be prevented by thorough analysis of adequate geological information prior to injection. Careful management of pressure during injection, appropriate sealing procedures for well closure, and efficient methods for measure-monitoring and verification (MMV), during and after CO_2 injection, are the other components of the security chain. In saline aquifers and depleted hydrocarbon fields, once CO_2 is trapped by capillary, solubility, or mineral trapping it will be practically immobile. But we have seen that these processes require a long period of time to proceed. Hence, during and immediately after injection operations, CO_2 that is mostly kept in place by stratigraphic or structural trapping can potentially migrate out of the storage formation. Leakage could occur through poorly designed injection wells, or badly sealed old abandoned wells penetrating the reservoir. Another leakage mechanism can exist if injection operations provoke an excessive pressure build-up in the reservoir that could damage the cap rock integrity. In unconfined aquifers leakages could happen if previously undetected fractures or faults in the cap rock are encountered, when the CO_2 plume migrates at the top of the reservoir.

These events are rather unlikely, however, MMV procedures must be implemented in order to verify the effectiveness and safety of storage activities both during and after the period of CO_2 injection. The objectives of MMV will be first to verify the quantities of CO_2 injected and stored for accounting purposes. During the storage operations, the first task is to monitor the injection well integrity to ensure that the operation does not result in leaks. It will also ensure that the CO_2 remains in the reservoir and detects leakage from the storage zone early enough for remediation to be effective. Finally it will monitor the effectiveness of any necessary remediation procedures. Established techniques currently in use in the oil and gas industry can be used to monitor the integrity of CCS injection wells. Logging techniques are very useful to assess the bond and continuity of cement around the well casing, which could constitute a preferential leakage pathway, if improperly completed.

Other methods for CO_2 MMV are adapted from technologies existing in oil and gas production or in underground natural gas storage projects. A number of direct and indirect techniques for monitoring the subsurface distribution of CO_2 have been tested and validated in the framework of the first three industrial projects in Sleipner, Weyburn, and In Salah, whereas specifically designed research projects have been supported by private and public funding mostly in Europe, the United States, Canada, and Australia.

Once CO_2 injection has stopped and the injection wells are sealed, postinjection monitoring will continue to detect any CO_2 migration that might occur afterward. Among all the techniques utilized in the first phase, the most cost effective and easy to deploy will be permanently installed. After closure, the pressure will decrease in the reservoir, and more secure storage mechanisms will tend to have more effect over time. Consequently, the monitoring needs will most likely decrease over time.

Should CO_2 leakage from a reservoir occur during or after remediation, methods must be implemented rapidly. Advantageously, unlike other such stored materials, CO_2 is not explosive or flammable. It should not be toxic at the concentration levels caused by very slow leakage. In the case of an improperly sealed well resulting in failure, the resulting concentrations will be well below harmful levels. Particular care, however, needs to be taken in low-lying or enclosed areas where CO_2, which is heavier than air, may tend to accumulate.

Once they are detected, leaks can generally be eliminated by reducing injection or storage reservoir pressures. Ultimately, leaks may also be eliminated by stopping injection at the current site.

Leaks from abandoned wells may present the most significant vulnerability in some areas with a low-intensity operation of oil and gas. Abandoned wells are sealed with cement. When needed, restoring cement plug integrity can be done, either directly from the surface or through another well drilled nearby to intercept the leaking well. If the injection well itself leaks, it can be repaired or properly sealed and abandoned. These standard techniques that have been established in many countries for abandonment of oil, natural gas, and other mineral extraction wells can be applied to CO_2 injection and storage.

6.5 Present State of Development

6.5.1 Policy Support and Regulations

CCS enters the demonstration stage through large-scale integrated projects. Today they still require policies and substantial financial support available only in a few countries. Moreover, policy uncertainty could be perceived in many countries as a major risk for project deployment. The first step to strengthen the process is to install a regulatory framework that allows financial support to exist. The importance of regulation has been recognized by the European Union which in 2005 has set up the European Emission Trading Scheme (EU ETS). In this system, emission quotas are allocated to large emitters for a given period of time. Emitters can use and combine trading and technological investment to achieve CO_2 emission reduction. Another major step has been taken through the implementation of the European Directive on Geologic storage of CO_2 in 2009. The Directive, which is currently being transposed in the member states' national laws, enables this new industrial practice and exposes the rules and methodologies for the selection and monitoring of storage sites.

Australia, the United States, and Canada have also delivered a number of laws, regulations, and initiatives that play a significant role in the development of CCS projects. But in some cases project proponents have identified a number of issues that have yet to be addressed. The adequacy and timing of the review exercises undertaken by regulators and policymakers to address these issues, will greatly affect the CCS deployment agenda in the forthcoming years.

The importance of effective regulation has also been recognized by many emerging countries. Some actions have already been undertaken to integrate CCS into the future climate change mitigation strategies, but today, many of these countries have yet to pass legislation or finalize their regulatory frameworks.

The Conference of the Parties (COP) of the United Nations Framework Convention on Climate Change (UNFCCC) held in Durban in 2011, has decided to include CCS in the Clean Development Mechanism (CDM: see Chaps. 11, 12, and 13), or in any future mechanism post the Kyoto protocol. This could ease the future demonstration of the technology in developing countries.

Government funding to support large-scale CCS demonstration projects is the second key mechanism indispensable for allowing CCS demonstration projects to exist. In 2011, approximately US$ 23.5 billion has been made available by governments worldwide. The projects are selected via competitive funding programs and are financially viable. The European Union is utilizing the NER300 program in which 13 CCS projects have been identified to go forward to the final decisions stage, expected in the second half of 2012.

6.5.2 Ongoing and Planned Projects

There are presently a great number of research and demonstration projects focused on some aspects of capture and/or storage. They are indispensable for economic and technological improvements but they often deal with small quantities of CO_2 for short or moderate time periods. But the industrial deployment of CCS will be based on the realization of large projects, typically on a scale greater than 0.8–1 Mt per year of carbon captured, optionally transported and stored. It is then very important to count the implementation of large-sized industrial CCS projects.

Among the eight active projects identified by Table 6.2, the first three are mostly EOR-oriented, but large amounts of CO_2 are captured from industrial plants and stored during operations allowing much experience to be gained. The fourth and following are considered as the pioneer project for CCS. Compared to the four other EOR projects, Weynburn–Midale comprises a highly valuable monitoring research program.

Looking at the list of large projects engaged in the construction phase (Table 6.3), note that the EOR objective is frequently a key element of the global decision, but on three occasions, power production is the source of the CO_2 emission and these projects will form a decisive step in the general learning phenomenon. The projects planning to store CO_2 in saline aquifers will mostly capture CO_2 from gas processing, and the ultimate demonstration based on power production and aquifer storage still needs to be finalized.

Table 6.2 Characteristics of active CCS projects

Name (location)	Capture type (industrial CO_2 source)	CO_2 volume (Mt/y)	Storage type	Starting date
Shute Creek Gas Processing Facility (USA)	Precombustion (gas processing)	7	EOR	1986
Val Verde Natural Gas Plants (USA)	Precombustion (gas processing)	1.3	EOR	1972
Enid Fertilizer Plant (USA)	Precombustion (fertilizer)	0.7	EOR	1982
Century Plant (USA)	Precombustion (gas processing)	5 (and 3.5 in construction)	EOR	2010
Sleipner CO_2 Injection (Norway)	Precombustion (gas processing)	1	Deep saline formation	1996
Great Plains Synfuels Plant and Weyburn-Midale Project (USA/Canada)	Precombustion (synfuels)	3	EOR with MMV	2000
In Salah CO_2 Storage (Algeria)	Precombustion (gas processing)	1	Deep saline formation	2004
Snøhvit CO_2 Injection (Norway)	Precombustion (gas processing)	0.7	Deep saline formation	2008

Table 6.3 Characteristics of CCS projects under construction

Name (location)	Capture type (industrial CO_2 source)	CO_2 volume (Mt/y)	Storage type	Starting date
Lost Cabin Gas Plant (USA)	Precombustion (gas processing)	1	EOR	2012
Boundary Dam with CCS Demonstration (Canada)	Postcombustion (power)	1	EOR	2014
Agrium CO_2 Capture with ACTL (Canada)	Precombustion (fertilizer)	0.6	EOR	2014
Kemper County IGCC Project (USA)	Precombustion (power)	3.5	EOR	2014
Illinois Industrial Carbon Capture and Sequestration (ICCS) Project (USA)	Industrial (ethanol production)	1	Deep saline formation	2013
Gorgon Carbon Dioxide Injection Project (Australia)	Precombustion (gas processing)	3.4–4	Deep saline formation	2015

6.5.3 Public Perception and Social Acceptance

For almost 15 years, research projects and opinion surveys have been conducted in several countries to evaluate public awareness and acceptance of CO_2 capture and storage. Some studies relate how, in a concrete way, the dialogue with the public took place around planned large integrated CCS, with in some cases poorly conducted relationships that contributed to halt the project. Very recently, a major inquiry has been conducted in 12 European countries (EUROBAROMETER 364 2011) and looks truly representative of many previous studies.

The number of people saying they have heard of CCS and knew what it was rarely exceeds 10 %. Another 20 % have heard of it but did not really know what it was. Note that these values have remained practically unchanged in 10 years. Even when there is a major EU cofinanced CCS project in their country, close to 90 % said they had not heard of the project. Among informed people, half of the respondents agreed that CCS could help to combat climate change, and up to 60 % agreed that "capturing and storing CO_2 should be compulsory when building a new coal-fired power plant." "For the implementation of CCS, more than three quarters (77 %) felt that public authorities should be able to monitor CCS operations" and close to 70 % felt that "harmonized and consistent methodologies should be developed internationally to manage the capture and storage of CO_2".

For dissemination of information about CO_2 storage, the most trusted bodies are universities and research institutions (45 %), followed by NGOs (31 %), whereas journalists (24 %), regional and local authorities (23 %), national governments (20 %), and the European Union (14 %) are considered trustworthy by less than a quarter of the people.

Risk perception has always been a major subject of most studies, and it mostly concentrates on the storage part, and to a lesser extent on CO_2 transport. Generally, up to slightly above 50 % of the respondents feel that CO_2 storage represents a risk. When asked where CO_2 storage sites should be located, 24 % of people preferred onshore storage in areas of low population density (20 %), close to the source where the CO_2 is produced. Only a fifth (21 %) preferred CO_2 to be stored offshore under the seabed, in spite of a lower risk of exposure for populations. It is not surprising to see in many studies that almost 60 % of the people would be worried of which only 25 % were "very worried," if an underground CO_2 storage site were to be located within a few km of their home. In such a situation, 40 % said they would like to be directly consulted and to participate in the decision-making processes, with 19 % willing to see NGOs participate in the decision-making process.

These results show that the major difficulty today is the very low public awareness of CCS. They also demonstrate that there is a general demand to see public entities playing a great role in regulation, research, and dissemination of information. When a real industrial storage project starts to fulfill its first administrative requirements, it is essential to organize a real dialogue with the public and both local and national NGOs. On some occasions, the dialogue has proceeded smoothly (Lacq, Ketzin), although it has taken quite a long time. In some other cases seemingly poorly organized public dialogue has contributed to halt the project (Carson, United States).

6.6 Conclusions

In conjunction with many other efforts, including the continuous search for energy efficiency, reforestation, and adequate agricultural practices, and the growth of renewable and nuclear energies share in the global energy mix, CCS can contribute significantly to the limitation of GHG emissions in the twenty-first century. Considering the very recent emergence of CCS around 1990, it can truly be considered that many steps have been rapidly accomplished on the road of industrial deployment. However, in order to play a significant role in the years 2050 and beyond, the cost-competitive deployment of CCS technologies in power plants and large emitting industries must be accomplished by 2020–2025.

The first key point is to establish international treaties and regulations to voluntarily fix goals in emission reductions at the national and international scales. Public support will be needed to finance the first large integrated demonstration projects, especially for electricity production plants.

Meanwhile more efficient and cost-competitive CCS technologies than are available at present, have to be developed through ongoing R&D. Improvement of power plant efficiency, development of new materials and of innovative and more cost-effective capture processes, are considered feasible and essential to reduce the costs of capture. Assessing the safety of CO_2 storage is also a priority topic inasmuch as the greatest public concern with long-term geological storage is its security. In parallel, a better assessment of storage potential and site characterization, especially of saline aquifers, is needed. Although CO_2 transport has been demonstrated on a commercial scale, concepts for transportation in populated areas are needed, and some issues related to CO_2 stream composition with respect to impurities are still of concern and need to be addressed.

Building public awareness and strengthening a confident dialogue with nonexpert people will also be a key to successful and rapid development.

References

Audigane P, Gaus I, Czernichowski-Lauriol I, Pruess K, Xu T (2007) Two-dimensional reactive transport modeling of CO_2 injection in a saline aquifer at the Sleipner Site, North Sea. Am J Sci 307:974–1008

Ennis-King JP, Paterson L (2003) Role of convective mixing in the long-term storage of carbon dioxide in deep saline formations. Presented at Society of Petroleum Engineers annual technical conference and exhibition, Denver, 5–8 Oct 2003, SPE paper no. 84344

EUROBAROMETER 364 (2011) Public Awareness and Acceptance of CO_2 capture and storage. http://ec.europa.eu/public_opinion/archives/ebs/ebs_364_en.pdf

Finkelrath M (2011) Cost and performance of carbon dioxide capture from power generation. IEA Working Paper

Gunter GW, Finneran JM, Hartmann DJ, Miller JD (1997) Early determination of reservoir flow units using an integrated petrophysical method. SPE annual technical conference and exhibition, San Antonio, TX, 5–8 Oct 1997

Hesse MA, Orr FM, Tchelepi HA (2008) Gravity currents with residual trapping. J Fluid Mech 611:35–60

Ide SK, Jessen K, Orr FM (2007) Storage of CO_2 in saline aquifers: effects of gravity, viscous, and capillary forces on amount and timing of trapping. Int J Greenh Gas Control 4:481–491

Koschel D, Coxam JY, Rodier L, Majer V (2006) Enthalpy and solubility data of CO_2 in water and NaCl (aq) at conditions of interest for geological sequestration. Fluid Ph Equilib 247:107–120

Lagneau V, Pipart A, Catalette H (2005) Reactive transport modelling of CO_2 sequestration in deep saline aquifers. Oil Gas Sci Technol 60:231–247

Metz B, Davidson O, Loos M, de Coninck H, Meyer L (2005) Carbon dioxide capture and storage, IPCC. Cambridge University Press, Cambridge

Morbee J, Serpa J, Tzimas E (2010) The evolution of the extend and the investment requirements of the trans-European CO_2 transport network. JRC scientific and technical reports. Publication Office of the European Union, Luxembourg

Oelkers EH, Schott J, Gauthier JM, Herrero-Roncal T (2008) An experimental study of the dissolution mechanism and rates of muscovite. Geochim Cosmochim Acta 72:4948–4961

Oldenburg C (2007) Migration mechanisms and potential impacts of CO_2 leakage and seepage. In: Wilson EJ, Gerard D (eds) Carbon capture and sequestration: integrating technology, monitoring, regulation. Blackwell Publishing, Ames, pp 127–146

Spycher N, Pruess K, Ennis-king J (2003) CO_2-H_2O mixtures in the geological sequestration of CO_2. I. Assessment and calculation of mutual solubilities from 12 to 100°C and up to 600 bar. Geochim Cosmochim Acta 67:3015–3031

ZEP (2009) The cost of CO_2 transport. Post demonstration CCS in the EU European Technology Platform for Zero Emission Fossil Fuel Power Plants

ZEP (2011) The cost of CO_2 storage. Post demonstration CCS in the EU European Technology Platform for Zero Emission Fossil Fuel Power Plants

Chapter 7
New Energy Sources and CO_2 Treatment

Siglinda Perathoner and Gabriele Centi

Abstract The conversion of CO_2 through the use of renewable energy sources offers new possibilities to develop innovative new approaches to improve sustainability of chemical and energy production. After introducing the topic, this chapter discusses the reuse of CO_2 as a valuable carbon source and an effective way to introduce renewable energy in the chemical industry value chain, to improve resource efficiency, and to limit greenhouse gas emissions. The specific challenge of using CO_2 for the production of light olefins (ethylene, propylene) is discussed. The recycling of CO_2 back to fuels using sunlight (solar fuels) is also briefly analyzed to demonstrate a relevant opportunity to develop effective energy vectors for the storage of solar energy integrated into the existing energy infrastructure and allowing a smooth but fast transition to a more sustainable energy future. The role of these topics for the future strategies of chemical and energy industries, especially in terms of resource efficiency, is also highlighted.

Keywords CO_2 valorization • Renewable energy • Solar fuels • Olefin from CO_2 • Sustainable chemistry and energy • Resource efficiency

List of Acronyms

ATP	Adénosinetriphosphate
BUA	Business as usual
C2, C3…	Hydrocarbons with two or with three atoms of carbon
>C1 (>C5)	Hydrocarbons, or alcohol with more than one (5) atom(s) of carbon
CCR	Carbon capture and recycling

S. Perathoner (✉) • G. Centi
Dipartimento di Chimica Industriale ed Ingegneria dei Materiali, University of Messina and INSTM/CASPE, V.le F. Stagno D'Alcontres 31, 98166 Messina, Italy
e-mail: perathon@unime.it

J.B. Saulnier and M.D. Varella (eds.), *Global Change, Energy Issues and Regulation Policies*, Integrated Science & Technology Program 2, DOI 10.1007/978-94-007-6661-7_7,

CCS	Carbon capture and sequestration
DME	Dimethylether
DNV	Det Norske Veritas, company involved in CCS and in ERC research
DoE	US Department of Energy
DR	Dry reforming
EJ	Exajoule (1 EJ = 10^{18} J)
ERC	Electrochemical reduction of CO_2
FT	Fischer–Tropsch
GHG	Greenhouse gas
Gt/y	Gigaton per year
Gteq CO2	Gigaton of equivalent CO_2
Gtoe	Gigaton of oil equivalent
GW	Gigawatt (= 10^9 W)
IEA	International Energy Agency
IPCC	Intergovernmental Panel on Climate Change
LCA	Life-cycle assessment
MTG	Methanol-to-gasoline
MtCO2/y	Megaton CO_2 per year
Mt/y	Megaton per year
NADPH	Nicotinamide adenine dinucleotide phosphate-oxidase
OECD	Organisation for Economic Co-operation and Development
PEM electrolyzer	Proton electrolyte membrane electrolyzer
PV	Photovoltaic
RCO2 AS	RCO2 AS is a small research company located in Norway, developing technologies for recycling CO_2
RE	Renewable energy
rWGS	Reversewater gas shift
SNG	Substituted natural gas
SOEC	Solid oxide electrolysis cells
TPES	Global total primary energy supply
TW	Terawatt (= 10^{12} W)
UCLA	University of California, Los Angeles
UNFCCC	United Framework Convention on Climate Change
ZSM-5	Zeolite Socony Mobil – 5

7.1 Introduction

The public perception of the relevance of protecting the environment, mitigating climate changes, and saving resources, particularly energy, has been significantly raised over the last decade and in combination with the enormous pressures that global competition puts on strategic resources has changed the future scenario. The necessity of improving the overall efficiency of resource consumption for industrial production and increasing the speed of innovation, particularly in Europe, caused a change in the priorities in both the social and industrial dynamics. Resource efficiency has being raised as the key topic of R&D activities. For example, a resource-efficient Europe is the flagship initiative of the Europe 2020 Strategy the objective of which is a radical shift towards a resource-efficient, low-carbon economy to achieve sustainable growth. New energy sources, particularly renewable, and addressing CO_2 emissions are two of the key elements to achieving this goal, and are closely interconnected as discussed in this chapter.

The European Union is at the forefront of these initiatives and is making a real effort to reduce its greenhouse gas emissions, but still a change in speed is necessary. With its "Roadmap for moving to a competitive low-carbon economy in 2050" (European Commission 2011) the European Commission is setting out a plan to meet the long-term target of reducing domestic emissions by 80–95 % by 2050. Although this target requires the harmonized effort of all the sectors responsible for Europe's emissions—power generation, industry, transport, buildings, and construction, as well as agriculture—to make the transition to a low-carbon economy over the coming decades effective, the chemical and petrochemical sectors are particularly relevant, because they have a more than 30 % share of total industrial energy use worldwide (including feedstocks) and thus are the sector with the largest energy use in industry (Saygın et al. 2009).

The introduction of renewable energy in the chemical chain, and the reuse of CO_2 to improve resource efficiency and limit greenhouse gas (GHG) emissions are thus key factors, together with other elements (increase bio-based feedstock, improve recycling and materials life cycle, introduce new processing technologies) for the new strategy of the chemical industry to be more innovative, competitive, and boost resource efficiency.

CO_2 valorization (Centi and Perathoner 2009) and producing solar fuels from CO_2 (Centi and Perathoner 2010, 2011; Centi et al. 2011b) are two of the relevant components in this strategy which integrates renewable energy and the chemical production chains, as shown in this chapter. The production of H_2 using renewable energy (Centi et al. 2011b) is a further element to implement this general objective and is thus also briefly discussed in this chapter.

7.2 Sources of CO_2

Electricity, heat generation, and transport sectors account for two thirds of global CO_2 emissions. According to the International Energy Agency (IEA), the generation of electricity and heat is by far the largest producer of CO_2 emissions (about 40 % of the world CO_2 emissions) and relies, worldwide, heavily on coal: emissions from coal will grow to about 18.6 Gt CO_2 in 2030 (IEA 2010). Although coal represented only 25 % of the world TPES (global total primary energy supply), it accounts for 43 % of global CO_2 emissions. As compared to gas, coal has on average nearly twice the emission intensity.

Suitable sources for CO_2 capture are large stationary sources that emit large volumes of CO_2 (about +0.1 $MtCO_2$/y). Large point sources are present mainly in fuel combustion activities, industrial processes (cement production, refineries, iron and steel industry, and petrochemical industry), and oil and gas processing. In terms of the global distribution of large stationary sources of CO_2 and of areas in sedimentary basins where suitable saline formations, oil or gas fields, or coal beds are present for carbon capture and sequestration[1] (CCS; see Chap. 6), there is a mismatch in some geographical areas (Europe, e.g.) that determines the need to transport CO_2 to long distances (higher than 100–150 km). Note also that in these areas, the construction of new pipelines dedicated to CO_2 transport is often not possible, and thus road/naval transport is necessary, with large cost/energy penalties.

Typical gas stream CO_2 concentrations are lower than 15 % (of the total gas volume). The higher the content of CO_2 from a source, the more suitable CO_2 capture becomes. A key variable in the selection of the separation method is partial pressure. High partial pressure makes CO_2 recovery easier than low partial pressure and therefore keeps the cost for capture lower. Depending on the method of oxidation in a power cycle, the percentage of CO_2 concentrations varies from about 0.03 to over 0.90, and different concentrations of other components (H_2O, SO_x, NO_x, etc.) are also present (see Chap. 6).

7.3 CO_2 Reuse Versus Storage

The agreement gathered under the United Framework Convention on Climate Change (UNFCCC) for the environmental goal of keeping the global temperature increase below 2 °C implies a very large reduction in CO_2 emissions (about 30 Gt CO_2 were emitted in 2010, at world level). The World Energy Outlook 2010 (IEA 2011) report prepared by the International Energy Agency discusses different scenarios to control GHG emissions and indicates a reduction of the CO_2 emissions in the 2/4 Gt/y range in a decade (2021) and 11./15 Gt/y range in two decades (2031), with respect to the business-as-usual (BUA) scenario. About 20 % would derive from (CCS), that is, about 400–800 Mt/y of CO_2 in a decade and about 2,200–3,000 Mt/y of CO_2 in two decades will be available. Large amounts of CO_2 as raw material at zero cost or even negative cost (the reuse of CO_2 avoids the costs of sequestration and transport, up to about 40–50 % of the CCS cost, depending on the distance between the sequestration site and the place of emission of CO_2) will soon be available. This creates clear opportunities for the utilization of CO_2.

However, the reuse of CO_2 goes well beyond that of a possible economic opportunity for some specific chemical synthesis, such as that of producing

[1] **Carbon capture and storage (CCS)**, is a technology to prevent large quantities of CO_2 from being released into the atmosphere from the use of fossil fuels in power generation and other industries. The process is based on capturing carbon dioxide (CO_2) from large point sources, such as fossil fuel power plants, and storing it in such a way that it does not enter the atmosphere.

CO_2-incorporating polymers (Quadrelli et al. 2011). We may roughly distinguish two main routes for reusing CO_2 to produce commercially valuable products, apart from routes involving bacteria and microorganisms:

- Those reactions incorporating the whole CO_2 moiety in organic or inorganic backbones
- Those involving the rupture of one or more of the C–O bonds.

This subdivision is crucial in terms of energy balance and applications. The first type of reaction (both organic and inorganic) is not energy intensive and sometimes may also occur spontaneously (although with low kinetics, as in the production of inorganic carbonates), however, the reactions of C–O cleavage require the use of reducing agents (typically H sources, e.g., H_2) and are energy intensive.

In the context of CO_2 management, this energy necessary for the reaction should derive from renewable ones (solar, wind, geothermal energy, etc.), or at least from noncarbon-based sources (nuclear energy) or eventually waste-energy sources. Although significant R&D development is still necessary to achieve this possibility technically, as discussed later, converting CO_2 to molecules for chemical or energy applications, via involvement of renewable energy sources is a relevant opportunity to introduce renewable energy into the chemical and energy chains effectively, thus realizing a relevant step forward in the achievements of resource efficiency discussed before. Conversely, the first class of reactions (e.g., organic or inorganic carbonates) yields stable material and becomes relevant in a carbon management strategy for ensuring long-lasting storage.

The IPCC (Intergovernmental Panel on Climate Change) report on CO_2 capture and storage (Metz et al. 2006) remarked the time lapse between the moment of CO_2 conversion into a product and CO_2 release back into the atmosphere, as the critical parameter to consider. A long lifetime of the CO_2-based product will fix carbon dioxide for a long time, thus preventing its reintroduction into the atmosphere. Most product lifetimes of the second class of reactions will range between months to a few years with the exception of inorganic carbonates and polymers issued from organic carbonates (first type of reaction) which store CO_2 from decades to centuries. Excluding from these bases the second type of reaction, and considering the effective market potential of the first type of reaction products as limited, IPCC reports did not consider reuse of CO_2 as a valuable option for GHG control. Furthermore, a general comment was added, estimating that in all scenarios developed to address GHG emissions, a portfolio of technologies is always necessary.

Yet these are not the proper considerations, because CO_2 recycling via incorporating renewable energy introduces a shorter path (in terms of time) to close the carbon cycle compared to natural cycles and an effective way to introduce renewable energy sources into the chemical/energy chain. In addition, it will reduce the use of fossil fuels for these chemical/energy uses. Although sequestration has an effective factor lower than one in reducing GHG emissions (considering the energy required to capture, transport, and sequester CO_2, the effective factor is 0.5 or below), the effective factor for converting carbon dioxide to CO_2-based polymers is higher than one, considering that it reduces the use of fossil fuels to produce equivalent polymers and that some indirect factors are also present (e.g., foams produced from CO_2-based polymers may be used for better thermal insulation of buildings, thus allowing an energy saving).

For the second type of reaction, on a constant time-horizon (20 years, e.g.), the chemicals/fuels produced from CO_2 conversion by incorporating renewable energy are cycled several times (depending on the lifetime of the product, shorter for fuels). The effective factor for reduction of GHG emissions is thus at least one order of magnitude higher than that for CCS. Thus, the effective potential of CCR (carbon capture and recycle) technologies in GHG control is at least similar to that of CCS technologies (Quadrelli et al. 2011). The potential reduction is equivalent to 250–350 Mt/y in the short to medium term (Quadrelli et al. 2011). This amount represents about 10 % of the total reduction required globally; that is, it is comparable to the expected impact of CCS technologies, but with additional benefit in terms of

- Fossil fuel savings
- Additional energy savings (e.g., the cited insulating effect of polyurethane foams)
- Accelerating the introduction of renewable energy into the chemicals and energy chain

7.4 Solar Fuels from CO_2

Different sources of renewable energy (solar, wind, tides, hydro, etc.) are currently used and have great potential to grow further, even with some disparities. The actual fraction of renewable energy sources (year 2011) is still often limited to ecologically inefficient options (such as wood combustion) or to routes where further potential increase is close to the limit (hydroelectricity). For other types of renewable energies, particularly solar energy, however, the potential is large.

The actual world average global energy consumption is about 12 Gtoe, leading to a mean annual power of 16 TW, which is estimated to increase to about 25 TW by the year 2050 (IEA 2010). A conservative estimate of the potential for solar energy is at least five to ten times higher than this estimated consumption, although significantly lower for other renewable sources: 2–4 TW for wind, 2–3 TW for tides, 5–7 TW for biomass, and 3–6 TW for geothermal energy (Lewis et al. 2005). Of these different renewable energy sources, only biomass can be converted to liquid fuels, whereas almost all the others produce electrical energy. Biomass, however, is quite complex, and producing liquid fuels in a sustainable and economical way is still a challenge, even considering the fast developments in this area. In addition, several question marks exist on the effective contribution to GHG reduction of the use of biomass to produce fuels for the transport sector (see also Chap. 14).

Relevant progress has been achieved over the last 7 years (2006–2013) in the introduction of renewable energy (RE). The IPCC Special Report on Renewable Energy Sources and Climate Change Mitigation (IPCC 2011) indicates that renewable energy capacity grew significantly in the last 2 years, despite the global financial crisis. On a global basis, it is estimated that RE accounted for 12.9 % of the total 492 Exajoules (EJ) of the primary energy supply in 2008, but the largest RE contributor was associated with burning biomass (10.2 %). Hydropower represented 2.3 %, whereas other RE sources accounted for 0.4 %. RE capacity continued to grow rapidly in 2009 compared to the cumulative installed capacity from the previous year, including wind power (32 % increase, 38 gigawatts (GW) added), hydropower (3 %, 31 GW added), grid-connected photovoltaics (53 %, 7.5 GW added), geothermal power (4 %, 0.4 GW added), and solar hot water/heating (21 %, 31 GW added). Although in some cases the rate of increase in the last 2 years was quite high (wind and solar), the share still remains low.

In these conditions, the electrical energy produced from RE technologies could be given to the grid, but in the future, with an increasing share, it will be necessary to provide continuity of the supply. On the other hand, even with the expected increase in the storage efficiency of advanced batteries (Garcia-Martinez 2010), direct electrical energy storage of the batteries will continue to be a major limit in many applications.

The actual society and energy scenario is largely based on the use, as in the energy vector, of liquid hydrocarbons, derived mainly from the refining of oil. Of the world energy final consumption, about 43 % is accounted for by oil and derived liquid fuels (gasoline, diesel, jet fuels, gasoil, etc.) (IEA 2010). In contrast, only about 17 % is accounted for by electrical energy, also due to the limit in energy storage. The specific energy of batteries, that is, their capacity for storing energy per kilogram of weight or the unit volume, is still only 1 % of the specific energy of gasoline.

The distribution of energy, matching time and geographical demand with production capacities (wind and solar energy, the two RE with the higher rate of increase, are typically discontinuous), as well as use of energy in mobile applications (from cars to laptops) are all pushing the development of effective solutions for energy storage and transport. The further development of a sustainable energy scenario requires finding efficient solutions to store and transport RE.

The conversion of electrical to chemical energy still remains the preferable option and the issue is thus the development of optimal energy vectors for RE. This is the base concept of developing solar fuels (Centi and Perathoner 2010, 2011; Gray 2009; Nozik 2010; Roy et al. 2010; Morris et al. 2009).

Suitable energy vectors must fulfill a number of requirements. They should:

- Have both a high energy density by volume and by weight
- Be easy to store without the need for high pressure at room temperature
- Be of low toxicity and safe to handle, and show limited risks in their distributed (nontechnical) use
- Show good integration in the actual energy infrastructure without the need for new dedicated equipment
- Have a low impact on the environment in both their production and use

H_2 could be an ideal energy vector (Züttel et al. 2008; Sartbaeva et al. 2008; Muradov and Veziroglu 2008) regarding many of these requirements, but the energy density of H_2, even considering future possible developments in storage materials, will still be a main issue for practical large-scale use. A clear gap exists between both H_2 and electrical storage with respect to liquid fuels based on fossil or renewable (biomass) sources. In addition, its use requires large costs for a new energy infrastructure, and it does not integrate with the actual devices, thus not allowing a smooth transition. In terms of sustainability, the cost parameter is one of the critical elements. When a novel technology requires high costs to be introduced, it will take a long time to be eventually applied. A solution that is better integrated in the actual infrastructure is thus preferable, because it has a lower economical barrier for the introduction and may be applied in a shorter term.

Solar–hydrogen may be a better and more sustainable alternative when combined with the possibility of forming easily transportable liquid fuels with high energy density (Bockris 2008; Olah et al. 2009b; Ferenc 2008; Leitner 1995). In addition, hydrogen is unsafe, in particular in transporting energy over long distances. The transformation of hydrogen into a safe transportable chemical is thus desirable for bulk energy transport.

Current energy vectors based on liquid fuels fulfill all the requirements for suitable energy vectors, except the last point, particularly regarding the emissions of greenhouse gases, because the available catalytic technologies can effectively minimize the emissions of the other pollutants produced during combustion (particularly NO_x).

Although challenging, finding an efficient solution to recycle the CO_2 produced during combustion, to form back fuels using RE will be thus preferable and more sustainable than to develop new energy vectors and a new energy infrastructure. The CO_2 conversion to liquid fuels (methanol or other liquid chemicals) forms safe chemicals with high energy density, with minimal (and well-established) risks in storage, and this route may be well integrated with the existing energy infrastructure with minimal investment.

Producing solar fuels via recycling CO_2 is thus a carbon-neutral approach to store and transport solar energy (and of other renewable or noncarbon-based energy sources) that can be well integrated into the current energy infrastructure.

7.5 CO_2 as a Carbon Source to Introduce Renewable Energy in the Chemical Production Chain

Light olefins[2] (ethylene and propylene, about 200 Mt/y global market) are the building blocks of current chemical production, but their synthesis is also the single most energy-consuming process in the chemical industry (Ren et al. 2006, 2008). They are mainly produced by steam cracking. On a world scale the operation uses about 3 EJ of primary energy (due to the combustion of fossil fuels and excluding the energy content of products) and produces large amounts of CO_2 emissions (Neelis 2007).

[2] *Olefin*, also called *alkene*, is an unsaturated hydrocarbon containing one or more pairs of carbon atoms linked by a double bond.

The pyrolysis section of a naphtha steam cracker alone consumes approximately 65 % of the total process energy and accounts for approximately 75 % of the total energy loss. The specific emission factors (CO_2 Mt/Mt light olefin) depend on the starting feedstock, but range between 1.2 and 1.8. Therefore, about 300 Mt/y of CO_2 derive from the production of these building blocks of the chemical production chain. It is thus a relevant objective to look at the possibility of using CO_2 as the carbon source for producing light olefins, with the multiple benefit of potentially reusing large amounts of carbon dioxide, avoiding large volume emissions of CO_2 in producing these raw chemicals, and introducing RE at the beginning of the chemical production chain.

There is a combination of factors, between which the lowering of fossil fuel resources, the time requested to change the energy system and infrastructure, the increase of fossil fuel cost and need of cleaner fuels, the increase of social pressure on sustainability and environmental protection issues, and so on, indicating that in addition to a significant increase in the cost of carbon sources for chemical production in the next two decades, there are many constraints limiting the use of oil-alternative carbon sources, particularly to produce base raw materials such as C2–C3 olefins. On the other hand, it is necessary to widen the possible sources to produce these base chemicals, because it will moderate the increase in their price, while maintaining the actual structure of value chain. This is an important element to keep production competitive and avoid large investments related to a value change in petrochemistry. Within this scenario, producing light olefins from CO_2, in order to avoid wasting this relevant source of carbon, is an interesting option (Centi et al. 2011a).

The energy necessary for the endothermic[3] process of producing light olefins from CO_2 must clearly derive from RE sources, because otherwise the process is not sustainable. It was previously remarked that converting CO_2 to fuels is a great opportunity to store and transport solar or other RE sources efficiently. Storing the energy in the form of chemical energy is still the preferable method to have a high energy density in the energy vector, with easy storage and transport. Due to their high energy of formation, C2–C3 olefins represent an excellent opportunity to store solar energy and incorporate it in the value chain for chemical production instead of that for energy. The high value of the energy of olefin formation also explains why the actual formation process is the most energy-consuming process in the chemical industry, with a large impact on CO_2 emissions.

The largest part of ethylene and propylene is currently used to produce polymers, directly (polyethylene and polypropylene; polypropylene production, e.g., accounts for more than 60 % of the total world propylene consumption) or indirectly (e.g., the main products of propylene are acrylonitrile, propylene oxide, acrylic acid, and cumene, which at the end are also mainly used to produce polymers). Producing ethylene and propylene from CO_2 can thus also be viewed as a way to capture CO_2 for a longer term with respect to the much shorter life cycle of CO_2-based fuels.

[3] An endothermic reaction needs external heat to be provided to the system.

7.5.1 Producing Light Olefins from CO_2

The synthesis of light olefins from CO_2 requires the availability of H_2. Ethylene and propylene have a positive standard formation energy with respect to H_2, but water forms in the reaction and thus the process essentially does not need extra energy with respect to that required to produce H_2. From the energetic point of view, the energy efficiency of the process is thus related to the energy efficiency of the production of H_2. The process for olefin synthesis from CO_2 may be described as the combination of a stage of reverse water gas shift (rWGS):

$$CO_2 + H_2 \rightarrow CO + H_2O \tag{7.1}$$

and a consecutive stage of Fischer–Tropsch (FT) synthesis

$$CO + H_2 \rightarrow C_nH_{2n} + C_nH_{2n+2} + H_2O + CO_2 \ (n = 1,2,\ldots) \tag{7.2}$$

The Fischer–Tropsch catalyst should be modified in order to minimize the formation of alkanes (especially CH_4), and increase selectivity to C2–C3 olefins. The two stages above may be combined, but water should preferably be removed in situ to shift the equilibrium, and avoid F–T catalyst reversible inhibition.

A number of recent studies have been dedicated to the effect of CO_2 on FT synthesis. For example, Sai Prasad et al. using a Fe/Cu/Al/K catalyst operating at 300 °C find a selectivity to alkene in the C2–C4 fraction (about 40 % of the whole hydrocarbon fraction) of about 85 % (Sai Prasad et al. 2008). Jiang et al. used a zeolite capsule catalyst with a core (Fe/SiO_2)-Shell (Silicalite-1) structure for the direct synthesis of light alkenes from syngas[4] (Jiang et al. 2011). This zeolite capsule catalyst exhibited excellent selectivities compared with the traditional FT catalyst, suppressing formation of the undesirable long-chain hydrocarbons.

Light olefins may also be obtained via conversion of methanol/DME (dimethylether) on multifunctional catalysts instead of through the Fischer–Tropsch mechanism. Kang et al. used Fe–Cu–K catalysts supported on ZSM-5 (Zeolite Socony Mobil–5) to improve selective olefin production (Kang et al. 2010). Park et al. used instead a dual-bed reactor approach, with a Fe–Cu-Al-based FT catalyst in the first stage and a ZSM-5 cracking catalyst in the second stage (Park et al. 2009). The formation of olefins is enhanced by doping the first catalyst with K. A 52 % selectivity to C2–C4 hydrocarbon rich in olefins (77 % selectivity) was reported.

Although further improvements in catalysts are possible, reactor design and process operations are also necessary and these results show that it is possible to modify FT or methanol catalysts to selectively form light olefins. It is reasonable to consider further optimization with a target of selective synthesis of light olefins over 80 % at higher productivity.

---

[4] *Syngas* (from *synthetic gas* or *synthesis gas*) is the name given to a gas mixture that contains varying amounts of carbon monoxide (CO) and hydrogen (H_2). Examples of production methods include steam reforming of natural gas or liquid hydrocarbons, the gasification of coal and biomass, and in some types of waste-to-energy gasification facilities.

7.5.2 H_2 Production Using Renewable Energy

Converting carbon dioxide to light olefins requires H_2 (cf. Eqs. 7.1 and 7.2) and the latter should be produced using RE sources to make the whole process sustainable (and effective in terms of GHG impact). For the same reason, only methods for producing hydrogen using RE sources directly are considered appropriate. Even if H_2 produced from biomass-derived products (via aqueous phase catalytic reforming, e.g.) is formally derived from renewable resources, the net impact on GHG emission per mole of H_2 is not very different from that of fossil fuels.

In methane steam reforming

$$CH_4 + 2H_2O \rightarrow CO_2 + 4H_2 \tag{7.3}$$

four moles of H_2 are produced per mole of CO_2, that is, 5.5 kg CO_2 per kg H_2. However, this value rises to about 8.9 kg CO_2/kg H_2 considering the life cycle of production (Ruether et al. 2005).

The life-cycle assessment (LCA) for producing H_2 from biomass is more complex, because one should include many factors, from the growing of biomass (including aspects related to GHG impact of the use of fertilizers, land change uses, etc.) to biomass harvesting/transport, to biomass conversion and related technologies/routes utilized, and in addition considering that H_2 is often a byproduct and not the primary product. Therefore, even if fully reliable estimations do not exist or are highly dependent on the specific case, on the average of 5–6 kg CO_2/kg H_2 could be indicated (Djomo and Blumberga 2011; Marquevich et al. 2002).

For the direct production of H_2 using RE, LCA data are also not reliable, because all technologies are still at the development stage. For wind/electrolysis, a value below 1 kg CO_2/kg H_2 was estimated (Spath and Mann 2004). Life-cycle CO_2 emissions for various hydrogen production methods (Utgikar and Thiesen 2006) indicate values for H_2 from hydroelectric/electrolysis and solar thermal around 2 kg CO_2/kg H_2, whereas values can reach around 6 kg CO_2/kg H_2 for photovoltaic/electrolysis. However, the latter value appears overestimated and does not include the latest developments in the field. It is thus a reasonable estimate that an average of 1–2 kg CO_2/kg H_2 could be indicated, that is, a value at least one third of that from biomass conversion routes.

Already well established is the production of H_2 by electrolysis, where the electrical energy derives from photovoltaic (PV) cells, wind turbines, or other RE sources. Today, the efficiency of the PV-electrolysis system has been optimized, and efficiencies up to about 12 % can be obtained (Gibson and Kell 2008). H_2 could be produced under pressure in modern electrolyzers, whereas other routes typically produce $H_{2\,\mathrm{at}}$ atmospheric pressure and its compression for use is necessary. Proton electrolyte membrane (PEM) water electrolysis technology provides a safe and efficient way to produce electrolytic hydrogen and oxygen from RE sources. Stack efficiencies close to 80 % have been obtained operating at high (1 A · cm^{-2}) current densities using low-cost electrodes and high operating pressures (up to 130 bar) (Millet et al. 2010). Among the advantages of PEM (polymer electrolyte membrane) electrolyzers over the well-established alkaline technology, the absence of corrosive electrolytes may be cited, as well as better integration with solar and wind power. Recent developments in terms of catalyst optimization, improved design of electrolyzer cells, and cost reduction of the membrane–electrode assembly have already led to significant stack capital cost reductions.

The DoE (US Department of Energy) targets for distributed water electrolysis indicate a hydrogen cost ranging in the \$2–4/kg H_2. These costs would increase if renewable energy were considered. Considering the incidence of electricity on H_2 production cost and that off-peak electrical energy at competitive costs is available, it is possible to estimate a future renewable H_2 around \$8–10/kg H_2 decreasing to \$5–6/kg H_2 in the year 2020. This cost may be significantly lowered when cheap excess electrical energy were available, for example, produced by nuclear plants at night. The cost of hydrogen produced from water splitting using nuclear technologies is around \$2/kg H_2 (Elder and Allen 2009).

Further improvements in the H_2 technology could be achieved by operating electrolysis at high temperature (above 200 °C), in particular using solid oxide electrolysis cells (SOEC; temperatures around 950–1,000 °C) (Hauch et al. 2008). For low-temperature electrolysis a larger quantity of electrical energy is necessary to overcome the endothermic reaction heat, whereas at high temperatures the primary electrical energy demand is considerably reduced and the electrical losses in the cell decrease due to lower ohmic resistance in the electrolyte and lower polarization losses from the electrode reactions, due to the faster kinetics of the electrolysis reactions. H_2 costs from high-temperature electrolysis have been estimated in the \$1.5–2.6/kg H_2 range (Hauch et al. 2008).

There are three alternative routes to produce renewable H_2 (Centi et al. 2011b).

- Bio-route using cyanobacteria or green algae (bioH_2)
- High-temperature thermochemical using concentrated solar energy
- Photo(electro)chemical water splitting or photoelectrolysis using semiconductors

All these routes are still under development and thus many uncertainties are present regarding their technoeconomic feasibility. Productivities are still quite low.

In conclusion, H_2 production from RE sources may be estimated to have a cost in the US\$2–10/kg H_2 range. H_2 using PEM electrolyzers can already be produced on a small to medium scale. There are relevant possibilities for a further decrease in the cost, particularly in terms of optimization of the electrodes and device configuration. H_2 production from biomass transformation products (bio-oil, byproducts, etc.) is also falling in the same range, but as discussed before, the impact on GHG emissions is significantly higher. BioH_2 using cyanobacteria or green algae, and photo(electro)chemical water splitting or photoelectrolysis have a larger potential for the future large-scale production of renewable $H_{2\,at}$ competitive costs, but intense R&D effort is necessary to realize this potential.

7.5.3 *Assessing the Production of Light Olefins from CO_2*

The process flow-sheet in light olefin production from CO_2 (Centi et al. 2011a) is based on a first step of H_2 production by electrolysis, either using PEM electrolyzers or high-temperature SOEC, and electrical energy deriving from RE. The core of the process is the combination of rWGS and modified Fischer–Tropsch reactions. The two steps are in separate stages for the optimization of the relative catalysts and reaction conditions. Inorganic membranes permeo-selective to water are integrated in the reactors to improve the performance of the process. H_2 is added in part between the two stages and the methane, light alkanes, and >C5 hydrocarbons produced in the process are recycled. The overall process, due to the formation of water, has slightly exothermic[5] results. After thermal recovery, the light olefins are separated in a sequence of columns similar to the steam cracking process, and CO and CO_2 are recycled to the rWGS unit.

A technoeconomic analysis of this possibility (Centi et al. 2011a) indicates that for a predicted renewable H_2 cost as a target for the year 2020 (US$2–3/kg H_2), the process is economically valuable. Current renewable H_2 cost is still higher, but not so much as to not consider in more detail the possibility to produce light olefins from CO_2.

7.6 Routes to Solar Fuels

Although the concept of solar fuels is often broad and typically includes H_2 production using solar energy, we do not consider hydrogen strictly a solar fuel here, but an intermediate to produce the true solar fuels in the form of liquid (easily storable and with high energy density) chemicals. Between the possible energy vector, those deriving from the reduction of CO_2 are preferable for the reasons discussed above and which are also discussed in more detail elsewhere (Centi and Perathoner 2010, 2011).

There are different possible energy vectors that derive from the hydrogenation of CO_2, either directly or through the intermediate stage of the rWGS reaction (Eq. 7.1) to produce syngas (mixture of CO/H_2) which can then be converted through the already established (and commercially applied) routes:

- Formic acid,[6] which may be used in formic acid may be used in formic acid fuel cells or as a vector to store and transport H_2. The reaction of synthesis is reversible and formic acid can be catalytically decomposed in mild conditions to form back H_2 and CO_2 (Ferenc 2008; Leitner 1995)

[5] An exothermic reaction creates heat that must be evacuated to the external ambient.

[6] *Formic acid* (also called *methanoic acid*) is the simplest carboxylic acid. Its chemical formula is HCOOH or HCO_2H.

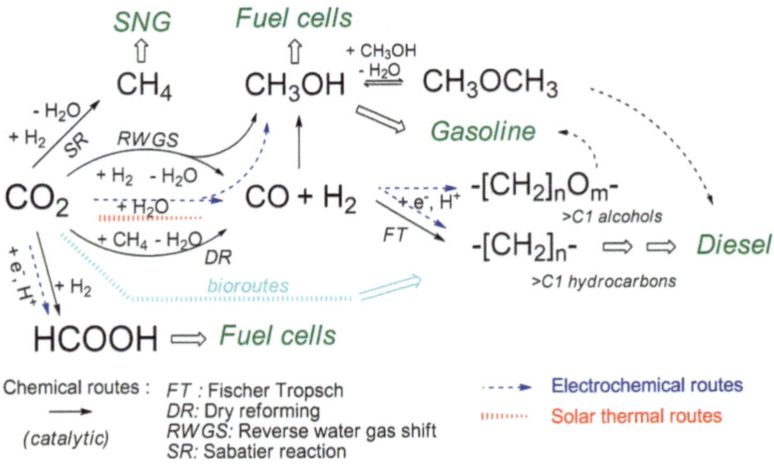

Fig. 7.1 CO_2 routes to solar fuels. The renewable energy is used either directly (in the solar thermal production of syngas) or indirectly, according to two main possibilities: (i) production of renewable H_2 or (ii) production of electrons, or electrons/protons (by water photo-oxidation), used in the electrochemical routes

- Methanol and dimethylether (DME)
- Methane (substituted natural gas, SNG)
- >C1 alcohols or hydrocarbons

An overview of the different possibilities and routes is presented in Fig. 7.1. The main routes are chemical (catalytic), but the possibility of using electrochemical routes or solar thermal routes is also shown. Syngas may also be produced by reaction with hydrocarbons (particularly methane) through so-called dry reforming (DR). The main potential advantage of this route is that it can be applied directly to flue gases[7] (even if technical problems exist), whereas all the other routes require a first step of separation of CO_2 from the flue gases. However, DR is an endothermic reaction occurring at high temperature (about 900–1,000 °C) and with the formation of carbon (which deactivates the catalyst) as a side reaction. The reaction may be combined with the wet reforming of methane (with H_2O) and its partial oxidation (so-called try-reforming) (Song 2006).

The catalytic chemistry of the rWGS reaction and the following transformation to methanol/DME, or hydrocarbons via Fischer–Tropsch, and the subsequent production of gasoline (methanol-to-gasoline, MTG) or of diesel via hydrocracking of

[7] *Flue gas* is the gas exiting to the atmosphere via a flue, which is a pipe or channel for conveying exhaust gases from a fireplace, oven, furnace, boiler, or steam generator. Quite often, the flue gas refers to the combustion exhaust gas produced at power plants. Its composition depends on what is being burned, but it will usually consist of mostly nitrogen (typically more than two thirds) derived from the combustion air, carbon dioxide (CO_2), and water vapor as well as excess oxygen.

the alkanes produced in the Fischer–Tropsch process (using Co-based catalysts) is well established, even if there is still need of development, due to the change of feed composition starting from CO_2 instead of from syngas. Also in terms of process development most of the knowledge necessary is available. Minor technological barriers to develop these routes are thus present. Only for the synthesis of formic acid, either catalytically or electrocatalytically, is there still need of development in terms of productivity and stability.

The main gap in the catalytic routes of CO_2 conversion to fuels is economic, with the cost of production of renewable H_2 as the key factor. However, opportunities already exist in terms of available (low-cost) sources of renewable H_2 which make the production of fuels from CO_2 interesting. Mitsui Chemicals and Carbon Recycling International are two companies that are running pilot plant projects to exploit the conversion of CO_2 to methanol, and Mantra Venture Group and DNV (Det Norske Veritas) are exploring at pilot-plant scale the electro reduction of CO_2 to formic acid. RCO2 AS has instead developed a pilot-scale process based on recovery of CO_2 from flue gas and its conversion to methane using renewable H_2. Details on these processes and the related chemistry and catalysis are discussed elsewhere (Centi and Perathoner 2009; Quadrelli et al. 2011). A book discussing the various possible routes for CO_2 conversion was recently published from Aresta (2010). Other authors have also recently published reviews on this topic (Dorner et al. 2010; Jiang et al. 2010; Olah et al. 2009a; Sakakura et al. 2007; Schaefer et al. 2010).

The catalytic synthesis of higher alcohols from CO_2 is an interesting route, but still not competitive. New interesting catalysts, however, have been developed. Conversion of CO_2 to higher alcohols and hydrocarbons (≥C2) using biocatalysis or electrocatalysis methodologies is also an interesting route, but is still at the preliminary stage (Centi and Perathoner 2010; Centi et al. 2011b). Genetically modified cyanobacteria have recently been reported to consume carbon dioxide in a set of steps to produce a mixture of isobutyraldehyde (primarily) and isobutanol. Using a gas-phase electrocatalysis approach CO_2 may be reduced to a mixture of C2 hydrocarbon and alcohols, mainly isopropanol (Ampelli et al. 2010). Artificial metabolic pathways involving enzymes or cyanobacteria have been proposed to use NADPH (nicotinamide adenine dinucleotide phosphate-oxidase) and ATP (adenosinetriphosphate) from photosynthesis for the synthesis of n-butanol (UCLA) or isobutene (Global Bioenergies) directly from carbon dioxide and water.

The conversion of CO_2 to hydrocarbons and alcohols using radiation and/or electrons (photo-, electro-, or photoelectrocatalytic methods) is actively being investigated worldwide with the aim of developing nonbiomimetic artificial leaves that use solar light, CO_2, and water to make solar fuels directly. Even if impressive results have been reported recently, it is still a long-term solution for still low productivity. It has also been shown that the intrinsic barriers for practical applications on the investigated routes often are unclear, and it is thus necessary to focus the research better. Details are discussed elsewhere (Centi and Perathoner 2009, 2010; Centi et al. 2011a, b).

7.7 Conclusions

This short overview has presented the concept of how the conversion of CO_2 through the use of renewable energy sources offers new possibilities to develop innovative new approaches to improve sustainability of chemical and energy production. After introducing the general aspects and the motivations for this R&D topic, as well as how these aspects are a key component of the chemical and energy industries' strategies (particularly in Europe), especially in terms of addressing resource efficiency, this chapter discusses two aspects in more detail:

- The challenge of using CO_2 for the production of light olefins (ethylene, propylene) as an example of the possible reuse of CO_2 as a valuable carbon source and an effective way to introduce renewable energy in the chemical industry value chain, improve resource efficiency, and limit greenhouse gas emissions
- The conversion of CO_2 to fuels using sunlight (solar fuels), with a concise presentation of the possible routes, also with some indication of the industrial developments in the field, in order to demonstrate this relevant opportunity to develop effective energy vectors for the storage of solar energy which are integrated into the existing energy infrastructure and allow a smooth but fast transition to a more sustainable energy future.

These examples show that it is possible to create CO_2-based and resource-efficient chemical and energy production. It is thus possible to expand the current view of considering CO_2 as a valuable resource only for target high-value products, which, however, cannot significantly contribute to the recycling of carbon dioxide and the creation of a low-carbon economy. It is also evident that the process of producing light olefins from CO_2 can be an enabling factor for an effective introduction of renewable energy into the chemical production chain, and the push towards solar fuels an effective driving force towards more sustainable energy.

References

Ampelli C, Centi G, Passalacqua R, Perathoner S (2010) Synthesis of solar fuels by novel photoelectrocatalytic approach. Energy Environ Sci 3:292–301

Aresta M (2010) Carbon dioxide as chemical feedstock. Wiley, Weinheim

Bockris JO'M (2008) Hydrogen no longer a high cost solution to global warming: new ideas. Int J Hydrog Energy 33:2129–2131

Centi G, Perathoner S (2009) Opportunities and prospects in the chemical recycling of carbon dioxide to fuels. Catal Today 148:191–205

Centi G, Perathoner S (2010) Towards solar fuels from water and CO_2. ChemSusChem 3:195–208

Centi G, Perathoner S (2011) CO_2-based energy vectors for the storage of solar energy. Greenh Gases Sci Technol 1:21–35

Centi G, Iaquaniello G, Perathoner S (2011a) Can we afford to waste carbon dioxide? Carbon dioxide as a valuable source of carbon for the production of light olefins. ChemSusChem 4:1265–1273

Centi G, Perathoner S, Passalacqua R, Ampelli C (2011b) Solar production of fuels from water and CO_2, Chapter 4. In: Muradov NZ, Veziroğlu T (eds) Carbon-neutral fuels and energy carriers. CRC Press, Boca Raton, pp. 291–324

Djomo SN, Blumberga D (2011) Comparative life cycle assessment of three biohydrogen pathways. Bioresour Technol 102:2684–2694

Dorner RW, Hardy DR, Williams FW, Willauer HD (2010) Heterogeneous catalytic CO_2 conversion to value-added hydrocarbon. Energy Environ Sci 3:884–890

Elder R, Allen R (2009) Nuclear heat for hydrogen production: coupling a very high/high temperature reactor to a hydrogen production plant. Prog Nucl Energy 51:500–525

European Commission (2011) A roadmap for moving to a competitive low carbon economy in 2050. In: COM (2011) 112 final, Brussels, 8 Mar 2011

Ferenc J (2008) Breakthroughs in hydrogen storage – formic acid as a sustainable storage material for hydrogen. ChemSusChem 1:805–808

Garcia-Martinez J (ed) (2010) Nanotechnology for the energy challenge. Wiley, Weinheim

Gibson TL, Kell NA (2008) Optimization of solar powered hydrogen production using photovoltaic electrolysis devices. Int J Hydrog Energy 33:5931–5940

Gray HB (2009) Powering the planet with solar fuel. Nat Chem 1:7

Hauch A, Ebbesen SD, Jensen SH, Mogensen M (2008) Highly efficient high temperature electrolysis. J Mater Chem 18:2331–2340

IEA (2010) Energy technology perspectives 2010. OECD/IEA, Paris

IEA (2011) World energy outlook 2010. OECD/IEA, Paris

IPCC (2011) Special report on renewable energy sources and climate change mitigation, May 2011. Available at: srren.ipcc-wg3.de

Jiang Z, Xiao T, Kuznetsov VL, Edwards PP (2010) Turning carbon dioxide into fuel. Philos Trans R Soc Math Phys Eng Sci 368:3343–3364

Jiang N, Yang G, Zhang X, Wang L, Shi C, Tsubaki N (2011) A novel silicalite-1 zeolite shell encapsulated iron-based catalyst for controlling synthesis of light alkenes from syngas. Catal Commun 12:951–954

Kang SH, Bae JW, Woo KJ, Sai Prasad PS, Jun KW (2010) ZSM-5 supported iron catalysts for Fischer-Tropsch production of light olefin. Fuel Proc Technol 91:399–403

Leitner W (1995) Carbon dioxide as a raw material: the synthesis of formic acid and its derivatives from CO_2. Angew Chem Int Ed 34:2207–2221

Lewis NS, Crabtree G, Nozik A, Wasielewski M, Alivisato P (2005) Basic research needs for solar energy utilization. US Department of Energy Pub, Washington, DC

Marquevich M, Sonnemann GW, Castells F, Montane D (2002) Life cycle inventory analysis of hydrogen production by the steam-reforming process: comparison between vegetable oils and fossil fuels as feedstock. Green Chem 4:414–423

Metz B, Davidson O, de Coninck H, Loos M, Meyer L (eds) (2006) Carbon dioxide capture and storage, IPCC special reports. Cambridge University Press, Cambridge

Millet P, Ngameni R, Grigoriev SA, Mbemba N, Brisset F, Ranjbari A, Etievant C (2010) PEM water electrolyzers: from electrocatalysis to stack development. Int J Hydrog Energy 35:5043–5052

Morris AJ, Meyer GJ, Fujita E (2009) Molecular approaches to the photocatalytic reduction of carbon dioxide for solar fuels. Acc Chem Res 42:1983–1994

Muradov NZ, Veziroglu TN (2008) "Green" path from fossil-based to hydrogen economy: an overview of carbon-neutral technologies. Int J Hydrog Energy 33:6804–6839

Neelis M (2007) Approximation of theoretical energy-saving potentials for the petrochemical industry using energy balances for 68 key processes. Energy 32:1104–1123

Nozik AJ (2010) Nanoscience and nanostructures for photovoltaics and solar fuels. Nano Lett 10:2735–2741

Olah GA, Goeppert A, Prakash Surya GK (2009a) Chemical recycling of carbon dioxide to methanol and dimethyl ether: from greenhouse gas to renewable, environmentally carbon neutral fuels and synthetic hydrocarbons. J Org Chem 74:487–498

Olah GA, Goeppert A, Surya Prakash GK (2009b) Beyond oil and gas: the methanol economy, 2nd edn. Wiley, Weinheim

Park JY, Lee YJ, Jun KW, Bae JW, Viswanadham N, Kim YH (2009) Direct conversion of synthesis gas to light olefins using dual bed reactor. J Ind Eng Chem 15:847–853

Quadrelli EA, Centi G, Duplan JL, Perathoner S (2011) Carbon dioxide recycling: emerging large-scale technologies with industrial potential. ChemSusChem 4:1194–1215

Ren T, Patel M, Blok K (2006) Olefins from conventional and heavy feedstocks: energy use in steam cracking and alternative processes. Energy 31:425–451

Ren T, Patel MK, Blok K (2008) Steam cracking and methane to olefins: energy use, CO_2 emissions and production costs. Energy 33:817–833

Roy SC, Varghese OK, Paulose M, Grimes CA (2010) Toward solar fuels: photocatalytic conversion of carbon dioxide to hydrocarbons. ACS Nano 4:1259–1278

Ruether J, Ramezan M, Grol E (2005) Life-cycle analysis of greenhouse gas emissions for hydrogen fuel production in the United States from LNG and coal, DOE/NETL-2006/1227 report, Nov 2005. Available at: www.netl.doe.gov/energy-analyses/pubs/H$_2$_from_Coal_LNG_Final.pdf

Sai Prasad PS, Bae JW, Jun KW, Lee KW (2008) Fischer–tropsch synthesis by carbon dioxide hydrogenation on Fe-based catalysts. Catal Surv Asia 12:170–183

Sakakura T, Choi JC, Yasuda H (2007) Transformation of carbon dioxide. Chem Rev 107:2365–2387

Sartbaeva A, Kuznetsov VL, Wells SA, Edwards PP (2008) Hydrogen nexus in a sustainable energy future. Energy Environ Sci 1:79–86

Saygın D, Patel MK, Tam C, Gielen DJ (2009) Chemical and petrochemical sector. Potential of best practice technology and other measures for improving energy efficiency. International Energy Agency (IEA) Pub, Paris

Schaefer M, Behrendt F, Hammer T (2010) Evaluation of strategies for the subsequent use of CO_2. Front Chem Eng China 4:172–183

Song C (2006) Global challenges and strategies for control, conversion and utilization of CO_2 for sustainable development involving energy, catalysis, adsorption and chemical processing. Catal Today 115:2–32

Spath PL, Mann MK (2004) Life cycle assessment of renewable hydrogen production via wind/electrolysis, NREL/MP-560-35404 report, Feb 2004. Available at: www.nrel.gov/docs/fy04osti/35404.pdf

Utgikar V, Thiesen T (2006) Life cycle assessment of high temperature electrolysis for hydrogen production via nuclear energy. Int J Hydrog Energy 31:939–944

Züttel A, Borgschulte A, Schlapbach L (2008) Hydrogen as a future energy carrier. Wiley, Weinheim

Chapter 8
Introduction to Hydrogen and Fuel Cell Technologies and Their Contribution to a Sustainable Energy Future

Deborah J. Jones

Abstract Research and development of hydrogen and fuel cell technologies are motivated by the same drivers as for other new energy production/conversion/storage options, in particular the increase in greenhouse gas emissions and in sea and land mass temperatures, and peaking of oil production capacity and the technical difficulties and safety issues associated with extracting oil from offshore deep drilling below the seabed, which together lead towards a global requirement for use of lower fossil carbon energy sources. In this context, this chapter outlines actual and potential roles for hydrogen and fuel cell technologies. It provides a short historical perspective of fuel cells and describes fuel cell types and their applications, in particular automotive and stationary fuel cell uses. Directions in fuel cell materials research on electrocatalysts and their supports and electrolyte membranes are described in a final section.

Keywords Electrolysis • Energy conversion • Fuel cell • Hydrogen

List of Acronyms

AFC	Alkaline fuel cell
CO	Carbon monoxide
CO_2	Carbon dioxide
DLR	Deutsche Zentrum für Luft- und Raumfahrt (German Aerospace Centre)
e-	Symbol of electron
EU	European Union

D.J. Jones (✉)
Institut Charles Gerhardt Montpellier, UMR CNRS 5253,
Université Montpellier 2, 34095 Montpellier, Cedex 5, France
e-mail: Deborah.Jones@univ-montp2.fr

J.B. Saulnier and M.D. Varella (eds.), *Global Change, Energy Issues and Regulation Policies*, Integrated Science & Technology Program 2, DOI 10.1007/978-94-007-6661-7_8,

EUCAR	European Council for Automotive R&D
FCHV	Fuel cell hydrogen vehicle
FCV	Fuel cell vehicle
Fe/N/C	Iron, nitrogen, carbon catalyst
H+	Proton (hydrogen nucleus)
H_2	Hydrogen molecule
HOPG	Highly oriented pyrolytic graphite
Hyfleet CUTE	This is a European project comparing the advantages and disadvantages of hydrogen internal combustion engine (ICE) buses with fuel cell buses. CUTE stands for Clean Urban Transport for Europe and the goal of the project is to test and demonstrate hydrogen buses in 10 different cities in Europe, Asia, and Australia to reduce CO_2 emissions and move away from fossil fuels.
IPCC	Intergovernmental Panel on Climate Change
IPHE	International Partnership for Hydrogen and Fuel Cells in the Economy.
kW	Kilowatt (SI power unit)
mb/d	Million barrels per day
MCFC	Molten carbonate fuel cell
MEA	Membrane electrode assembly
OPEC	Organization of the Petroleum Exporting Countries
PAFC	Phosphoric acid fuel cell
PCFC	Proton ceramic fuel cell
PEM	Polymer electrolyte membrane
PEMFC	Proton exchange membrane fuel cells, also known as polymer electrolyte fuel cells (PEFC)
PFSA	Perfluorosulfonic acid
Pt/M	Alloy of platinium with a metal (M)
Pt	Symbol of platinum
PURE	Promoting Unst Renewable Energy (Unst is one of the North Isles of the Shetland Islands, Scotland)
SI	Système International d'unités (International System of units)
SOFC	Solid oxide fuel cell
WHFS	Worldwide hydrogen fueling stations

8.1 Sustainable Energy Future

From what could have been considered as recently as 15–20 years ago, as bleating in the wind, to a recognized urgent problem with huge geopolitical implications, the means to assure a sustainable energy future are in the forefront of international considerations. Without undue exaggeration, it is something of a race against time in an overall context where the unpredictable or at least unexpected can occur and change international opinion in one fell stroke, with possible significant repercussions on future energy supply options, as has happened following the tsunami and earthquake in Japan in March 2011 and (although to a lesser extent) the deepsea oil rig explosion in the Gulf of Mexico in May 2010. The main driver for research and implementation of new energy technologies is their contribution to an overall mix of energy sources, where renewables play a much more important role than in the past, and which require means for energy conversion, and hugely increased storage capacity. The trilogy of declining fossil reserves, increasing emissions, and climate change concerns, and the need for energy security motivate accelerating efforts, against a background where the current infrastructure, developed for high-carbon energy, "locks-in" introduction of new energy technologies, and makes the 2 °C climate change goal more challenging and expensive to meet.

Data logged at the Mauna Loa Observatory (Hawaii) indicates an increase in atmospheric carbon dioxide from ca. 320 to ca. 390 ppm in the 50 years to 2011. The annual mean growth rate is also instructive, inasmuch as it shows a steady rise from 0.8 ppm average annual increase in CO_2 in the decade 1950–1960, to 1.9 ppm annual increase in the decade 2000–2010 (Earth Systems Research Laboratory 2011). Despite economic difficulties at the end of this decade, global primary energy demand rebounded by 5 %, pushing CO_2 emissions to a new high (World Energy Outlook 2011). In its fourth Assessment Report, the Intergovernmental Panel for Climate Change (IPCC 2007) provides charts that show that global land mass and ocean temperature increases since the beginning of the last century are modeled correctly only when anthropogenic impact is included, those models considering only natural forcings (forcing factor within the climate system) providing poor agreement. It was concluded that anthropogenic warming over the last three decades has likely had a discernible influence on the global scale on observed changes in many physical and biological systems (IPCC 2007). Improving technical aspects of detection and attribution, especially harmonizing terms and definitions, is an important goal to advance this topic, with emphasis on impact-relevant changes in the climate system and impacts on natural and human systems (IPCC 2009). In its most recent World Energy Outlook (World Energy Outlook 2011) the International Energy Agency predicts that although production of conventional crude oil will remain at current levels before declining slightly to around 68 mb/d (million barrels per day) by 2035, to compensate for declining crude oil production at existing fields, 47 mb/d of gross capacity additions are required, which corresponds to twice the current total oil production of all OPEC countries in the Middle East (Fig. 8.1).

Sustained development is therefore needed, mainly to combat the decline in output at existing fields, which will drop by around two thirds by 2035, with pressure from emerging economies on rising oil use by the transport sector. A fundamental change in transport technologies is required through use of plug-in hybrids and electric vehicles (fuel cells and battery vehicles) to bring about a drop in CO_2 emissions from this sector. Renewable energy sources are predicted increasingly to occupy center stage, with an increase in power production from nonhydropower renewables from a current 3–15 % by 2035, with China and the European Union expected to drive this expansion. The challenges are well documented—their intermittent character, energy density, and their integration into

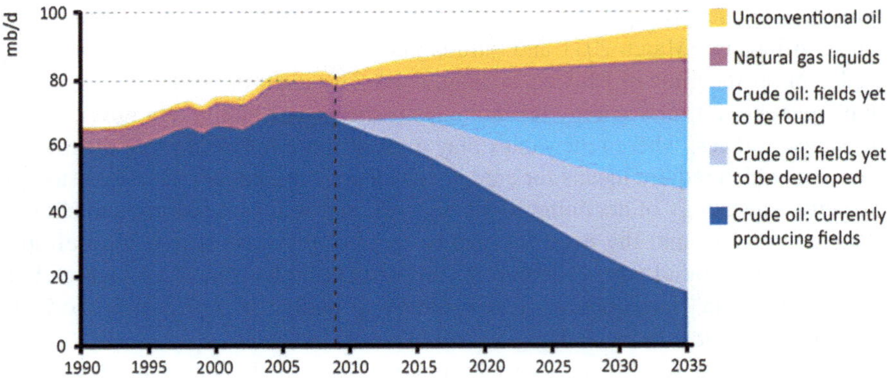

Fig. 8.1 World oil production by type (World Energy Outlook Executive Summary © OECD/IEA 2011)

the grid—nevertheless these are outweighed by the environmental protection and energy security that they ensure, and data estimating that photovoltaic cells covering 1 % of the usable land mass would meet world energy demand, or that eight 100 × 100 km offshore wind farms could produce the energy demand of the European Union are sobering. Accommodating more electricity from renewable sources, sometimes in remote locations, will require additional investment in transmission networks amounting to 10 % of total transmission investment: in the European Union; 25 % of the investment in transmission networks is needed for this purpose (World Energy Outlook 2011).

8.2 Hydrogen as a Future Energy Carrier

Hydrogen is the simplest element, comprising one proton and one electron. It is the lightest and most abundant of all elements. About 90 % of all atoms in the universe are hydrogen. Its gravimetric energy density is more than twice that of carbon-containing fuels such as gasoline or natural gas, but its volumetric energy density is low, and is improved by either high pressures (compressed gas) or extremely low temperatures (cryogenic liquid), although the low boiling point of hydrogen makes liquefaction very energy intensive. The energy storage capability of hydrogen is increasingly being considered as a means of buffering fluctuations in energy from renewable sources.

Hydrogen does not exist in gaseous form in a natural state on earth, but is always mixed with other elements: combined with oxygen in water; combined with carbon in methane, coal, petroleum, and with both these elements in methanol and ethanol. A primary energy source is therefore needed to release hydrogen from such resources. In the hydrogen cycle, energy from a renewable source (wind, solar) dissociates water to hydrogen and oxygen. The latter is generally released to the atmosphere (although it could be dried and purified and used for other applications), and hydrogen stored and transported as required, and used as fuel in a fuel cell, where it produces electricity, heat, and water, thus closing the cycle. In this cycle, renewable energy is stored as hydrogen for use as and when required and, although the number of single processes (hydrogen production, conditioning, storage, distribution, transportation) reduces the overall energy efficiency (Züttel 2008), these considerations are to be weighed against the opportunity of large-capacity renewable energy storage. In general, hydrogen technologies favor decentralized systems and local and individual solutions for hydrogen generation rather than central generation and delivery via a pipeline. The hydrogen network is estimated at around 1,600 km in Europe and 1,100 km in North America, so that supply from a central production site has to take place via road transport, either a compressed gaseous hydrogen or liquid hydrogen.

The most common process for hydrogen production currently is by steam reforming using hydrocarbons (e.g., natural gas, liquid petroleum gas, and naphtha) as feed. Until recently, steam reforming plants were designed for production capacity ranging from 200 up to 100,000 Nm^3/h. By using a newly developed type of reformer it is now possible to serve ranges of 50 up to 200 Nm^3/h economically by compact, small-scale hydrogen generation plants based on steam reforming of natural gas. This capacity range is well suited for an on-site supply of small vehicle fleets with hydrogen (Global Hydrogen Bus Platform). Hydrogen supply pathways in cities operating fuel cell buses and hydrogen internal combustion engine buses developed in the European Commission's HyFleet:Cute are shown in Fig. 8.2 (Global Hydrogen Bus Platform).

8.3 Fuel Cells

Fuel cells convert the chemical energy of the fuel to electricity by the electrocatalyzed reaction of hydrogen and oxygen. Using hydrogen and air, water is the only by-product, thus eliminating all local production of greenhouse gas emissions. The fundamental principles underlying water electrolysis and hydrogen and oxygen recombination in a fuel cell have been known for almost 200 years, since the publications of Schönbein (1839) and Grove (1839, 1842). Fuel cell history is characterized by developments in fits and starts, thus following these first observations, any progress slumbered for more than a century, to accelerate again with the advent of the space program, where the fuel cells provided not only electricity but also water. From this same period date the first stationary and terrestrial transportation applications, as well as portable machinery tools such as welding machines. Following a further period of relative dormancy, broken only by a brief revival of interest during the time of the energy crises of the 1970s, fuel cell technologies have attracted sustained attention since the end of the 1980s.

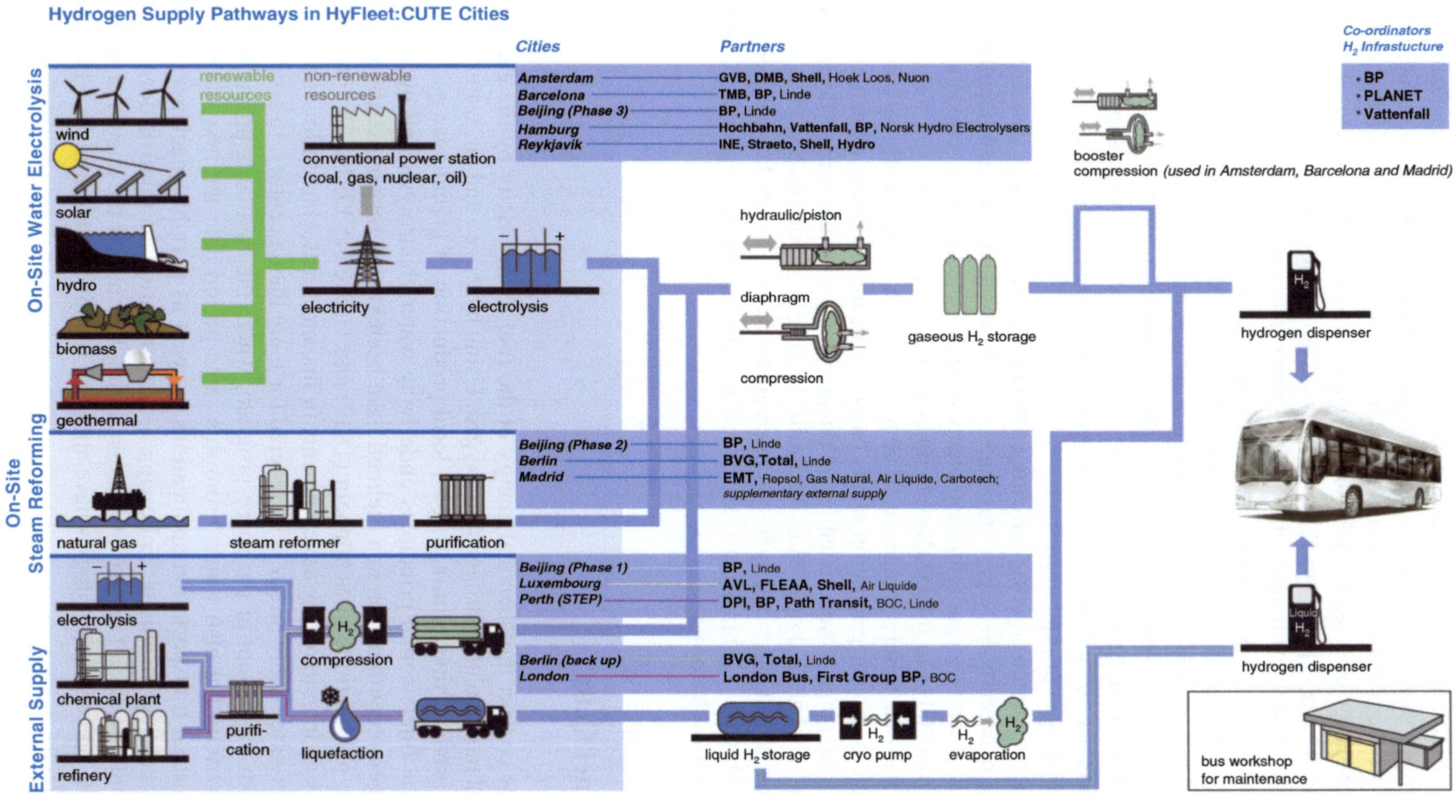

Fig. 8.2 Hydrogen supply pathways by water electrolysis on site, on-site steam reforming, and external supply in Hyfleet:CUTE cities (*Source*: Hyfleet:CUTE 2012)

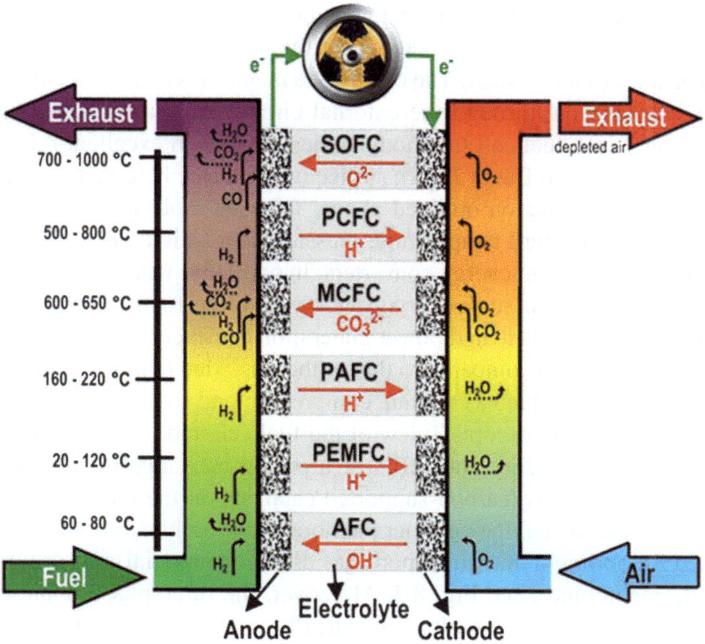

Fig. 8.3 Schematic representation of different fuel cell types, temperature range of operation, and charge carrier species. *AFC* alkaline fuel cell, *PEMFC* polymer electrolyte membrane fuel cell, *PAFC* phosphoric acid fuel cell, *MCFC* molten carbonate fuel cell, *PCFC* proton ceramic fuel cell, *SOFC* solid oxide fuel cell (Figure courtesy Erich Erdle, efceco)

There exist different types of fuel cell using different electrolytes and operating at different temperatures. Figure 8.3 illustrates this schematically, and provides operating temperature ranges for alkaline fuel cells, proton exchange membrane fuel cells (20–160 °C), phosphoric acid fuel cells (160–220 °C), molten carbonate fuel cells (600–650 °C), proton ceramic fuel cells (500–800 °C), and solid oxide fuel cells (700–1,000 °C). Electrolytes in these fuel cell types range from liquid (predominant in AFC, although alkaline anion exchange membranes are under intensive research development, and PAFC), through polymer (PEMFC) to ceramic (PCFC, SOFC). In addition to hydrogen fuel, low-temperature fuel cells can also be fed with methanol, ethanol, or other hydrogen-rich fuels (where the challenge lies in the development of highly active anode catalysts for methanol , ethanol etc. oxidation), whereas the operating temperature of solid oxide fuel cells (>700 °C) is sufficient for internal reforming of natural gas, for example. With pure hydrogen as fuel, low-temperature (to 200 °C) fuel cells exhibit major advantages regarding power density and durability in dynamic operation, and the PEMFC can be taken as an example to outline fuel cell operation briefly. As shown in Fig. 8.3, the basic structure of a fuel cell is that of two electrodes, separated by an electrolyte, and connected by an external circuit. Hydrogen gas (H_2) is oxidized at the anode on catalyst particles that disassociate H_2 to $2H^+$ (protons) and 2 e- (electrons).

The protons (associated with water molecules) are transferred across the polymer electrolyte membrane to the cathode, where they combine with oxygen at electrocatalyst particles to form water. The electrolyte material is electronically insulating, and the electrons flow through the external circuit. It is not the purpose of this chapter to enter into fuel cell thermodynamics, for which excellent references are available (Friedrich and Büchi 2008); suffice it to indicate here that the equilibrium cell voltage for the reaction outlined above, for each molecule of hydrogen converted to water at standard temperature, pressure, and activity, is 1.23 V. Fuel cells can be extremely efficient energy converters, in particular when use is also made of the heat produced, as in stationary systems where the total (electric + heat) efficiency of the combined heat and power generation can exceed 80 %. The efficiency of the system is reduced compared to that of the stack due to power requirements of peripheral components such as the fan, compressor, and so on.

Figure 8.3 is a "macroscopic" view of the fuel cell and its membrane electrode assembly (MEA). In reality, the MEA of a proton exchange membrane fuel cell comprises nanomaterials (carbon-supported catalysts) at the interface between the gas diffusion material and the polymer membrane, the architecture of both of which is designed for optimum mass transport (gas diffusion material) and proton transfer (electrolyte membrane); see Fig. 8.4. Moreover, the three-phase boundary at the anode and cathode interface regions is critically important in determining catalyst reaction kinetics and fuel cell performance.

8.4 A Case Study: The Vestenskov Hydrogen Community in Denmark

Vestenskov, Lolland in Denmark (population 540) provides an excellent example of how hydrogen can be used for the storage of energy from a renewable source. Lolland's wind turbines produce some 50 % more electricity than the area can use. Because wind energy as such cannot be stored, and in the absence of an adequately dimensioned means of electricity storage, when insufficient wind power is available, households must revert to the use of fossil sources for heating and power. In this 6-year project, the Danish consortium Dansk Mikrovarme is testing the potential of storing wind energy by converting it to hydrogen. Wind power from the turbines is used to supply power to alkaline electrolyzers, where water is split into its hydrogen and oxygen components. The hydrogen is stored until needed, and is piped to individual houses and used in microcombined heat and power units. These units provide both the electricity supply for the homes, and heating, and they have replaced former gas-fired central heating boilers. The Vestenskov hydrogen community project is taking place in three phases, the first of which was inaugurated in November 2006. In September 2008 the first five households were connected to the underground hydrogen pipeline which enabled assessment of the system's operational stability and safety. In the current phase,

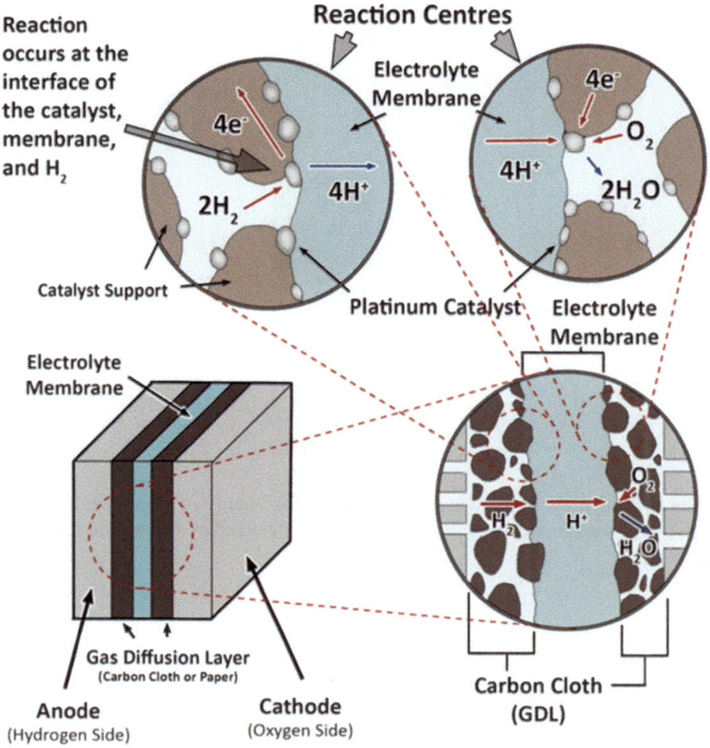

Fig. 8.4 Membrane-electrode assembly of a proton exchange membrane fuel cell and its constituent interfaces

35–40 households are being equipped with a fuel cell module the size of a small central heating unit, with an output of approximately 2 kW. Projects such as this are important demonstrations of how hydrogen can be used as storage medium for renewable electricity (see Fig. 8.5).

A second case study is provided by the PURE (Promoting Unst Renewable Energy) project, on Unst, one of the North Isles of the Shetland Islands, Scotland. Unlike Vestenskov, the issue for this implementation was the lack of electrical grid connection, hence the need for an innovative solution for integrating green technologies on the island. Therefore the main aim of the project was to provide a complete off-grid renewable energy solution for powering an industrial estate. The proposed renewable energy solution was to deliver a reliable source of electrical and heating energy through the use of renewables and energy storage technologies. This led to the selection of storing wind power as the primary energy source into hydrogen, and of hydrogen fuel cell technology to deliver an uninterrupted power supply (PURE project 2007). The PURE project comprises two 15 kW wind turbines, an electrolyzer, high pressure hydrogen storage system, and hydrogen dispensing

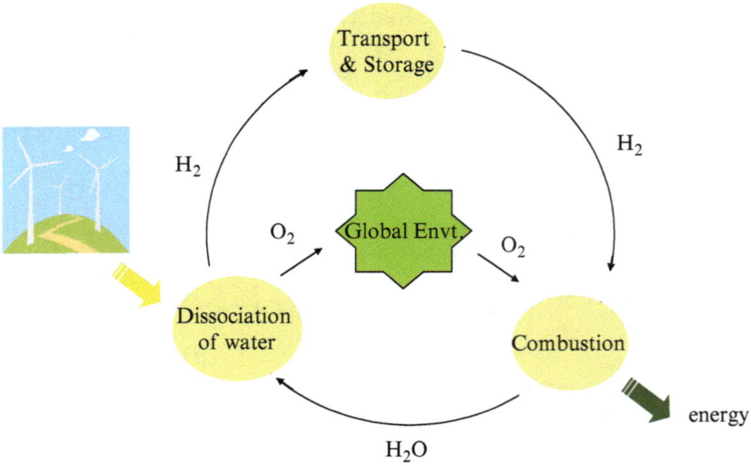

Fig. 8.5 The hydrogen cycle, showing the storage of the energy of a renewable source in hydrogen, through use of a water electrolyzer (to dissociate water) and hydrogen-fed fuel cells (to generate heat and power)

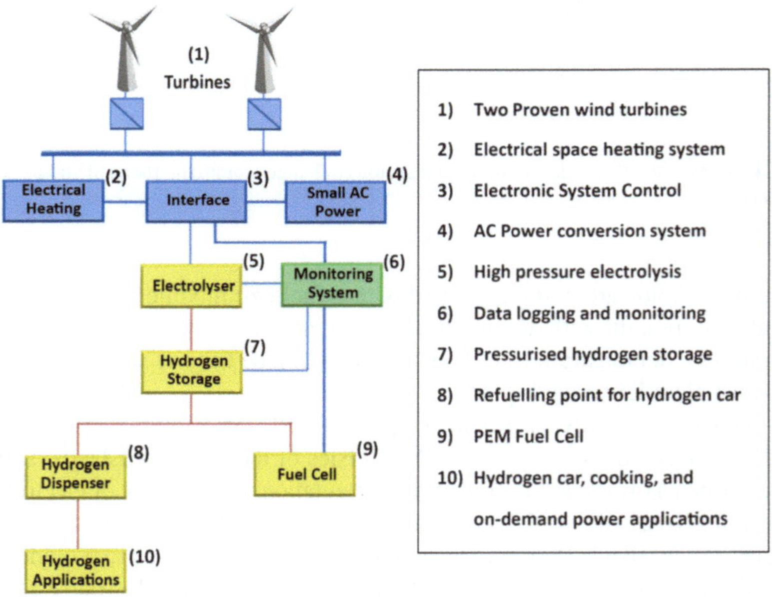

Fig. 8.6 Overview of the PURE project system in Unst, Scotland (*Source*: PURE project 2007, http://www.pureenergycentre.com/pureenergycentre/Pureprojectcasestudy.pdf)

facility. The facility is used to fill cylinders in a fuel cell/battery hybrid vehicle, and is used in a back-up power supply based on a 5 kW fuel cell to supply an industrial estate with electrical power when there is no wind. An overview of the PURE project system is shown in Fig. 8.6.

8.5 Fuel Cells in the Transport Sector

The first fuel cell vehicles date from the 1960s with the Electrovan of General Motors (1966) being the first hydrogen fuel cell car on record. It was powered by a 5 kW Union Carbide fuel cell, and the vehicle had a range of ca. 190 km, which rivals that of current battery electric vehicles. The Electrovan is displayed at the General Motors Heritage Center near Detroit. The Electrovan never ran on public roads, unlike the passenger car based on an Austin A40 built by Karl Kordesch (1970). The K. Kordesch was a fuel cell hybrid car using seven lead acid batteries and a 6 kW Union Carbide alkaline fuel cell. Compressed hydrogen gas was contained in six tanks strapped to the top of the car and the K. Kordesch ran on public roads for 3 years.

Fuel cells can be used directly for vehicle propulsion, or used as a range extender in a battery/fuel cell hybrid vehicle. Compared with battery electric vehicles, fuel cell powered passenger cars are well adapted for long distances and, just as battery electric vehicles, they provide for clean city operation. In heavy-duty applications and public transport, fuel cells are well adapted for buses for city operation, and several demonstration and implementation projects are underway. In the Hyfleet project, 33 hydrogen fuel cell powered buses are operated in nine cities around the world, Amsterdam, Barcelona, Beijing, Hamburg, London, Luxembourg, Madrid, Perth, and Reykjavik, and hydrogen internal combustion engine powered buses are operated in Berlin (Hyfleet:CUTE project 2012). In the aviation sector, hydrogen derived on-board from jet fuel is under consideration for fuel cells for auxiliary power units (e.g., for on-ground taxiing, in-flight environment control), and fuel cells have also been used in demonstration projects for aviation propulsion. The Antares DLR-H2 was the world's first piloted aircraft capable of performing a complete flight powered by fuel cells alone, with a range of 700 km and an altitude of 2,560 m. Current developments are underway to extend the range (6,000 km) and endurance (50 h; DLR 2010).

The real impact of the potential in terms of reduced emissions and alternative fuels allowed by fuel cell technology awaits the introduction of a mass market application, and arguably the greatest bearing will be brought by fuel cell powered cars, an essential driving force for hydrogen in the economy. With this as a market driver, research efforts directed at widespread use of fuel cells have accelerated in recent years, and have been supported by industrial, governmental, and international initiatives. The challenges for fuel cell passenger vehicles around 10 years ago embraced the need to increase volumetric power density, to improve the ability to start up at subzero temperatures, to extend the driving range, to improve durability and reliability, and to reduce costs. Research has advanced over the intervening period such that currently essentially only the last two challenges remain. For example, the driving range has been extended both by improvement in the efficiency of the fuel cell stack, and by means of increasing the amount of hydrogen stored on-board by the use of high-pressure tanks. Thus Toyota's FCHV-adv fuel cell vehicle has a range of 830 km compared to the 330 km of the previous generation of FCHVs, enabled by the use of 70 MPa hydrogen tanks, compared to 35 MPa formerly used. Honda's Clarity has a range of 620 km, compared to the 360 or 470 km range of FCX generations of 2003 and 2005, respectively. Marked progress has also been made in increasing the output power density per volume, by a factor of two for both of Nissan and Honda fuel cell vehicles, for example. Top speed in current passenger cars is around 150–160 km/h, with fuel cell stacks in the power range of 90–100 kW. These improvements are very considerable, giving range and top speeds that are comparable to current diesel powered engines, for identical capacity in terms of number of passengers that can be carried.

Comparison of the carbon dioxide emissions between various possible power train options must take into account a complete analysis from "well to wheel." Several studies have been made, and recent data show that fuel cell vehicles fed with hydrogen from wind power, biomass, or compressed natural gas have respectively 95, 93, and 46 % lower CO_2 emissions than from a reference gasoline internal combustion engine (JRC-EUCAR-CONCAWE 2008). These values can be compared with those from battery electric vehicles: 100, 96, and 56 % lower emissions when charged using electricity from wind power, biomass, and EU-mix, respectively,

Fuel cell vehicles fed with hydrogen, and battery vehicles based on renewable electricity have the highest potential for CO_2 emission reduction. Both of these options represent technology step changes compared to internal combustion engines, and the costs for transformation of technology also have to be considered. Battery electric vehicles fueled by 100 % electricity from EU-mix or fuel cells fueled by hydrogen from fossil sources are intermediate solutions: they have higher emissions of CO_2 equivalent per km, higher well-to-wheel energy consumption, and relatively lower costs for transformation of technology than fuel cell or battery electric vehicles fueled by renewable hydrogen or electricity.

Infrastructure is a vitally important factor for market introduction of fuel cell vehicles and several recent H_2 mobility initiatives point the way forward whereby industrial participants from the utility, gas, infrastructure, and global car manufacturing sectors jointly evaluate the potential for hydrogen as a fuel in a given country before developing an action plan for an anticipated roll-out of fuel cell vehicles and appropriate hydrogen infrastructure in a concerted implementation program. The first such initiative in Europe was that in Germany (2009), followed by the United Kingdom (2012). A similar initiative has been undertaken in Japan (2011). Various electronic sources are available that provide an up-to-date vision of hydrogen fueling stations worldwide, the partners involved, and the hydrogen production technologies adopted (IPHE 2009).

8.6 Directions in Fuel Cell Materials Research

Increase in fuel cell durability and reduction in cost are closely related to materials improvements, in terms of performance, robustness, and composition. Space does not allow anything approaching a comprehensive analysis of the huge advances that have been made in fuel cells materials research over the past decade and a half in particular, and excellent reviews (Rozière and Jones 2003; Jones and Rozière 2008; Gasteiger et al. 2010; Peckham et al. 2010; Thompsett 2010; Xie et al. 2010) and handbooks (Vielstich et al. 2003, 2009) are available to complete this necessarily summary overview, which is restricted to some considerations of catalyst support materials, catalysts, and membranes for proton exchange membrane fuel cells (PEMFC).

In general, catalyst support materials for PEMFC have a high surface area (required to disperse the catalyst particles) and electronic conductivity, and to resist electrochemical corrosion they must be electrochemically stable in the voltage range of fuel cell operation. A specific interaction between catalyst and support enhances activity and durability, and Vulcan XC-72 carbon black is among the most-used fuel cell catalyst supports. Under steady-state fuel cell operation, carbon support corrosion is minor, but is considerably accelerated during stop/start cycles or local hydrogen starvation. Conventional carbon supports can be replaced by fully graphitized carbon supports that are more corrosion resistant. Fundamental study of the influence of argon and nitrogen treatment of highly oriented pyrolytic graphite (HOPG) has shown that Ar-treatment and N-doping promote nucleation and a smaller metal particle size, with a narrower platinum particle size distribution (Zhou et al. 2009). Furthermore, nitrogen doped HOPG has increased resistance to agglomeration on cycling and enhanced intrinsic catalyst activity. In recent work, mesoporous carbons (Joo et al. 2001), conducting diamonds (Wang and Swain 2003), carbon nanotubes (Wu et al. 2011), and carbon nanofibers (Tsuji et al. 2007) have been used as PEMFC catalyst support materials.

Oxide, nitride, and carbide materials are alternative supports to carbon for PEM fuel cells. Recent examples include development of support materials based on tungsten (Saha et al. 2009) and molybdenum (Elezovic et al. 2009) oxides, tungsten carbide (d'Arbigny et al. 2011), and titanium oxide (Bauer et al. 2010) among others. Such support materials generally have the advantage of higher electrochemical stability than carbon under the conditions outlined above and, due to their composition, the metal/support interaction may be stronger. In the case of tungsten carbide, the material has been elaborated in the form of microspheres with high surface area, that are adapted for the deposition of platinum nanoparticles. The tungsten carbide supported platinum has an electrochemical surface area comparable to that of Pt supported on carbon, but significantly higher electrochemical stability on voltage cycling (d'Arbigny et al. 2011). Nanofiber supports provide the opportunity for the laying down of alternative catalyst layer structures and for reductions in precious metal loadings that are a main contributor to fuel cell materials cost. Among methods for the preparation of nanofiber oxide (nitride, carbide) supports, electrospinning is attracting attention as a versatile method leading reproducibly to nanofiber supports or targeted architecture that can be catalyzed during or after the fiber preparation step (Cavaliere et al. 2011).

Fuel cell vehicle durability is currently limited by catalyst degradation, and significant academic and industrial research efforts are currently devoted to resolving the challenges of cost and durability issues associated with state-of-the-art PEM fuel cell catalysts. For a fuel cell vehicle fitted with a 100 kW stack, in which membrane electrode assemblies giving a power density of 1 W/cm^2 are loaded with 0.5 mg/cm^2 of platinum catalyst, 50 g of platinum are required. Improvement in the mass activity, enabling a reduction in platinum loading, is required to reduce costs, and current research directions include routes both to catalysts with ultra-low platinum loadings, and nonplatinum group metal catalysts. Developments over the past 5 years have embraced methods to tailor the platinum nanoparticle size or shape to enhance electrochemical activity (Cavaliere-Jaricot et al. 2007), by forming a monolayer of Pt on a nanoparticle of another metal or alloy (Zhao et al. 2011), through the dealloying of transition metals from bimetallic Pt/M alloys, where compressive strain in platinum-rich nanoparticles following electrochemical dealloying of the transition metal and ion exchange is considered to increase the strength of oxygen adsorption (Neyerlin et al. 1009), and the elaboration of platinum catalysts as continuous layers (Gancs et al. 2008), in the groundbreaking work on oriented "whiskers" of perylene derivatives encapsulated with platinum developed at 3 M over the past decade (Debe et al. 2006). In this last approach, contiguous platinum grains are considered to impart high voltage stability. Ultra-low loadings of platinum have been achieved by a plasma sputtering technique (Rabat and Brault 2008). On the other hand, replacement of platinum by catalysts comprising only nonprecious metals has long been considered a mere scientific curiosity with limited prospective for practical fuel cells with high power density due to the historically low activity for the oxygen reduction reaction. Recently, however, several breakthroughs have occurred in the field of Fe/N/C-catalysts (Jaouen et al. 2011), making them worthy competitors to platinum (Proietti et al. 2011). However, before these Fe/N/C-catalysts can be introduced in commercial polymer electrolyte fuel cells, the durability of the most active Fe/N/C-catalysts reported to date must be greatly improved compared to the current state.

In PEM fuel cell membrane research, efforts are oriented in three directions: further improving the chemical and mechanical stability of fully fluorinated sulfonic acid functionalized membranes (perfluorosulfonic acid, PFSA, membranes), increasing the conductivity of hydrocarbon-based proton conducting membranes (and their chemical and mechanical stability), and in the development of membrane materials for medium- (110–130 °C) and high- (160 °C) temperature PEMFC where water-assisted proton conductivity plays a diminishing role. Computational simulations (Liu et al. 2010) are greatly supporting experimental observations and contribute to the current state of understanding of the microstructure of PFSA membranes, and the importance of microphase separation into the hydrophobic region (structural function) and hydrophilic regions, the latter being characterized by well-identified ion-conducting channels. Such understanding provides guidelines for the development of sulfonic acid functionalized hydrocarbon membranes where particular polymer architectures are required to facilitate interconnected ion channel formation and high conductivity (Park et al. 2011). High conductivity is essential, but far from the only property required of PEM fuel cell membranes that must also be dimensionally stable in the wet and dry states, water insoluble, of high mechanical and thermal stability, chemically stable in the fuel cell environment, without competing ion transport or electronic conductivity, and having a sufficiently robust and scaleable preparation route to keep cost low. In addition, the particular system constraints of automotive fuel cell application add the challenging requirement for membranes to demonstrate sufficiently high proton conductivity at conditions of low relative humidity and at adequately high temperature (Wieser 2004).

Advances have been made in PFSA membranes that have led to new generations of materials in fuel cell applications (Aricò et al. 2010), for example, PFSA polymers having shorter side chain length (Arcella et al. 2003) than in the long side chain Nafion®-type ionomer and lower polymer equivalent weight. Such higher ion exchange capacity materials have a greater tendency to swell in water, and can undergo large dimensional change in a fuel cell operating over a range of different temperatures and relative humidities, where the induced mechanical stress can lead to membrane failure. Mitigation strategies include the use of mechanical reinforcements (Grohs et al. 2010) or the development of covalent cross-linking between polymer chains (Zhang et al. 2009).

Finally, the role played by phosphoric acid-doped basic polymers is increasingly being recognized for its importance in membrane electrode assemblies for PEM fuel cells operating in the medium and high temperature ranges (Li et al. 2009), above the temperature at which PFSA membranes can be used. Benefits include operating features such as no humidification, high CO tolerance, better heat utilization, and possible integration with fuel processing units.

8.7 Outlook

Fuel cells have few commercial applications to date, partly because they must compete with existing technologies, but also because they must be competitive in terms of durability and cost. Recent progress on understanding of degradation/ageing processes in fuel cells has enabled valuable insight into viable operation windows and the need for new and improved materials. Strong research directions in materials research must be continued alongside large demonstration projects. Parallel development of hydrogen infrastructure is essential, and hydrogen mobility initiatives are currently driving the way forward to the concerted introduction of hydrogen infrastructure and fuel cell vehicles. Huge progress has been made in FCV, with regard to cruising ranges, acceleration, and top speeds that cannot so far be achieved by battery vehicles. Finally, the energy storage capability of hydrogen when coupled with renewable energy sources is a means of overcoming the drawbacks arising from the intermittent character of such sources with several landmark demonstration projects in operation using the hydrogen cycle.

Acknowledgments The author thanks Surya Subianto for his assistance with the graphics of this chapter. The research leading to these results has received funding from the European Community's Seventh Framework Programme (FP7/2010-2013) under the call ENERGY-2010-10.2-1: Future Emerging Technologies for Energy Applications (FET) under contract 256821 QuasiDry.

References

Arcella V, Ghielmi A, Tommasi G (2003) High performance perfluoropolymer films and membranes. Ann N Y Acad Sci 984:226–244

Aricò AS, Di Blasi A, Brunaccini G, Sergi F, Dispenza G, Andaloro L, Ferraro M, Antonucci V, Asher P, Buche S, Fongalland D, Hards GA, Sharman JDB, Bayer A, Heinz G, Zandonà N, Zuber R, Gebert M, Corasaniti M, Ghielmi A, Jones DJ (2010) High temperature operation of a solid polymer electrolyte fuel cell stack based on a new ionomer membrane. Fuel Cells 10:1013–1023

Bauer A, Song C, Ignaszak A, Hui R, Zhang J, Chevallier L, Jones D, Roziere J (2010) Improved stability of mesoporous carbon fuel cell catalyst support through incorporation of TiO_2. Electrochimica Acta 55:8365–8370

Cavaliere S, Subianto S, Savych I, Jones DJ, Roziere J (2011) Electrospinning: designed architectures for energy conversion and storage devices. Energy Environ Sci 4:4761–4785

Cavaliere-Jaricot S, Etcheberry A, Herlem M, Noel V, Perez H (2007) Electrochemistry at capped platinum nanoparticle Langmuir Blodgett films: a study of the influence of platinum amount and of number of LB layers. Electrochimica Acta 52:2285–2293

Joint Research Centre – EUCAR – CONCAWE (2008) Well to wheels analysis of future automotive fuels and powertrains in the European context. Retrieved 26 Jan 2012, from http://ies.jrc.ec.europa.eu/uploads/media/V3.1%20TTW%20Report%2007102008.pdf

d'Arbigny JB, Taillades G, Marrony M, Jones DJ, Roziere J (2011) Hollow microspheres with a tungsten carbide kernel for PEMFC application. Chem Commun 47:7950–7952

Debe MK, Schmoeckel AK, Vernstrom GD, Atanasoski R (2006) High voltage stability of nanostructured thin film catalysts for PEM fuel cells. J Power Sources 161:1002–1011

DLR (2010) Antares H3: DLR and Lange aviation develop the next generation of fuel-cell powered aircraft. Retrieved 26 Jan 2012, from http://www.dlr.de/en/desktopdefault.aspx/tabid-6216/10226_read-26189/

Earth Systems Research Laboratory (2011) Trends in atmospheric $CO_{2\ at}$ Mauna Loa, Hawaii, 2011. Retrieved 26 Jan 2012, from http://www.esrl.noaa.gov/gmd/ccgg/trends/mlo.html

Elezovic NR, Babic BM, Radmilovic VR, Vracar LM, Krstajic NV (2009) Synthesis and characterization of MoOx-Pt/C and TiOx-Pt/C nano-catalysts for oxygen reduction. Electrochimica Acta 54:2404–2409

Friedrich KA, Büchi FN (2008) Fuel cells using hydrogen. In: Züttel A, Borgschule A, Schlapbach L (eds) Hydrogen as a future energy carrier. WILEY-VCH, Weinheim, pp 335–364

Fuel Cells 2000 (2009) Worldwide hydrogen fueling stations. Retrieved 26 Jan 2012, from http://www.fuelcells.org/info/charts/h2fuelingstations.pdf

Gancs L, Kobayashi T, Debe MK, Atanasoski R, Wieckowsk A (2008) Crystallographic characteristics of nanostructured thin-film fuel cell electrocatalysts: a HRTEM study. Chem Mater 20:2444–2454

Gasteiger HA, Baker DR, Carter RN, Gu W, Liu Y, Wagner FT, Yu PT (2010) Electrocatalysis and catalyst degradation challenges in proton exchange membrane fuel cells. In: Stolten D (ed) Hydrogen and fuel cells. WILEY-VCH, Weinheim, pp 1–14

Global Hydrogen Bus Platform. On-site steam reforming. Retrieved 26 Jan 2012, from http://www.global-hydrogen-bus-platform.com/Technology/HydrogenProduction/reforming

Grohs JR, Li Y, Dillard DA, Case SW, Ellis MW, Lai Y-H, Gittleman CS (2010) Evaluating the time and temperature dependent biaxial strength of Gore-Select series 57 proton exchange membrane using a pressure loaded blister test. J Power Sources 195:527–531

Grove WR (1839) On voltaic series and the combination of gases by platinum. Lond Edinb Philos Mag J Sci Ser 3 14:127–130

Grove WR (1842) On a gaseous voltaic battery. Lond Edinb Philos Mag J Sci Ser 3 21:417–420

HyFleet – Clean Urban Transport for Europe. Consulted 26 Jan 2012, from http://www.global-hydrogen-bus-platform.com/Home

International Partnership for Hydrogen and Fuel Cells in the Economy (2009) http://www.iphe.net/resources/demonstration.html

IPCC (2007) IPCC fourth assessment report: climate change 2007 section 2.4: causes of climate change. Retrieved 26 Jan 2012, from http://www.ipcc.ch/publications_and_data/ar4/syr/en/mains2-4.html

IPCC (2009) IPCC expert meeting on detection and attribution related to anthropogenic climate change. Retrieved 26 Jan 2012, from http://www.ipcc.ch/pdf/supporting-material/expert-meeting-detection-anthropogenic-2009-09.pdf

Jaouen F, Proietti E, Lefevre M, Chenitz R, Dodelet J-P, Wu G, Chung HT, Johnston CM, Zelenay P (2011) Recent advances in non-precious metal catalysis for oxygen-reduction reaction in polymer electrolyte fuel cells. Energy Environ Sci 4:114–130

Jones DJ, Rozière J (2008) Advances in the development of inorganic/organic membranes for fuel cell applications. Adv Polym Sci 215:219–264

Joo SH, Choi SJ, Oh I, Kwak J, Liu Z, Terasaki O, Ryoo R (2001) Ordered nanoporous arrays of carbon supporting high dispersions of platinum nanoparticles. Nature (Lond) 412:169–172

Li QF, Jensen JO, Savinell RF, Bjerrum NJ (2009) High temperature proton exchange membranes based on polybenzimidazoles for fuel cells. Prog Polym Sci 34:449–477

Liu J, Suraweera N, Keffer DJ, Cui S, Paddison SJ (2010) On the relationship between polymer electrolyte structure and hydrated morphology of perfluorosulfonic acid membranes. J Phys Chem C 114:11279–11292

Neyerlin KC, Srivastava R, Yu C, Strasser P (1009) Electrochemical activity and stability of dealloyed Pt-Cu and Pt-Cu-Co electrocatalysts for the oxygen reduction reaction (ORR). J Power Sources 186:261–267

Park C-H, Lee C-H, Guiver MD, Lee Y-M (2011) Sulfonated hydrocarbon membranes for medium-temperature and low-humidity proton exchange membrane fuel cells. Prog Polym Sci 36:1443–1498

Peckham TJ, Yang Y, Holdcroft S (2010) Proton exchange membranes. In: Wilkinson DP, Zhang JJ, Hui R, Fergus J, Li X (eds) Proton exchange membrane fuel cells. CRC Press, Boca Raton

Proietti E, Jaouen F, Lefevre M, Larouche N, Tian J, Herranz J, Dodelet J-P (2011) Iron-based cathode catalyst with enhanced power density in polymer electrolyte membrane fuel cells. Nat Commun 2:1427/1–1427/9

Pure Energy Centre (2007) Promoting Unst Renewable Energy (PURE) project – from wind to green fuel. Retrieved 26 Jan 2012, from http://www.pureenergycentre.com/pureenergycentre/Pureprojectcasestudy.pdf

Rabat H, Brault P (2008) Plasma sputtering deposition of PEMFC porous carbon platinum electrodes. Fuel Cells 8:81–86

Rozière J, Jones DJ (2003) Non-fluorinated polymer materials for proton exchange membrane fuel cells. Annu Rev Mater Res 33:503–555

Saha MS, Banis MN, Zhang Y, Li R, Sun X, Cai M, Wagner FT (2009) Tungsten oxide nanowires grown on carbon paper as Pt electrocatalyst support for high performance proton exchange membrane fuel cells. J Power Sources 192:330–335

Schönbein CF (1839) On the voltaic polarization of certain solid and fluid substances. Lond Edinb Philos Mag J Sci Ser 3 14:43–45

Thompsett D (2010) Recent developments in electrocatalyst activity and stability for proton exchange membrane fuel cells. In: Wilkinson DP, Zhang JJ, Hui R, Fergus J, Li X (eds) Proton exchange membrane fuel cells. CRC Press, Boca Raton

Tsuji M, Kubokawa M, Yano R, Miyamae N, Tsuji T, Jun M-S, Hong S, Lim S, Yoon S-H, Mochida I (2007) Fast preparation of PtRu catalysts supported on carbon nanofibers by the microwave-polyol method and their application to fuel cells. Langmuir 23:387–390

Vielstich W, Lamm A, Gasteiger H (eds) (2003) Handbook of fuel cells. Wiley, Chichester

Vielstich W, Harumi Y, Gasteiger HA (eds) (2009) Handbook of fuel cells. Wiley, Chichester

Wang J, Swain GM (2003) Fabrication and evaluation of platinum/diamond composite electrodes for electrocatalysis. J Electrochem Soc 150:E24–E32

Wieser C (2004) Novel polymer electrolyte membranes for automotive applications – requirements and benefits. Fuel Cells 4:245–250

World Energy Outlook (2011) World energy outlook executive summary. Retrieved 26 Jan 2012, from http://www.worldenergyoutlook.org/docs/weo2011/executive_summary.pdf

Wu B, Hu D, Kuang Y, Yu Y, Zhang X, Chen J (2011) High dispersion of platinum-ruthenium nanoparticles on the 3,4,9,10-perylene tetracarboxylic acid-functionalized carbon nanotubes for methanol electro-oxidation. Chem Commun 47:5253–5255

Xie Z, Song C, Wilkinson DP, Zhang JJ (2010) Catalyst layers and fabrication. In: Wilkinson DP, Zhang JJ, Hui R, Fergus J, Li X (eds) Proton exchange membrane fuel cells. CRC Press, Boca Raton

Zhang YM, Li L, Tang JK, Bauer B, Zhang W, Gao HR, Taillades-Jacquin M, Jones DJ, Rozière J, Lebedeva N, Mallant RKAM (2009) Development of covalently cross-linked and composite perfluorosulfonic acid membranes. ECS Trans 25:1469–1472

Zhao J, Jarvis K, Ferreira P, Manthiram A (2011) Performance and stability of Pd-Pt-Ni nanoalloy electrocatalysts in proton exchange membrane fuel cells. J Power Sources 196:4515–4523

Zhou Y, Pasquarelli R, Holme T, Berry J, Ginley D, O'Hayre R (2009) Improving PEM fuel cell catalyst activity and durability using nitrogen-doped carbon supports: observations from model Pt/HOPG systems. J Mater Chem 19:7830–7838

Züttel A (2008) Chapter 1: Introduction. In: Züttel A, Borgschule A, Schlapbach L (eds) Hydrogen as a future energy carrier. WILEY-VCH, Weinheim, pp 1–6

Chapter 9
Biomass for Energy: Energetic and Environmental Challenges of Biofuels

Jean Michel Most, Marie Thérèse Giudici-Orticoni, Marc Rousset, and Mireille Bruschi

Abstract Transportation is 94 % dependent on oil, represents around 20 % of global consumption of energy, and is responsible for 23 % of total emissions from fossil fuels. For several years, progress has been made to enhance the energy efficiency of the systems, but increasing the part of biofuel still seems irremediable both for environmental, economic, and energy independence reasons. Fuel production from biomass is clearly considered as an important substitute for liquid fossil fuels such as bioethanol for motor gasoline, biodiesel for diesel, jet fuel for biokerosene, and for gaseous fuels (hydrogen, natural gas for vehicles, biomethane, etc.). This chapter presents the main pathways for the production of biofuels, and classifies their degree of maturity:

- The first-generation processes that value the reserves of a plant (starch, sugar, oil) are now mature and industrially deployed.
- The second generation processes extend their resource to the whole plant tissues (agricultural, forest) or to organic waste, and are almost under scientific control but they still need more economic and energetic assessment before being commercially deployed.
- The last innovative pathway, the advanced or third biofuel generation, shows significant potential by using bioalgae or microorganisms capable of producing much more biomass oil convertible into biodiesel and gaseous fuels such as methane or hydrogen.

J.M. Most (✉)
Institut P', CNRS UPR 3346/ENSMA/Université de Poitiers, 1 Avenue Clément Ader, BP 40109, 86961 Futuroscope Chasseneuil, Cedex, France
e-mail: Jean-michel.most@ensma.fr

M.T. Giudici-Orticoni • M. Rousset • M. Bruschi
Laboratoire de Bioénergétique et Ingénierie des Protéines CNRS, 31 Chemin Joseph Aiguier, 13402 Marseille, Cedex 20, France
e-mail: giudici@imm.cnrs.fr; rousset@imm.cnrs.fr; bruschi@imm.cnrs.fr

J.B. Saulnier and M.D. Varella (eds.), *Global Change, Energy Issues and Regulation Policies*, Integrated Science & Technology Program 2, DOI 10.1007/978-94-007-6661-7_9,

Keywords First-generation biofuels • Second-generation biofuels • Algae for energy • Biohydrogen • Biodiesel • Bioethanol

List of Acronyms

ADEME	French Environment and Energy Management Agency
BtL	Biomass to liquid
CVO	Crude vegetable oils
Da	amu (atomic mass units)
DIREM	General Directorate for Energy and Raw Materials
DOE	US Department of Energy
E10	Gasoline with 10 % in volume of ethanol
E85	Gasoline with 85 % in volume of ethanol
ETBE	Ethyl tert-butyl ether
FAEE	Fatty acid ethyl ester
FAME	Fatty acid methyl ester
GHG	Greenhouse gases
GtL	Gas to liquid
HC	Hydrocarbon
IEA	International Energy Agency
MDa	$1\text{MDa} = 10^6$ Da
MJ	10^6 Joules
Mtoe	10^6 toe
NCV	Net calorific value
PAH	Polycyclic aromatic hydrocarbons
PCW	Plant cell wall
ppm	Part per million
PS II	Photosystem II (or water-plastoquinone oxidoreductase) is the first protein complex in light-dependent reactions.
TAG	Triacylglycerol
Toe	Ton of oil equivalent
v/v	Volume by volume

9.1 Introduction

Biomass potential is found in all living organisms on the continents and in the oceans; it is presently the main renewable energy resource on earth, providing 10 % of global primary energy demand (industrial combustion, transport, and traditional use, mainly wood for heating and cooking). This percentage has remained stable since the 1980s, and it should retain this value in the coming decades despite the increase in global energy demand. Consequently, biomass should clearly be considered in the energy mix for the years 2020–2050.

For decades, biomass conversion technologies have been developed to produce energy for transportation, agriculture, manufacturing, buildings, waste processing, and the like. In 2008, transport represented around 20 % of global energy consumption, corresponding to 2,300 Mtoe in comparison to a total of nearly 12,000 Mtoe. Transport currently is 94 % dependent on oil (IEA 2009a, p. 324), and is responsible for 23 % of global CO_2 emissions (IEA 2011, p. 9). Fuel production from biomass—bio fuels—could be an important substitute for liquid fossil fuels such as gasoline, diesel, kerosene, and jet fuel and for gaseous fuels (hydrogen, natural gas for vehicles, biomethane, etc.) simultaneously reducing the anthropogenic CO_2 emissions.

In 2008, 78 % of biomass energy was allocated for heating and cooking (IEA 2009b, p. 41). The new uses of biomass for electricity, heat, cogeneration, biofuels, and the like represented only 150 Mtoe, of which 46 Mtoe (2 % of fuel requirements, BP 2011, p. 39) dedicated to biofuels (ethanol for gasoline engines, biodiesel for diesel engines, etc.).

The projections (ADEME 2011), in the short and medium term, clearly cannot be strictly fixed and will depend on the energy options chosen for the future:

- For the "Blue Map" scenario, in 2050, the percentage of biomass could reach 3,500 Mtoe and biofuels could be around 750 Mtoe, which represents 27 % of energy consumption for transportation.
- For the same timeframe, the scenario trend (little change from the current energy policy) provides a more nuanced view: 1,900 Mtoe for the production of bioenergy and 160 Mtoe for biofuels (covering 4 % of demand).
- Finally, the "450 scenario" (see Chaps. 1 and 2), which proposes limiting the atmospheric concentration of CO_2 to 450 ppm, is a compromise and is already focused on the use of 280 Mtoe of biofuels by 2030.

It can be observed that, concerning aviation use, the impact is marginal inasmuch as the consumption of oil represents only 9 % of total hydrocarbons. The predicted increase in air traffic and the simultaneous decreasing supply of fossil fuels will induce a competitive use of biomass as BtL (second or third generation biofuel, biomass to liquid), natural gas, biogas, or even hydrogen if the limitations of its use related to the logistics and safety of its storage are solved.

Among the numerous reports available on biofuels (IEA 2009a, b; ADEME/DIREM 2002; ADEME 2011; DOE 2006, etc.), let us mention one of the last up-to-date reviews, an "advanced biofuels roadmap," available on the Web (ADEME 2011), including the context, issues, and worldwide prospective visions on biofuels over a timescale to 2050.

Even if biofuels have engendered widespread enthusiasm in recent years, presuming to be a good solution for the growing demand for energy and reducing greenhouse gas emissions, an overall balance of advantages and disadvantages (real energy gain, CO_2 mitigation, land uses, water requirements, etc.) of their development must be prepared.

Presently, many organizational and technological locks subsist at all levels in the different processes of biomass fuel production. The earth's surface, especially the area available for cultivation, is limited and it will be necessary, as the population grows, to control the social problems linked to food security (competition between crops), to recycle agricultural waste and by-products, and to rationally exploit the forests. On this last point, let us evoke the controversial effects of deforestation, as the production of biofuels could increase the impacts on the environment (water, soil, or air) and on biodiversity.

We refer the reader here to Chap. 14 of this book, "The triple A issue: agriculture, alimentation needs, agro-fuels,[1]" which deals with the following basic problem.

How relevant is it to build up agrofuel production in terms of a partial solution to energy self-sufficiency, in terms of the reduction of greenhouse gas (GHG) emissions, in terms of its impact on meeting humanity's alimentation needs?

The present chapter is devoted more to scientific and technical considerations dealing with the state of the art of the different ways of producing biofuels from agriculture and with the main issues for research.

In the following paragraphs, an updated review is performed on the different processes for the production of:

- First-generation biofuels (the only currently operational ones) considering fermented sugars and oils, using the noblest part of the plant (seeds, grains, roots, etc.)
- Second-generation, based on the thermochemical processing of wood (lignocelluloses), implying the utilization of the whole plant (seeds and stems) as well as utilization of the wood and waste wood (lignocellulosic matter)
- Third generation, with, for instance, the development of biochemistry of algae

9.2 First-Generation Biofuels

First-generation biofuels, already present on the market and manufactured using mature technologies, only use part of the plant by processing the sugar or the starch that it contains or by extracting the oil from its seeds. The lignocellulosic part of the plant (stem, root) is generally not used for energy purposes (except for direct combustion). Two classical production pathways are currently being exploited: the ethanol pathway that leads to the production of ethyl tert-butyl ether (ETBE), shown in Fig. 9.1, and the diesel pathway that produces fatty acid methyl ester (FAME) or fatty acid ethyl ester (FAEE) fuel for diesel engines from processed vegetable oils. Figure 9.1 illustrates the main production steps of these two pathways.

9.2.1 Ethanol Pathway and ETBE (Ethyl Tert-Butyl Ether)

Three major families of alcohol-producing crops (see Fig. 9.1) are taken into account for the production of bioethanol:

[1]*Agro-fuel*: Biomass from agricultural crops.

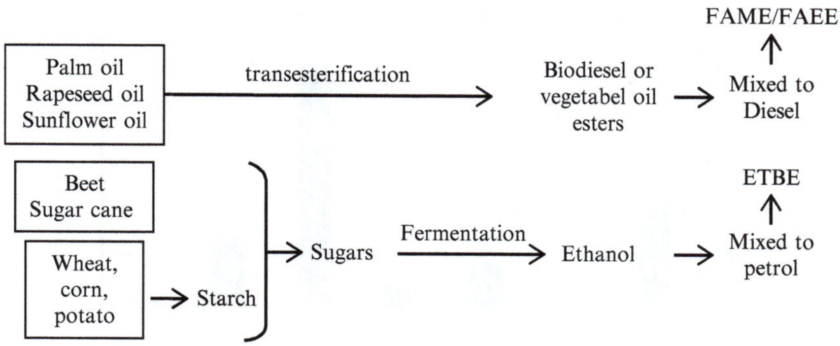

Fig. 9.1 Biofuel manufacturing principles diagram

- Sugar crops (beet, sugar cane, etc.), which contain sugars that are directly fermentable into ethanol.
- Cereal crops (corn, wheat, barley).
- Starchy crops and tubers (rich in fructans) such as potatoes, sweet potatoes, yams, manioc, and Jerusalem artichokes, among others, where sugars pre-exist in the form of starch. First of all, prior to fermentation, an hydrolysis of starch into sugar monomer must be carried out using enzymes.

The principle of alcoholic fermentation has been known for over 25 centuries. The sugars are converted into ethanol (C_2H_5OH) by the fermenting action of micro-organisms, yeasts, and bacteria, in accordance with Gay Lussac's reaction:

$$C_6H_{12}O_6 \rightarrow 2C_2H_5OH + 2CO_2$$

Fermentation is influenced by the presence of ethanol and oxygen, by temperature, by minerals, and by the pH of the solution. The maximum efficiency of this reaction establishes that 100 kg of glucose will produce 48.4 kg of ethanol, which results in industrial efficiency of around 95 % of the theoretical efficiency of Guy Lussac's reaction (Ballerini 2006). This fermentation reaction is exothermic (1.2 MJ/kg of ethanol produced) and produces ethanol along with a release of gaseous CO_2, which is a greenhouse gas.

To attain concentrations of more than 99.5 %, the ethanol needs to be purified. This is carried out industrially in two steps: a classic distillation to attain content close to that of azeotrope,[2] a second azeotropic distillation carried out with absolute ethanol at 99.8 % v/v; this step is generally performed with cyclohexane. With regard to the distillation step, energy savings are made by recovering the heat from the exothermic reactions.

[2]*Azeotrope*: A liquid mixture for which boiling occurs at a fixed temperature and composition.

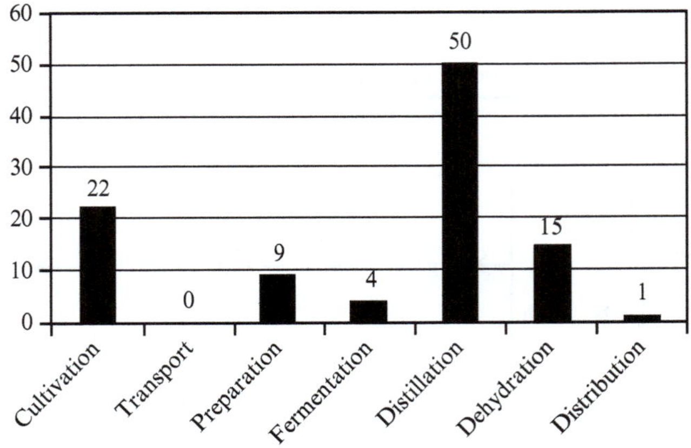

Fig. 9.2 Energy breakdown as a percentage based on wheat or beet ethanol production steps (After ADEME/DIREM 2002)

The production of bioethanol from cereals or starchy crops for manufacturing biofuel (ETBE) is initiated as soon as the agricultural resource is available. Examples of 6 m^3 bioethanol production units exist on the market at a modest cost (large farm or village scale). The bioethanol obtained can be mixed directly up to 20 % with gasoline, using pumps, without changing engine settings. For an 85 % ethanol (referred to, in Europe, as E85), vehicles must be equipped with an ethanol-flex fuel system and the country is required to implement a distribution network for this E85 biofuel.

Processes can be used for recovering the effluents and by-products from ethanol production, for example, dry pulps may be used for animal feed, the residues recovered at the bottom of the distillation column may be spread as an amendment on farm land (subject to prior authorization), solid residues from grinding steps may be utilized as a fuel for thermal power plants. As an additional bioethanol outlet, hydrogen may be produced by reforming to be used in fuel cells.

Figure 9.2 shows the energy breakdown required for wheat ethanol production. This process mobilizes 0.49 MJ of nonrenewable energy per MJ supplied (i.e., 13.1 MJ of nonrenewable energy mobilized per kg supplied) and overall energy efficiency (energy restored/nonrenewable energy mobilized) is, as such, 2.05 (ADEME/DIREM 2002). In comparison, 1.15 MJ of nonrenewable energy must be mobilized to supply 1 MJ from petrol, with an efficiency of 0.87.

We can also see that 65 % of the energy required for producing bioethanol comes from the distillation and dehydration steps. In addition, free heat, from power plants and from the metallurgical industry may be utilized in such a process

Around 100 kg of ethanol is produced from 300 kg of dry starchy plant matter, which corresponds to one ton of potatoes, sweet potatoes, or starchy plants. By contrast, around 100 kg of ethanol can be produced from 200 kg of corn (dry matter).

If we wish to get a mixture of gasoline incorporating 10 % (in NCV, net calorific value[3]) of bioethanol, and if we take account of the ratio between the NCVs, then 1 t (≈0.6 toe) of ethanol needs to be incorporated in 6 t (≈6 toe) of gasoline. Knowing (see Table 9.1) that the ethanol net production is 1.25 toe/ha/year for wheat and 2.9 toe/ha/year for beets (even better for sugar cane ≈3.5–4 toe/ha/year and less for starchy plants that comprise 70 % water), then the dedicated cultivation surface area for the production of 1 ton E10 (0.86 t gasoline and 0.14 t ethanol, containing 10 % bioethanol, according to the NCV criterion) will be 0.11 ha for wheat and 0.05 ha for beets. In intertropical regions, the use of sugar cane leads to slightly higher efficiency (≈10 %), therefore to a dedicated surface of around 0.045 ha.

9.2.2 *Crude Vegetable Oils and Vegetable Oil Esters (FAME, Fatty Acid Methyl Ester, FAEE; Fatty Acid Ethyl Ester)*

On a worldwide scale, soya oil represents the most important global production of crude vegetable oils (CVO), followed by palm oil, then rapeseed oil, sunflower oil, peanut oil, coconut oil (copra), and olive oil. The oils consist mainly of triglycerides, around 95–98 %. These latter are made up of an esterified glycerol molecule with three fatty acid molecules that may be either similar or different. Each oil has its own fatty acid composition, comprising between 6 and 21 carbon atoms.

There are three steps in the process for producing CVOs: trituration (seed crushing, cooking, and pressing), extraction (extraction of the oil contained in the oil cakes using solvent (n-hexane) at a temperature lower than 70 °C), and refining.

These CVOs are characterized by a molecular chain length that is longer than that of diesel, with high distribution and a significant number of unsaturations, a molecular weight of around 0.88 kg/mol and a density of more than 910 kg/m^3. They are low-volatile oils and crack from 300 °C. The composition of the oil affects their viscosity, their resistance to cold, their ignition time, and their temporal stability. If used in an engine, the high viscosity that affects atomization, the low volatility that increases the ignition time, the deposits on the walls of the chamber, and the observed congealing problems at low temperature impair diesel engine running. Their dilution in lubrication oils should also be mentioned as this causes the engine to wear more quickly. All these drawbacks prove that direct use of such oils in modern engines will be increasingly critical.

Moreover, the use of these oils increases emissions of CO (+100 %), of unburned hydrocarbons HC (+400 %), which are frequently considered as carcinogenic particles (+90 to +400 %), of aldehydes and ketones (+30 to 330 %) and of polycyclic aromatic hydrocarbons PAH (+20 %); on the other hand, the formation of NO_x

[3]The *Net Calorific Value* (NCV) of a substance, is the amount of heat released during the combustion of 1 kg, the water produced being still considered as a vapor. The NCV of ethanol and of gasoline are, respectively, 26.8 and 42.7 MJ/kg, and their ratio is #0.6.

Table 9.1 Crop efficiency for producing agrofuels in temperate-region countries (first and second generations)

	Mass production (t/ha/year)	Energy gross production (toe/ha/year)	Energy net production (toe/ha/year) inputs deducted	Theoretically usable surface areas (nonaccumulable)	Industrial deadlines	Optimized provision cost (excluding taxes)
Wheat ethanol	~2.8 tons of ethanol/ha/year	~1.8	~1.25	~10 M ha (100 % foodstuffs)	Immediate	~0.5 €/l (NCV coefficient 0.64)
Beet ethanol	~6.5 tons of ethanol/ha/year	~4.1	~2.9	~1 M ha (including 50 % foodstuffs)	Immediate	~0.4 €/l (NCV coefficient 0.64)
FAME (rapeseed)	~1.5 tons of FAME/ha/year	~1.35	~1	~4 M ha (including 50 % foodstuffs)	Immediate	~0.5 €/l (NCV coefficient 0.9)
Cellulose ethanol by fermentation	~12 tons of DM/ha/year	~2	~1.5	~10 Equivalent M ha (agricultural and forest)	~2015	(NCV coefficient 0.64)
BtL–thermo–chemical autothermic (green diesel)	Idem	~2	~1.5	Idem	2010–2015	~0.7 €/l (NCV coefficient 1)
BtL–thermo –chemical allothermic (green diesel)	Idem	~4	~3.5	Idem	2020	~0.7 €/l (NCV coefficient 1)

Source: After Roy, 2008, Interministerial coordinator for biomass energy, personal communication

BtL Biomass to liquid; Autothermic and Allothermic: with or without external energy contribution (see Sect. 9.3.2); *DM* is for dry matter; *NCV* is for natural calorific value; NCV coefficient 0.64 means the NCV ratio ethanol/petrol is 0.64

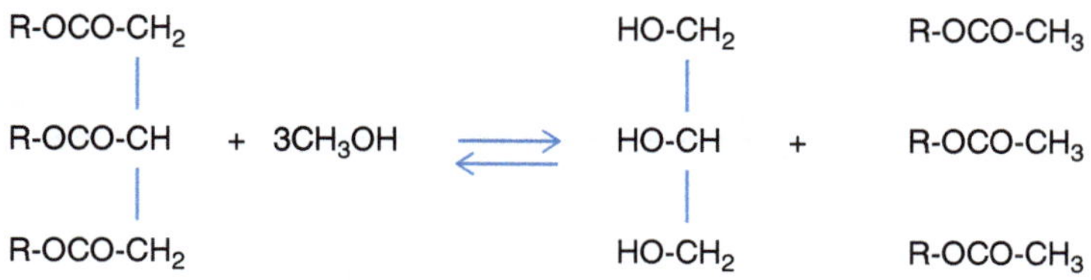

Triglycérides Méthanol Glycérol Esters Méthyliques

Fig. 9.3 The transesterification reactions

is reduced (–25 %) as a result of the lower combustion temperature. If such oils could be used directly in tractors or power generators, their use in modern engines would come up against the same limitations as their use in automobile engines.

Direct CVO use in an engine (especially direct injection engines) is, therefore, limited and modifies engine performances. Only the few engines remaining that are fitted with a precombustion chamber and have an extremely low revolution speed (marine engines, old agricultural equipment, certain power generators) can use these oils. There seem to be fewer drawbacks but wall deposits are liable to form over the long term.

Once purified, crude vegetable oils are sometimes used as fuel for specific diesel engines, but the technology and the performance of new engines have requirements that can no longer accept these fuels. These oils are now transesterified with alcohol (generally methanol or ethanol) to produce fatty acid methyl ester (FAME) and fatty acid ethyl ester (FAEE). As such, this ester may be used in any diesel engine by mixing the diesel, of any concentration, even pure, which helps reduce the consumption of fossil fuels.

9.2.3 Processes for Producing FAME (Fatty Acid Methyl Ester) and FAEE (Fatty Acid Ethyl Esters)

When the CVOs are transesterified the triglycerides are transformed by the catalytic reactions of trans-esterification in a basic medium by methanol to form methyl esters and to form a subproduct: glycerol. One hundred kilograms of alcohol need to be added to a triglyceride cake to obtain one ton of biodiesel and 100 kg of glycerol that must be eliminated or utilized. This reaction requires the reagents to be slightly heated (see Fig. 9.3).

Transesterification may also be carried out using ethanol via a similar process; ethyl esters FAEE with similar properties to FAME are produced. Oil ester and glycerine is formed once the excess alcohol evaporates.

These esters are fatty acids used in human foodstuffs and animal feed and, they have a low impact on the environment. They are totally biodegradable which, unfortunately, results in their rapid oxidation; they are, as such, not very stable and it is compulsory to add additives for storing them.

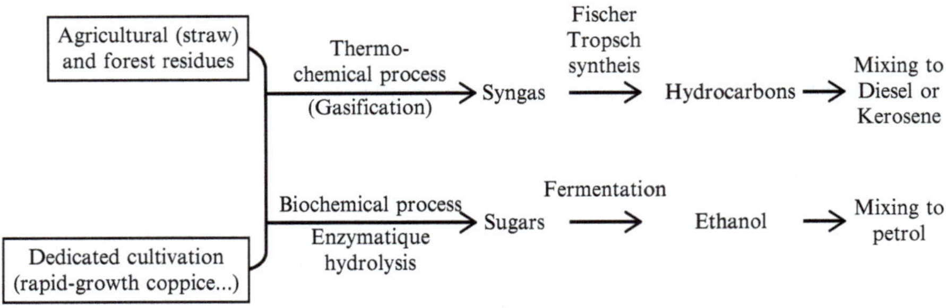

Fig. 9.4 Future production pathways for manufacturing second-generation biofuels

Direct FAME or FAEE use in diesel engines does not lead to the drawbacks related to CVO use. These esters are even beneficial for the lubrication of fuels with very low sulphur content. They are miscible with diesel (they may be used mixed with diesel or even pure) but their calorific value is slightly lower and leads to a 7 % higher consumption (Ballerini 2006).

FAEEs have a positive impact on the environment with, generally, a decrease in CO, HC, and particle emissions but with an increase in NO_x.

Similar to first-generation fuels, two pathways are used for utilizing this lignocellulosic resource: a thermochemical production pathway (dry) and a biological one (moist). The general principle is illustrated in Fig. 9.4.

9.3 Second-Generation Biofuels

This type of fuel is still in the research phase and has a number of aims: the utilization of the whole plant (seeds and stems) as well as the utilization of the wood and waste wood (lignocellulosic matter) and a reduction in the competition with food crops.

9.3.1 Lignocellulosic Biomass Resources

This resource comes from agricultural, forest, and wood transformation subproduct waste as well as from dedicated cultivation, whether it may be woody plants or herbaceous plants. Fuelwood represents half of the gross world production of this resource; the other half covers industrial uses (materials). Transforming wood generates a considerable quantity of waste recyclable in the form of energy or of raw material for trituration; dry agricultural and forest waste that also provides significant quantities of lignocellulosic biomass. Dedicated cultivation is an important source of lignocellulose (poplars, eucalyptus, willows, short rotation coppice and other herbaceous species, such as miscanthus) that is energetically utilizable.

The characteristics of the raw material influence the energy conversion process. For example, the concentration of inorganic matter disadvantages thermochemical conversion, the concentration of ash conditions the temperature of the reaction, as such, and the efficiency of the process, the concentration of sulphur and of nitrogen are responsible for the formation of pollutants that must be purified. The percentage of lignin, moisture, and its granulometry also play an important role. By contrast, in biochemical processes, it is the lignin that controls the conversion.

The lignocellulosic biomass resource is real or potential in different parts of the world, but the degree of mobilization of this resource depends greatly on the constraints of continuity of supply and of the land policy of the country, of standardization of the quality, and the cost of the cultivation (plantation, maintenance, exploitation, transport). This resource has two major interests, and this is one of the major challenges for the pathway: not to be in competition with food crops, a choice that might lead to the starvation of the world, and to make forest maintenance profitable (carbon storage protection).

Furthermore, these processes lead to clean synthetic fuels without aromatic molecules that are liable to affect health.

9.3.2 The Thermochemical Pathway

The thermochemical conversion of the lignocellulosic biomass will represent a major challenge to the world for 2020. It is currently referred to as BtL (biomass to liquid). Two main production pathways exist for producing energetically utilizable products: pyrolysis and gasification (Ballerini 2006).

9.3.2.1 Pyrolysis

Pyrolysis is thermochemical transformation of a solid fuel (coal, biomass, etc.) which is carried out in an inert medium (no oxygen). Two pyrolysis processes have been developed for biomass processing: a slow and a fast one.

In slow pyrolysis, the heating is performed at a slow rate (<50 °C/min) up to 600 °C and with a residence time of some hours. The result is essentially vegetal charcoal.

On the contrary, during fast pyrolysis the heating is performed at a much higher rate (100 °C/s) up to 500/550 °C and with much smaller residence times (on the order of 1 s). As a result of the action of heat and the lack of oxygen, lignocellulose constituents are pyrolyzed and may be converted here in three phases: the solid (vegetable coal), the condensable organic compounds (oils rich in oxygen), and the gases, where the main ones are CO_2, CO, H_2, and CH_4. By controlling process parameters, these reactions may be directed towards the preferential production of one of these three phases and efficiency may be maximized.

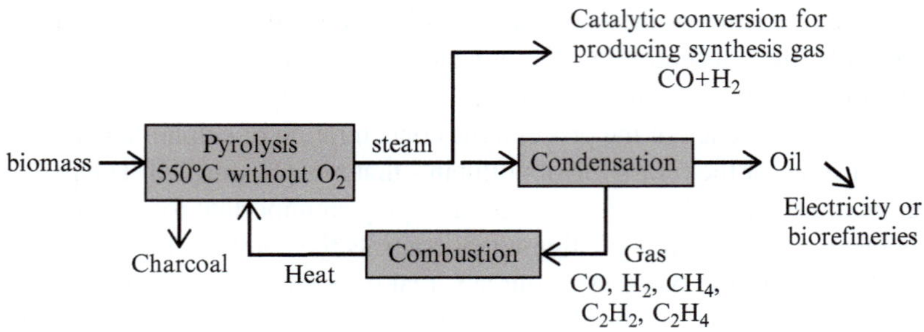

Fig. 9.5 Diagram of a fast pyrolysis principle

Fast pyrolysis is an homogenization stage and, as such, it is useful for preconditioning the biomass. This process has two advantages:

- To produce, from heterogeneous lignocellulosic composites (which are difficult to use or to transport in their common state) homogeneous and transportable solid or liquid intermediates such as charcoal and pyrolysis oils.
- As such, to decouple temporally biomass production and utilization. In this way, the production unit is established close to the resource so that raw material transportation costs are minimized; biofuel transformation is centralized and benefits from the scale effect on investment.

Another advantage of biomass pyrolysis concerns the versatility of the technology which, depending on operating conditions (heating speed, temperature, residence time in the reactor, and pressure), can direct the reaction to liquid, solid, or gas and thus recover or concentrate the undesired fractions in one of these phases.

Fluidized-bed, ablative, and vacuum processes generally adhere to the diagram displayed in Fig. 9.5. They generate gaseous compounds, which, following cooling and condensing, are recovered as a liquid phase, the noncondensable gases and the charcoal are recovered to be burned and to be used as auxiliary energy for pyrolysis. Pyrolysis oils contain several hundred chemical compounds in highly variable molecular proportions and weight. They are rather unstable (decomposition above 100 °C) and present a rather high acidity (risk of corrosion)

The technology is mature but still lacks opportunities. Pyrolysis oils may only be used in a few engines that are specially designed for electricity production but they must be excluded from use in a gas turbine or an automobile engine (oils cracking is difficult if they only spend a short time in the combustion chamber, which thus explains the high amount of unburned particles in the smoke and of destructive deposit of oils on the walls). The most promising operation concerns the gasification of oils and/or of charcoal to produce syngas which is then converted into methanol or hydrocarbon using the Fischer–Tropsch process (see the following section, "Gasification"). The gasifier could, thus, work with oils with similar characteristics but from different biomasses.

9.3.2.2 Gasification

Gasification is the thermochemical transformation of a solid fuel (coal, biomass, etc.) using a reactive gas (oxygen, water vapor, hydrogen, etc.) compared to pyrolysis, which is carried out in an inert medium. The aim is to convert the solid into a gaseous mixture, referred to as "syngas" or synthetic gas containing H_2 and CO with other minor species CO_2, H_2O CH_4, in proportions that depend on the process used.

The gaseous mixture obtained may be used here for different applications:

- Engine combustion
- Heat and power cogeneration, with an improved efficiency compared to direct biomass combustion, given the increase in combustion temperature (efficiency up to 40 % of the biomass NCV)
- Biofuel synthesis, including liquid hydrocarbons including synthetic gasoline, kerosene, or diesel (because of their linear chains they must be mixed with fossil hydrocarbons to ignite the fuel correctly in the chamber).

With regard to combustion, syngas may contain methane and heavier compounds. But concerning biofuels synthesis, it is better to get a syngas with a maximum conversion into CO and H_2, and more precisely a ratio H_2/CO, in volume, close to 2. The process is normally autothermic.[4] But, to increase gasification efficiency, an allothermic[5] process is preferable and the introduction of external hydrogen can favor the CO/H_2 ratio in syngas.

Lignocellulose gasification (syngas) or pyrolysis oils process must be followed by a Fischer–Tropsch synthesis so that the liquid fuel produced corresponds to a mixture of hydrocarbons such as gasoline, diesel, or kerosene.

The Fischer–Tropsch process consists in establishing the catalytic reaction, at a temperature of 220 °C, of syngas, with a H_2/CO ratio close to 2.05, so that saturated paraffin (C_nH_{2n+2}) may be produced, ranging from ethane to $C_{80}H_{162}$, with cobalt as the catalyst.

$$(2n+1)H_2 + nCO \Leftrightarrow C_nH_{2n+2} + nH_2O \quad \text{where } \Delta H = -165\ \text{kJ/mol}$$

This reaction is highly exothermic and one of the technological barriers is the cooling of the multiple reactors. By controlling the temperature, the H_2/CO ratio, the gas flow rate (residence time), and the type of catalyst, the optimal hydrocarbon may be selected that has a large number of carbons (between C_{10}[6] and C_{20}, similar to diesel) and oil-refining chemistry is then applied. These Fischer–Tropsch units

[4]*Autothermic*: The reactions involved in the process are globally endothermic, but some of them are exothermic and can provide an internal energy contribution to the process. An endothermic reaction needs an external heat to be provided to the system. An exothermic reaction creates heat that must be evacuated to the external ambient.

[5]*Allothermic*: With contribution of external energy.

[6]*C_{10}* and *C_{20}*: Hydrocarbon molecules with, respectively, 10 or 20 carbon atoms.

should be set up close to a refinery: the natural gas to liquid fuel (GtL) process is carried out in the same way as that of coal or of biomass to biofuels and such a pilot unit exists in the United Arab Emirates. Let us also mention that the diesel obtained has high qualities: no sulphur, no nitrogen, no metal, no aromatic compounds, and a good cetane index (70/75 compared to 50 for commercial diesel).

Biomass gasification using this thermochemical process still lacks maturity, as it is still better adapted to coal gasification than to that of the lignocellulosic biomass. In addition, a continual and significant flux of biomass and/or coal resource is required to be able to establish an installation big enough to ensure efficiency of 30–40 % (allothermic process).

Efficiency not only depends on the gasification technology, but also on the choices made for the overall integrated process, in upstream steps (using the biomass directly or products from slow or fast pyrolysis), as well as in downstream steps, particularly as regards developing subproducts using Fischer–Tropsch synthesis, such as gases and naphtas: gases may be burned and supply part of the energy required for the industrial unit to operate.

To retain some figures, carbon efficiency from autothermic processes is less than 30 %. A supply of external energy, for example, as electricity injected into the gasifier or during the high-temperature gas reforming step, or a supply of external mass hydrogen, may increase efficiency significantly up to 40/50 %.

9.3.3 *The Biochemical Pathway*

The use of biomass to supply renewable fuel for transportation was first integrated in our economy in the first-generation biofuels by converting the edible part of crop biomass (stored sugars or lipids) into ethanol or biodiesel (first generation). With first-generation biofuels coming under increasing criticism for driving up food prices and putting biodiversity at risk, politicians and scientists turned their gaze towards next-generation biofuels (advanced biofuels), the production of which use the nonedible part of biomass (lignocellulose; second-generation biofuels), and new energy crops (third-generation biofuels; for instance, H_2, microalgae). However, the technologies behind these new forms of bioenergies are not yet mature owing to biological and technological barriers. Major efforts in R&D are therefore required to create innovative processes for the production of advanced biofuels. In the following pages we present an overview of the different biological energy alternatives that are currently the subject of intense research; we also identify the major obstacles that need to be overcome for each of them.

9.3.3.1 Recall of the Main Steps to Accomplish

Although the process for producing ethanol from the sucrose of sugary plants or from the starch of starchy plants is mature (first generation), the process of ethanol production from lignocellulosic matter (second generation) comes up against two barriers: the hydrolysis of cellulose to fermentable sugars and the conversion of pentoses from hemicellulose to ethanol. As lignin cannot be fermented in ethanol, the lignocellulosic matrix must be preprocessed to make the cellulose and the hemicellulose hydrolyzable to obtain potential sources of fermentable sugars

9.3.3.2 An Enzymatic Approach to the Production of Biofuels from Lignocellulose

The production of second-generation fuels from lignocellulosic biomass displays many benefits (energetic, economic, and environmental) that are associated with the use of enzymes or microorganisms as catalysts.

The process includes an enzymatic saccharification (breakdown to C_5 and C_6 sugars[7]) of the pretreated plant biomass, performed by the enzymes of a cellulolytic fungus, *Trichoderma reesei*. This step accounts for approx. 50 % of the total cost of production, because a huge amount of enzyme (about 1 % of the mass of substrate) is needed, especially for the hydrolysis of the main polymer, cellulose. Use of microbes for fermentation is the most common method for converting sugars produced from biomass into liquid fuel. To develop commercially viable processes for cellulose bioconversion to ethanol, an organism is needed that uses all sugars (C_5 and C_6) produced from biomass saccharification, at rates and in high alcohol concentrations that match or surpass current yeast-based glucose fermentation.

Trichoderma reesei is the primary producer of cellulase and hemicellulase because it secretes high amounts of cellulase (100 g/L). Analysis of its genome revealed a surprising small set of genes encoding enzymes for the breakdown of plant cell wall polysaccharides thereby suggesting this cocktail could be improved by additional enzymes. Furthermore, it was recently reported that some proteins devoid of measurable activity can significantly improve the activity of plant cell wall (PCW) degrading enzymes or cocktails, thus showing that our understanding of the mechanisms underlying the enzymatic degradation of biomass is far from complete. In an effort to identify novel enzymes or synergistic factors that would efficiently boost the enzymatic cocktail of *T. reesei*, Henrissat's group has surveyed a large number of fungal and bacterial genomes to identify putative novel proteins involved in plant cell wall digestion (Weiner et al. 2008; Martinez et al. 2009; Yang et al. 2009; Ma et al. 2010). Another promising approach lies in the exploration of atypical cellulolytic systems in which the catalytic subunits are gathered in huge complexes of 1 MDa[8] range called cellulosomes. Cellulosome engineering exhibits an important potential for biotechnological applications including biofuel production (Mingardon et al. 2007). Furthermore, anaerobic

---

[7] Hydrocarbon molecule with respectively five or six carbon atoms.

[8] 1MDa = 1M Dalton = 10^6 Da; 1 Da = 1 amu (atomic mass units).

bacteria represent a source of novel enzymes that can be mixed with fungal enzymes for improving the overall performance of the enzymatic cocktail. Another application is the introduction of an optimized cellulosome in a biofuel producer to establish a single-organism process enabling the direct conversion of biomass into biofuel.

9.4 Third-Generation Biofuels: Algae and Microorganisms

9.4.1 Using Microalgae for Biofuels

Microalgae are photosynthetic microorganisms that convert water and carbon dioxide in cell material using sunlight. Microalgae are receiving more and more attention as the search for sustainable and profitable biofuel feedstocks progresses. They exhibit several advantages, when compared to other biofuel crops such as wheat, corn, rapeseed, or sunflower. Microalgae show a very high surface productivity and they can be produced on noncultivable land surfaces, therefore not competing with food production by agriculture. Three different types of energy compounds can be produced by microalgae from which biofuels are made: lipids (transesterification will give biodiesel), starch (fermentation will give bioethanol), and biohydrogen (for reviews see Radakovits et al. 2010; Stephens et al. 2010; Larkum et al. 2011). When compared to major crops, microalgae show a very high productivity potential.

In spite of a high productivity potential, large-scale cultivation of microalgae shows relatively low sunlight conversion yields into biomass (from 0.5 to 3 %), much lower than the maximal photosynthetic conversion yield (~10 %). At a laboratory scale it is, however, possible to approach theoretical limits, but the existence of biological limitations considerably lowers the yields, especially when the metabolism is oriented towards the production of high-energy compounds. During growth in optimal conditions, the photosynthetic metabolism is oriented towards cellular growth and division. Cells accumulate high-energy compounds only in response to unfavorable conditions, such as nutrient deprivation (nitrogen, sulphur, etc.). This response allows optimizing survival and awaiting a return to better conditions. As a result, accumulation of lipids or starch relies on a phase of nutrient starvation which considerably lowers production yields.

9.4.1.1 Lipids for Biodiesel

The most abundant microalgae studied for biodiesel production are *Cyanophyceae Synechocystis* (blue green algae, Fig. 9.6), *Chlorophyceae* (green algae), *Bacillariophyceae* (diatoms), and *Chrysophyceae* (golden-brown algae). Some unicellular microalgae, such as *Chorella* or diatoms, accumulate triglycerides in response to

Fig. 9.6 Examples of microalgae: *Cyanophyceae Synechocystis* (*with permission Y. Tsukii*)

Table 9.2 Comparison of lipid productivity of major crops and microalgae

Plant source	Seed oil content (% oil by wt in biomass)	Oil yield (L oil/ha year)	Land use (m^2year/kg biodiesel)	Biodiesel productivity (kg biodiesel/ ha year)
Soybean	18	636	18	562
Jatropha C.	28	741	15	656
Camelina S.	42	915	12	809
Rapeseed	41	974	12	862
Sunflower	40	1,070	11	946
Palm oil	36	5,366	2	4,747
Microalgae (low oil content)	30	58,700	0.2	51,927
Microalgae (medium oil content)	50	97,800	0.1	86,515
Microalgae (high oil content)	70	136,900	0.1	121,104

nutrient shortage (nitrogen for most microalgae, silicium in the case of diatoms). Lipid accumulation can reach up to 60 % dry weight in some algal species. Table 9.2 presents the lipid production yield of main crops and microalgae. Lipid accumulation by algae is of considerable biotechnological interest in the perspective of producing renewable biodiesel.

However, at present there is a major discrepancy between predicted hydrocarbon yields based on laboratory experiments and the measured hydrocarbon yields of large-scale algae production systems. Algae possess a variety of lipid types ranging from the less commercially useful membrane-bound polar lipids to large droplets of unbound triacylglycerols, which are of prime interest as a biofuel source. Identifying those algal species capable of producing large stores of readily extractable neutral lipid droplets is the first step in the development of algal biofuels as an economically viable fuel source. On the other hand, the lipid content may vary according to the growth conditions. This means that it is theoretically possible to improve the lipid content in cells by adapting growth parameters but also by modifying the metabolic pathways responsible for lipid production. Several research programs around the world are currently devoted to exploring biodiversity to discover the best organism able to reach both a high growth rate and high lipid content. Global studies such as proteomic and transcriptomic are also being conducted in order to better understand the whole metabolic processes leading to lipid accumulation, and to identify the main enzymes as potential targets for metabolic engineering. Photobioreactors have allowed the production of such lipids at the lab scale (Pruvost et al. 2009).

9.4.1.2 Starch for Bioethanol

Carbohydrates are produced by microalgae mainly under nitrogen starvation conditions. They can be transformed by a subsequent fermentation step into a variety of biofuels, including ethanol, butanol, H_2, lipids, and/or methane. Glucans[9] are stored in microalgae in a variety of ways. The phyla *Chlorophyta*, *Dinophyta*, *Glaucophyta*, and *Rhodophyta* store glucans in linear α-1,4 and branched α-1,6 glycosidic linkages. In *Phaeophyceae* and *Bacillariophyceae*, water-soluble granules of laminarin and chrysolaminarin are synthesized, which are made up of β-1,3 linkages with branching at the C-2 and C-6 positions of glucose. In green algae, starch is synthesized and stored within the chloroplast, whereas it is stored in the cytoplasm in *Dinophyta*, *Glaucophyta*, and *Rhodophyta* and in the periplastidial space in *Cryptophyceae*.

Most microalgae do not contain lignin because they do not require the mechanical support that lignin provides. The fact that microalgae produce fermentable carbohydrates without containing lignin makes them an attractive feedstock for bioethanol production. This is especially true when the superior productivity of microalgae is taken into account. Table 9.3 illustrates the differences in productivity of three terrestrial crops and three microalgae. The microalgae exhibit considerably higher productivity. It is clear that they have the potential to produce more fermentable carbohydrates annually per hectare than terrestrial feedstock, despite containing a lower percentage of fermentable carbohydrates, because of their high rates of productivity. Table 9.3 also shows that the terrestrial feedstock produces a considerable amount of lignin annually, which inhibits their efficient conversion to bioethanol.

---

[9] *Glucans* are polysaccharides of D-glucose monomers, linked by glucosidic bonds.

Table 9.3 Comparison of lipid productivity of major crops and microalgae

Feedstock	Biomass productivity (dry T/ha·year)	Fermentable carbohydrate (%)	Lignin (%)	Fermentable carbohydrate productivity (T/ha·year)	Lignin productivity (T/ha·year)
Corn	7	80	15	5.6	1.05
Switchgrass	3.6–1.5	76	12	2.8–11.5	0.4–1.8
Woody biomass	10–22	70–85	25–35	7–18.7	4–7.7
Chlorella	127.8–262.8	33.4	0	42.7–87.8	0
Tatraselmis	38–139.4	11–47	0	4.2–65.5	0
Anthrospira	27–70	15–50	0	4.1–35	0

Fig. 9.7 Large-scale cultivation of microalgae in open ponds: Earthrise Farms, California (Aerial photo of Earthrise Nutritional Spirulina ponds with photoshop of microscopic spirulina algae in ponds by Robert Henrikson (2010), with permission)

9.4.1.3 Algae Cultivation, Harvest, and Extraction

Microalgal productivity depends on the culture system used. It is relatively low for open-pond systems (Fig. 9.7), and can be much higher for closed photobioreactors which allow better control of culture parameters and contamination with other species. Cultivating photosynthetic microorganisms in closed photobioreactors limits the environmental cost by recycling water and nutrients. Microalgae may also use industrial or urban waste as sources of CO_2, nitrogen, or phosphate. Considering the low productivity of the open-pond system and the high costs of closed photobioreactors, industrial applications of microalgae are presently limited. In this context, production of biofuel from microalgae can only be considered as a side product of an economically viable production. After separation of high-value compounds, residual biomass can be used for energy production through lipid extraction, hydrogen, or ethanol production from fermentation. It is estimated that productivity of microalgae actually obtained in production systems should be increased fivefold to meet the criteria of profitability.

It is believed that among the most costly downstream processing steps in fuel production using microalgae are the harvesting/dewatering steps and the extraction of fuel precursors from the biomass. Based on currently achievable productivities, most microalgae will not grow to a density higher than a few grams of biomass per liter of water. Although there are several possible low-cost solutions to concentrating the biomass, including settling and flocculation, these methods are slow and the resulting biomass may still require further dewatering. Alternative methods to concentrate algal biomass include centrifugation and filtration, which are faster, but are also typically much more expensive and energy intensive. In addition, many microalgal species have a very tough outer cell wall that makes extraction of fuel feedstock difficult, thereby requiring the use of harsh lysis conditions. The addition of all these steps dramatically raises the cost of the biofuel produced, meaning that the process is not economically viable.

One possible solution is to manipulate the biology of microalgal cells to allow for the secretion of fuels directly into the growth medium. There are, in fact, several pathways in nature that lead to secretion of hydrophobic compounds, including TAGs (triacylglycerol), free fatty acids, alkanes, wax esters, and soluble sugars. However, the introduction of metabolic pathways for the direct production of fuels faces many challenges. The product yields for pathways that lead to the accumulation of compounds that are not necessarily useful for the cell are unlikely to be economically viable without the comprehensive engineering of many aspects of microalgal metabolism. In addition, many types of fuel products have the potential to be toxic, and tolerant species of microalgae may have to be generated.

9.4.2 Biohydrogen Production by Microorganisms or Using Bacteria for Hydrogen Production

In the biosphere, H_2 (hydrogen) is generated by different metabolic processes in a wide range of microorganisms. It is also the energy source of many living species using hydrogenases as catalysts. The enzyme catalyzes the heterolytic splitting of H_2 ($H_2=H^+ + H^-$). The most significant discoveries on the structure/function of hydrogenases have been exclusively carried out by European research groups (Brugna-Guiral et al. 2003; Liebgott et al. 2009; Dementin et al. 2006, 2008, 2009; Rousset and Cournac 2008; Fernandez et al. 2007; Pandelia et al. 2010a, b, 2011). Hydrogenases could potentially replace platine as a first-row transition element in fuel cells, with the advantages of specificity, biodisponibility, and biodegradability (Lojou et al. 2008; Luo et al. 2009; Ciaccafava et al. 2010, 2011; Armstrong et al. 2009).

9.4.2.1 Biophotoproduction by Algae

Some microalgae, such as *C. reinhardtii*, harbor an hydrogenase interacting with the photosynthetic electron transfer chain. This property allows these organisms to produce hydrogen using light as the sole energy source. When an anaerobic adapted culture of *C. reinhardtii* is placed in the light, hydrogen production is observed for a few minutes at very high efficiency (close to the maximal photosynthesis yield ~10 %). However, hydrogen photoproduction rapidly stops, the hydrogenase being sensitive to oxygen produced at PSII (Photosystem II). Melis et al. (2000) proposed an experimental protocol to overcome this main limitation. Sulphur deprivation triggers PSII degradation. This results in installation of anoxic conditions because O_2 consumption by respiration becomes higher than its production. However, the general cellular metabolism is also severely affected by sulphur starvation, therefore considerably lowering the productivity of the algal culture. Two main research pathways are currently being explored for biohydrogen production by microalgae. The first is to control PS II activity and starch accumulation in nutrient-replete conditions. The second is to study at the enzyme level the mechanisms of hydrogenase inhibition by O_2. Once understood, the basic mechanism of natural algae has been reproduced by artificial means (biomimetism) and has led to the conception of photobioreactors (Fouchard et al. 2008; Shi et al. 2011).

9.4.2.2 Dark Fermentation by Bacteria

Dark hydrogen fermentation (or anaerobic digestion) is an attractive pathway to produce H_2. This metabolism is ubiquitous in nature under anaerobic conditions. The advantage of dark fermentation is that the H_2 production rate can be of a magnitude higher than those achieved by other means. Moreover, dark fermentation systems are relatively simple to construct and their functioning does not consume much energy. However, the amount of H_2 produced per molecule of fermented hexose (i.e., yield) is low. Instead of the expected 12 H_2 per hexose if 100 % conversion of hexose electron equivalents occurs, a yield of 2 H_2 per molecule of hexose is obtained (Nath and Das 2004) except in some thermophilic bacteria such as *Caldicellulosiruptor* that reach a molar yield close to 4 H_2 per hexose. Another advantage is that fermenting bacteria can utilize complex forms of organic substrate for H_2 production, that is, cellulose, food waste, and urban and industrial waste. These sources represented about 849 million tons worldwide in 2006. Pure cultures (i.e., containing a single microorganism species) have been intensively investigated in fermentation processes, including *B. coagulans*, *Thermoanaerobacterium* spp., *E. aerogenes*, and *C. butyricum*. In contrast, the use of mixed cultures (i.e., containing a mix of different microorganism species), which allows the degrading of more complex substrates, has received little attention. The major obstacle of mixed culture is the presence of three classes of microorganisms that establish close metabolic interactions, that is, H_2 producers, H_2 consumers (methanogenic bacteria producing methane), and metabolic competitors. The control of H_2 consuming pathways constitutes the main challenge for improving the stability of the fermentation process in bioreactors treating agricultural waste. Various studies have enabled the identification of major strains present in the bacterial consortia that develop spontaneously in fermenting wastes; yet few of these studies were completed by a metabolic characterization of the organism. Moreover, uncertainties remain regarding the identity of minor species as well as their role in the control and resilience of the process.

9.5 Conclusion

In a context of increasing world energy demand, biomass conversion into energy can propose a contribution to the mitigation of GHG emission and to the limitation of climate change. It has the potential to make a significant contribution to the energy mix, particularly biofuels for transportation. But the development of biofuels still requires awareness with regard to the complexity of the global problem where both the scientific and technological issues, but also the environmental, social, and economic impacts must be considered. The processes from well-to-wheel must be discussed in terms of a positive energy balance, GHS emission reduction, biodiversity and forest protection, environment degradation limitation, water availability, and food.

This chapter is essentially oriented towards the scientific and technological issues of the three biofuel pathways. An interesting complement can be found in Chap. 14 which gives a more worldwide and geopolitical analysis of first-generation biofuel potential and risks (environmental impacts, energy efficiency, world hunger, etc.).

Our chapter proposed a scientific state of the art on the three generations of biofuels and, when available, on the manufacturing processes.

Technologies, such as the first biofuel generation, that value the reserves of plants (starch, sugar, oil), are already fully mature and industrially controlled and they have still a fair potential of deployment.

The second biofuel generation that turns to the whole plant tissues (agricultural, forest, or organic waste resources), are rather operational for dry processes (thermochemical transformation of lignocellulose), and research is still underway for wet processes (enzymatic biological transformation of lignocellulose in sugars, and fermentation). Although these processes are already scientifically and technologically almost validated, or on the way to being so, the economical cost remains high as regards fossil fuel, the energetic and environmental balance must still be validated, and the indirect impacts of the pathways assessed.

Finally, very innovative processes have been developed on the third biofuel generation that show significant potential by transforming lipids, starch, and biohydrogen, thanks to bioalgae or microorganisms, into biofuel, which is a completely new concept in comparison with other biofuel crops such as wheat, corn, rapeseed, or sunflower. Fundamental research studies are still to be developed to describe the rather complex involved biological processes. Additional locks subsist on the cultivation, harvest, and extraction of algae to assess pathway productivity depending on the culture system and energy required for extracting a very low proportion of oil from a great water volume for producing biofuel.

References

ADEME (2011) *Feuille de route biocarburants avancés* (Advanced biofuels roadmap), Ademe. Retrieved from http://www2.ademe.fr/servlet/getDoc?cid=96&m=3&id=77923&p1=02&p2=10&ref=17597

ADEME/DIREM (2002) Bilan énergétique et gaz à effet de serre des filières de production de biocarburants en France (Greenhouse gas and energy balance in biofuel production pathways in France), ADEME/DIREM, Dec 2002. Retrieved from www.uclm.es/area/amf/Antoine/Energias/Informe_ADEME_fr.pdf

Armstrong FA, Belsey NA, Cracknell JA, Goldet G, Parkin A, Reisner E, Vincent KA, Wait AF (2009) Dynamic electrochemical investigations of hydrogen oxidation and production by enzymes and implications for future technology. Chem Soc Rev 38(1):36–51

Ballerini D (2006) Les biocarburants, Etats des lieux, perspectives et enjeux du Développement (Biofuels, inventories, development perspectives and challenges). IFP Publications/Editions Technip, Paris

BP (2011) British petroleum statistical review of world energy. Retrieved from http://www.bp.com/assets/bp_internet/globalbp/globalbp_uk_english/reports_and_publications/statistical_energy_review_2011/STAGING/local_assets/pdf/statistical_review_of_world_energy_full_report_2011.pdf

Brugna-Guiral M, Tron P, Nitschke W, Burlat B, Guigliarelli B, Stetter KO, Bruschi M, Giudici-Orticoni MT (2003) Biochemical, biophysical and phylogenetic characterisation of hyperthermostable hydrogenases from *Aquifex aeolicus*. Extremophile 7:145–157

Ciaccafava A, Infossi P, Giudici-Orticoni MT, Lojou E (2010) Stabilization role of a phenothiazine derivative on the electrocatalytic oxidation of hydrogen via *Aquifex aeolicus* hydrogenase at graphite membrane electrodes. Langmuir 26:18534–18541

Ciaccafava A, Infossi P, Ilbert M, Guiral M, Lecomte S, Giudici-Orticoni MT, Lojou E (2011) Electrochemistry, AFM and PM-IRRA spectroscopy of immobilized hydrogenase: role of a trans-membrane helix on enzyme orientation for efficient H2 oxidation. Angewandte 123:1–5

Dementin S, Belle V, Bertrand P, Guigliarelli B, De Lacey A, Fernandez V, Rousset M, Léger C (2006) Changing the ligation of the distal [4Fe4S] cluster in [NiFe] hydrogenase impairs inter- and intramolecular electron transfers. J Am Chem Soc 128:5209–5218

Dementin S, Belle V, Champ S, Bertrand P, Guigliarelli B, De Lacey A, Fernandez V, Léger C, Rousset M (2008) Molecular modulation of hydrogenase activity. Int J Hydrog Energy 33:1503–1508

Dementin S, Leroux F, Cournac L, De Lacey AL, Volbeda A, Léger C, Burlat B, Martinez N, Champ S, Martin L, Sanganas O, Haumann M, Fernandez V, Guigliarelli B, Fontecilla-Camps JC, Rousset M (2009) Introduction of methionines in the gas channel makes [NiFe] hydrogenase aero-tolerant. J Am Chem Soc 131:10156–10164

DOE (2006) Breaking the biological barriers to cellulosic ethanol. In: A research roadmap resulting from the biomass to biofuel workshop, sponsored by the US DOE, Rockville, 7–9 Dec 2005

Fernandez VM, De Lacey A, Rousset M, Cammack R (2007) Activation and inactivation of hydrogenase function and the catalytic cycle, spectroelectrochemical studies. Chem Rev 107:4304–4330

Fouchard S, Pruvost J, Degrenne B (2008) Investigation of H2 production using the green microalga *Chlamydomonas reinhardtii* in a fully controlled bioreactor fitted with on-line gas. Int J Hydrog Energy 33(13):3302–3310

Henrikson R (2010) Spirulina world food: how this micro algae can transform your health and our planet. Ronore Enterprises, Inc., Maui

IEA (2009a) World energy outlook 2009. International Energy Agency, Paris

IEA (2009b) Bioenergy – a sustainable and reliable energy source, Main Report, International Energy Agency. Retrieved from http://www.ieabioenergy.com/libitem.aspx?id=6479

IEA (2011) CO_2 emissions from fuel combustion, Highlights International Energy Agency statistics. IEA, Paris

Larkum AW, Ross IL, Kruse O, Hankamer B (2011) Selection, breeding and engineering of microalgae for bioenergy and biofuel production. Trends Biotechnol 30(4):198–205

Liebgott P-P, Leroux F, Burlat B, Dementin S, Baffert C, Lautier T, Fourmond V, Ceccaldi P, Cavazza C, Meynial-Salles I, Soucaille P, Fontecilla-Camps J-C, Guigliarelli B, Bertrand P, Rousset M, Léger C (2009) Hydrogenases: the relation between diffusion along the substrate tunnel and resistance to oxygen. Nat Chem Biol 6:63–70

Lojou E, Luo XJ, Brugna M, Candoni N, Dementin S, Giudici-Orticoni MT (2008) Biocatalysts for fuel cells: efficient hydrogenase orientation for H_2 oxidation at electrodes modified with carbon nanotubes. J Biol Inorg Chem 13:1157–1167

Luo XJ, Brugna M, Tron-Infossi P, Giudici-Orticoni MT, Lojou E (2009) Immobilization of the hyperthermophilic hydrogenase from *Aquifex aeolicus* bacterium onto gold and carbon nanotube electrodes for efficient H_2 oxidation. J Biol Inorg Chem 14:1275–1288

Ma LJ, van der Does HC, Borkovich KA, Coleman J-J, Daboussi M-J, DiPietro A, Dufresne D, Freitag M, Grabherr M, Henrissat B, Houterman PM, Kang S, Shim WB, Woloshuk C, Xie X, Xu JR, Antoniw J, Baker SE, Bluhm BH, Breakspear A, Brown DW, Butchko RAE, Chapman S, Coulson R, Coutinho PM, Danchin EGJ, Diener A, Gale L, Gardiner DM, Goff S, Hammond-Kosack KE, Hilburn K, Houterman PM, Hua-Van A, Jonkers W, Kazan K, Kodira CD, Koehrsen M, Kumar L, Lee Y-H, Li L, Manners JM, Miranda-Saavedra D, Mukherjee M, Park G, Park J, Park S-Y, Proctor RH, Regev A, Ruiz-Roldan CM, Sain D, Sakthikumar S, Sykes S, Schwartz DC, Turgeon BG, Wapinski I, Yoder O, Young S, Zeng Q, Zhou S, Galagan J, Cuomo CA, Kistler HC, Rep M (2010) Comparative genomics reveals mobile pathogenicity chromosomes in *Fusarium oxysporum*. Nature 464:367–373

Martinez D, Challacombe J, Hibbett D, Morgenstern I, Schmolls M, Kubicek CP, Martinez A, Ferreira P, Ruiz-Duenas F, Kersten P, Hammel K, Vanden WA, Gaskell J, Grigoriev I, Lindquist E, Sabat G, Splinter BS, Larrondo L, Canessa P, Yadav J, Doddapaneni H, Subramanian V, Pisabarro A, Lavín JL, Oguiza JA, Master E, Henrissat B, Coutinho PM, Harris P, Magnuson JK, Baker S, Bruno K, Keneally W, Hoegger P, Kues U, Ramiaiya P, Lucas S, Salamov A, Shapiro H, Tu H, Teter S, Yaver D, James T, Mokrejs M, Brettin T, Rokhsar D, Berka RM, Cullen D (2009) Genome, transcriptome, and secretome analysis of wood decay fungus Postia placenta supports unique mechanisms of lignocellulose conversion. Proc Natl Acad Sci USA 106:1954–1959

Melis A, Zhang L, Forestier M, Ghirardi ML, Seibert M (2000) Sustained photobiological hydrogen gas production upon reversible inactivation of oxygen evolution in the green alga *Chlamydomonas reinhardtii*. Plant Physiol 122(1):127–136

Mingardon F, Chanal A, Lopez-Contreras AM, Dray C, Bayer EA, Fierobe H-P (2007) Incorporation of fungal cellulases in bacterial minicellulosomes yields viable, synergistically acting cellulolytic complexes. Appl Environ Microbiol 73:3822–3832

Nath K, Das D (2004) Improvement of fermentative hydrogen production: various approaches. Appl Microbiol Biotechnol 65(5):520–529

Pandelia M-E, Tron-Infossi P, Giudici-Orticoni M-T, Lubitz W (2010a) The oxygen-tolerant hydrogenase I from *Aquifex aeolicus* weakly interacts with carbon monoxide: an electrochemical and time resolved FTIR study. Biochemistry 49(41):8873–8881

Pandelia ME, Fourmond V, Tron P, Lojou E, Bertrand P, Léger C, Giudici-Orticoni M-T, Lubitz W (2010b) The membrane-bound hydrogenase I from the hyperthermophilic bacterium *Aquifex aeolicus*: enzyme activation, redox intermediates and oxygen tolerance. J Am Chem Soc 132:6991–7004

Pandelia ME, Nitschke W, Infossi P, Giudici-Orticoni M-T, Bill E, Lubitz W (2011) Characterization of a unique [4Fe4S] cluster in the electron transfer chain of the oxygen tolerant NiFe hydrogenase of *Aquifex aeolicus*. PNAS 108(15):6097–6102

Pruvost J, Van Vooren G, Cogne G, Legrand J (2009) Investigation of biomass and lipids production with *Neochloris oleoabundans* in photobioreactor. Bioresour Technol 100(23):5988–5995

Radakovits R, Jinkerson RE, Darzins A, Posewitz MC (2010) Genetic engineering of algae for enhanced biofuel production. Eukaryot Cell 4:486–501

Rousset M, Cournac L (2008) Towards hydrogenase engineering for hydrogen production. In: Wall JD, Harwood CS, Demain A (eds) Bioenergy. ASM Press, Washington, DC, pp 249–257

Shi L, Belchik SM, Plymale AE, Heald S, Dohnalkova AC, Sybirna K, Bottin H, Squier TC, Zachara JM, Fredrickson JK (2011) Purification and characterisation of the [NiFe]-hydrogenase of *Shewanella oneidensis* MR-1. Appl Environ Microbiol 77(16):5584–5590

Stephens E, Ross IL, Mussgnug JH, Wagner LD, Borowitzka MA, Posten C, Kruse O, Hankamer B (2010) Future prospects of microalgal biofuel production systems. Trends Plant Sci 10:554–564

Weiner RM, Taylor LE, Henrissat B, Hauser L, Land M, Coutinho PM, Rancurel C, Saunders EH, Longmire AG, Zhang H, Bayer EA, Gilbert HJ, Larimer F, Zhulin IB, Ekborg NA, Lamed R, Richardson PM, Borovok I, Hutcheson S (2008) Complete genome sequence of the complex carbohydrate-degrading marine bacterium, *Saccharophagus degradans* strain 2–40 T. PLoS Genet 4(5):e1000087

Yang JC, Madupu R, Durkin AS, Ekborg NA, Pedamallu CS, Hostetler JB, Radune D, Toms BS, Henrissat B, Coutinho PM, Schwarz S, Field L, Trindade-Silva AE, Soares CA, Elshahawi S, Hanora A, Schmidt EW, Haygood MG, Posfai J, Benner J, Madinger C, Nove J, Anton B, Chaudhary K, Foster J, Holman A, Kumar S, Lessard PA, Luyten YA, Slatko B, Wood N, Wu B, Teplitski M, Mougous JD, Ward N, Eisen JA, Badger JH, Distel DL (2009) The complete genome of *Teredinibacter turnerae* T7901: an intracellular endosymbiont of marine wood-boring bivalves (shipworms). PLoS One 4:e6085

Chapter 10
Meeting Environmental and Energy Challenges: CNG Conversion of Motor Vehicles in Dhaka

Zia Wadud

Abstract Petroleum fuels used in motor vehicles are a major source of local air pollutants and greenhouse gas emissions in Dhaka, the capital of Bangladesh, which already has poor air quality. Lack of their own oil resources also presents a major challenge to the policymakers regarding the sustainability of the transportation system. Under this circumstance, conversion of petroleum vehicles to run on compressed natural gas (CNG) offers multiple benefits to the country. This chapter quantifies the social benefits due to a government initiative that led to widespread conversion of petroleum motor vehicles to CNG vehicles. An impact-pathway model has been developed to relate the changes in emissions resulting from the policy to changes in ambient air quality and resulting number of avoided premature deaths. It is estimated that around 11,100 premature deaths can be avoided in Dhaka annually as a result of a complete switch from petroleum to CNG vehicles. This amounts to a saving of USD 1.33 billion a year, which is around 1.3 % of the GDP of the country. For climate benefits, impacts of black carbon (BC), organic carbon (OC), and SO_2 have been considered, in addition to the traditional greenhouse gases (GHG), CO_2 and methane. Although CNG conversion was detrimental from a climate change perspective using the changes in CO_2 and methane only (methane emissions increased), after considering all the global pollutants (especially the reduction in black carbon) the conversion strategy was beneficial. Considering the damage costs of CO_2, we find a benefit of around USD 25 million in a year, which is small compared to the health benefits. The strategy also helps the country save around USD 620 million worth of foreign currency a year. This indicates that policies focusing on individual country's strengths can have large benefits in securing a sustainable energy future in the transportation sector.

Z. Wadud (✉)
Department of Civil Engineering, Bangladesh University of Engineering and Technology, Dhaka, Bangladesh
e-mail: ziawadud@yahoo.com

J.B. Saulnier and M.D. Varella (eds.), *Global Change, Energy Issues and Regulation Policies*, Integrated Science & Technology Program 2, DOI 10.1007/978-94-007-6661-7_10,

Keywords CNG • Alternate energy • Co-benefits • Climate change • Air pollution

List of Acronyms

BRTA	Bangladesh Road Transport Authority
GoB	Government of Bangladesh
CR	Epidemiological concentration-response
BC	Black carbon
BDT	Bangladesh currency (Bangladeshi Taka)
CNG	Compressed natural gas
EF	Respective emission factors
EU	European Union
EU-ETS	EU-Emissions Trading Scheme
GDP	Gross domestic product
GHG	Greenhouse gases
OC	Organic carbon
$PM_{2.5}$	Particulate matters, with a diameter less than 2.5 μm
PPP	Purchasing power parity
RPGCL	Rupantorito Prakritik Gas Company Limited
VSL	Value of statistical life

10.1 Introduction

Motor vehicles are a major source of greenhouse gas (GHG) emissions, which affect the climate system. In developed and rapidly developing countries such as China and India, growth in GHG emissions from motor vehicles is a major concern. Although concerns about local air pollutant emissions from motor vehicles were addressed in the developed countries decades ago, megacities in the developing countries are still suffering from poor air quality. Road transportation, especially motor vehicles, is a major source of air pollution in all these large cities. The air quality in the cities of the developing countries is often deteriorating, primarily because of increasing motor vehicle ownership resulting from high economic growth and relatively lax emissions control. This makes air quality management a priority issue for policymakers in these countries or cities.

Air quality in Dhaka, the capital of Bangladesh, is very poor, particularly during the dry seasons. Although vehicle ownership is still very low in Bangladesh, the number of vehicles has been growing at a rate of more than 6 % a year for the last few years. The growth in the number of motor vehicles not only worsens the air quality in large cities, but also places a large burden on foreign currency reserves because the country imports all its petroleum to keep its vehicles running. Although Bangladesh emits very little GHG as compared to the developed or the rapidly developing economies such as India or China, motor vehicles and their growth could be responsible for a major share of its GHG emissions in near future.

Under these circumstances, fueling vehicles using compressed natural gas (CNG) offers a four-way benefit to the country. CNG vehicles reduce emissions of local air pollutants, improving air quality of the large cities, and CNG vehicles also can emit fewer GHGs, improving the country's GHG accounts. At the same time, natural gas is an indigenous energy source, therefore, it saves foreign currency and improves the trade balance of the country. Using a local energy source also improves energy security.[1] Accordingly, the government of Bangladesh (GoB) actively pursued a policy to encourage the conversion of existing vehicles to run on dual fuels, one of them being CNG. This chapter quantifies (with an emphasis on modeling methods) the benefits of such a policy for Dhaka city, which represents around half of all registered vehicles in the country (Bangladesh Road Transport Authority, BRTA 2010a, b). Reduction in local air pollutants and GHG emissions reduction offer two very different types of benefits, therefore it is important to convert the benefits to a unit such that they can be compared. Conversion of these benefits to monetary units therefore has some appeal.

The chapter is organized as follows. Section 10.2 presents the background information on Bangladesh and Dhaka city and the policy initiatives for improving air quality. Section 10.3 describes the modeling approach to quantify health benefits from the CNG conversion policy. Section 10.4 presents the methods for determining climate change benefits (or costs) and the results. Section 10.5 describes the results of foreign currency benefits, with Sect. 10.6 presenting the uncertainties in such estimates. Section 10.7 draws conclusions of the exercise.

10.2 Background on Dhaka and Its CNG Program

Bangladesh is a low-income country with per capita GDP of around USD 550 (current dollar) in 2007, yet the growth rate is above 6 % (Bangladesh Bureau of Statistics, BBS 2009a). Dhaka, the capital of the country, is one of the most populous and densely populated cities in the world with a population of 12.3 million in 2007 (BBS 2009a). Until recently, there was a lack of (or too lax) emissions standards for industries or motor vehicles in Bangladesh. Even when emissions standards exist, enforcement of these standards is also poor. The city is also surrounded by brickfields which use coal for burning bricks. All these made Dhaka's air one of the most polluted in the world (Gurjar et al. 2008). The situation has been further deteriorating as a result of economic growth, with corresponding prosperity and increases in vehicle ownership, resulting in congested roads and higher vehicle emissions. A recent estimate concluded that air pollution in Dhaka alone can be traced to about 15,000 premature deaths a year (IRIN 2009).

[1] Although there can be a limit to energy security benefits if the resources are to be exhausted soon.

Responses to controlling air pollution have not been quick enough. Monitoring ambient air quality at the government level started only recently, with four continuous monitoring stations set up in 2002 in large cities with the help of the World Bank.[2] Some recent policy initiatives, however, were noteworthy. Bangladesh showed remarkable leadership in outright banning of leaded fuel in 1999, thus effectively reducing the lead content in the air. The ban was easier to implement as compared to other countries, inasmuch as the country had only one petroleum refinery. Emissions standards for motor vehicles were tightened in 2002, but these standards are still relatively relaxed compared to the developed countries (even as compared to China or India). Initiatives to regulate emissions from brickfields were undertaken. One major initiative that visibly improved the air quality in Dhaka city was banning the two-stroke three-wheeler autorickshaws from the city on January 1, 2003. Vehicles older than 20 years of age were also banned from the city during the same period. Also, Bangladesh is one of the few countries to have an ambient $PM_{2.5}$ standard, concerning particulate matters, with a diameter less than 2.5 μm (USEPA 2004).

Apart from the ban on leaded fuel, enforcement is very lax. There are no sophisticated vehicle emissions testing facilities, with only occasional tests carried out by smoke opacity meters. Certification for vehicle fitness can be a corrupt process, and vehicles without fitness certificates can run on the city streets due to a lack of enforcement. Enforcing the ban of older vehicles proved impossible by successive governments, inasmuch as a large number of those vehicles are trucks and buses, backed by a strong union and lobby. Although ambient air quality standards exist, there is no serious effort to achieve them.

As mentioned earlier, conversion of vehicles to CNG offers four benefits in Dhaka and Bangladesh: air quality improvement, GHG emissions reduction, foreign currency savings, and energy security. Natural gas is an indigenous source of energy in Bangladesh, which lacks resources in petroleum, and has only a small deposit of coal. Natural gas is also the largest source of energy in the country with around half of all energy use in the country coming from natural gas (Fig. 10.1). There is a fairly extensive gas network in major cities in the eastern part of the country, primarily because all the gas production fields and the largest cities lie to the east of the rivers Padma and Jamuna, which present a significant hurdle for the pipelines. The present gas network is about 21,110 km (end of December 2007; Petrobangla 2008), which is set for planned expansion to the western part of the country, especially to Khulna, a major industrial city, by 2014. Once completed, all the major cities in the country will be connected to a natural gas distribution network.

The first CNG refueling station in Bangladesh was established as early as in 1984 by the Rupantorito Prakritik Gas Company Limited (RPGCL), a government-owned subsidiary, but the next four came a decade later, in 1995. Between 1984 and 1997, there were only 174 vehicles running on CNG in Dhaka, clearly indicating that the introduction of CNG for vehicles was not a runaway success. In fact, until 2001, the number of CNG vehicles was very small (see Fig. 10.2). Refueling

[2] Some of these monitors are already nonoperational.

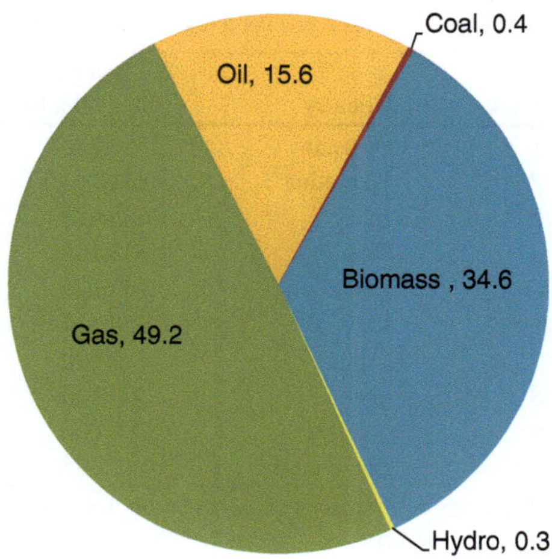

Fig. 10.1 Share of natural gas in the energy mix of Bangladesh in 2005

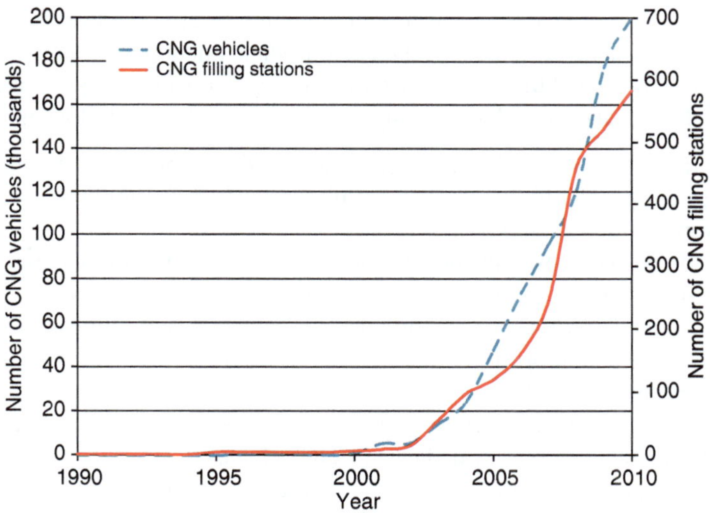

Fig. 10.2 Evolution of CNG vehicles and CNG filling stations in Bangladesh (data from RPGCL 2009)

facilities were even poorer. The CNG industry got some momentum during early 2001 when some 12,000 CNG-run taxis were gradually introduced in Dhaka city to alleviate its transportation problem.

The government's decision in late 2002 to remove old two-stroke petroleum-run autorickshaws (three-wheelers) from Dhaka city in order to improve the local air quality resulted in 9,000 new CNG-run autorickshaws in the city, almost overnight. These two initiatives helped the industry gain a critical mass, especially the chicken and egg (build CNG infrastructure first and wait for vehicles to convert later or

Table 10.1 Natural gas vehicles in different countries

Rank	Country	Natural gas (NG) vehicles[a]	NG refueling stations[a]	Stations per 1,000 NG vehicles
1	Pakistan	2,300,000	3,068	1.33
2	Argentina	1,807,186	1,851	1.02
3	Iran	1,665,602	1,021	0.61
4	Brazil	1,632,101	1,704	1.04
5	India	935,000	560	0.60
6	Italy	628,624	730	1.16
7	China	450,000	870	1.93
8	Colombia	300,000	460	1.53
9	Ukraine	200,000	285	1.43
10	Bangladesh	177,555	500	2.82

[a]Data from International Association for Natural Gas Vehicles (2010); data for 2009, except Ukraine (2006)

convert the vehicles first and wait for CNG filling stations to respond to demand later) problem was avoided by ensuring a minimum level of demand from the new vehicles. At the same time, the government took some important decisions to encourage the development of the CNG industry in the city. These decisions included boosting both the demand as well as the supply side and included the following.

- The private sector was allowed to participate in CNG conversion of vehicles and in setting up CNG filling stations, curbing the previous monopoly of RPGCL, a government-owned body.
- The private sector was encouraged to enter the industry by making available government land to them only for setting up CNG filling stations.
- Import duties on CNG conversion kits, storage tanks, and filling station-related equipment were dropped in order to bring down the conversion costs and expedite the conversion rate.
- Import duties on dedicated CNG buses were dropped and on autorickshaws were reduced to encourage more dedicated CNG vehicles.
- CNG and petroleum prices were restructured in order to make CNG a more lucrative choice as a transportation fuel, even after considering the conversion costs; especially subsidies on petroleum fuels were either reduced or removed.
- All government vehicles were asked to convert to CNG.
- Safety campaigns were run to ensure the use of proper CNG storage tanks (initially there were a few accidents with poor quality storage tanks in vehicles).

All these initiatives led a large number of different types of vehicles (private cars, SUVs, minibuses, buses) to gradually switch to CNG from petroleum.

Government statistics show that around 117,000 (more than 40 %) vehicles in Dhaka currently run on CNG, and nationwide, the number is around 200,000 (in year 2000). This puts Bangladesh among the top 10 countries in the world in terms of total number of natural gas vehicles (Table 10.1). However, the number of refueling stations per thousand CNG vehicles in Bangladesh is the highest among these 10 countries. A spot survey carried out at different locations in Dhaka indicates that almost 83 %

of all cars surveyed were running on CNG in 2010. The proportion of buses and minibuses running on CNG was around 75 %. The large difference with government statistics (40 %) is possibly due to the slow updating of the CNG vehicle register or the inaccurate vehicle registration data (no data on vehicle attrition, which deflates the share of CNG vehicles). Because of the discrepancy of the conversion numbers, instead of modeling the existing benefits, we model the potential benefits if all vehicles but trucks in Dhaka city are converted to CNG by 2012. For trucks, the conversion is assumed to stop at 50 %, inasmuch as long-distance trucks would continue to run on diesel until CNG infrastructure was available in every part of the country (e.g., southwest part of the country does not have access to natural gas pipeline yet).

10.3 Local Air Quality Benefits

The principal local benefits of using CNG as a transportation fuel are the reduced emissions of local or criteria air pollutants and resulting reductions in adverse health impacts. The US Environmental Protection Agency (USEPA 2010) reports that light-duty CNG vehicles can emit 90 % less CO, 35–60 % less NO_x and 50–75 % nonmethane HC compared to similar petroleum vehicles in use. Particulate emissions are also negligible in CNG vehicles. Heavy-duty CNG vehicles can emit 84–95 % less particulates and 25–50 % NO_x emissions (USEPA 2010). A recent study by the International Energy Agency (Nijboer 2010) finds that CNG vehicles offer a very large benefit in reduced particulate emissions for diesel vehicles, even for recent vehicle models. For newer vehicle models with stringent emissions control for petroleum and diesel vehicles, the local air quality benefits of CNG are diminished but they do not vanish. However, because the vehicle emissions standard and enforcement in Dhaka is lax, and in-use vehicles are converted to CNG, the potential benefits in Dhaka would be on the larger side.

In determining the benefits of a policy intervention that improves local air quality, the reduction in emissions is linked with well-defined improvements in damage endpoints and associated benefits through the impact-pathway approach, described graphically in Fig. 10.3 (European Commission 2003; ExternE 2005). The first step in an impact-pathway approach is to quantify the emissions (or changes in emissions for a policy intervention such as CNG conversion), which can be determined from a vehicle emissions inventory model. The changes in modeled emissions are then fed into an air quality model in order to determine the changes in ambient air quality (i.e., pollutant concentration).

In the third step the modeled improvements in ambient air quality are coupled with population distribution and epidemiological concentration-response (CR) functions of the health impacts to determine the avoided health impacts of different types.[3] Each of these health cases is then valued using the cost savings

[3]A CR function determines the mathematical relationship between the changes in an adverse health outcome with respect to changes in an ambient air pollutant concentration.

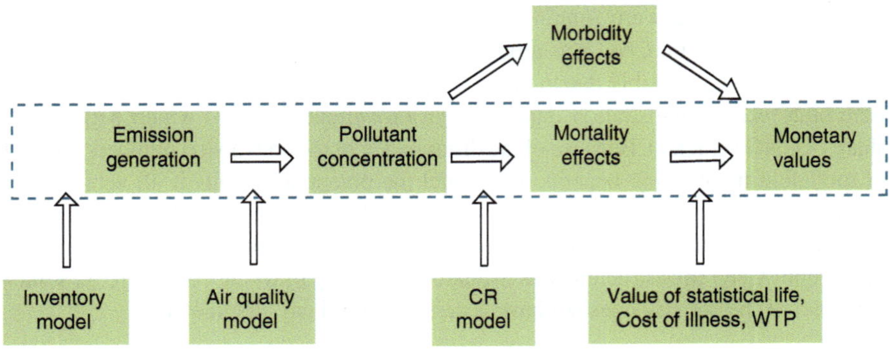

Fig. 10.3 Impact-pathway approach for air quality-related premature deaths

associated with those specific health impacts or willingness to pay to avoid those health cases (see Fig. 10.3) to determine the avoided costs, or benefits, of the policy intervention. Studies modeling the health impacts of air pollution using the impact-pathway model in the developed countries have found that the majority of the health impacts can be attributed to particulate matters, especially those with a diameter less than 2.5 μm, known as $PM_{2.5}$ (USEPA 2004 for a synthesis). Although $PM_{2.5}$ (or other local air pollutants) can have different effects on health (e.g., increase in mortality, asthma or respiratory troubles, eye irritation etc.), monetized health costs of air pollution are dominated by the premature mortality costs due to exposure to $PM_{2.5}$ (typically 85–95 % of total health costs, USEPA 2007). We therefore focus only on the reduced mortality impacts due to reduction in ambient $PM_{2.5}$ concentration (the dashed box in Fig. 10.3). The significant challenge lies in collecting all the relevant data, especially in the context of a developing country such as Bangladesh.

We model a spatially disaggregated emissions inventory of the vehicles following the well-known formulae

$$Emissions_{im} = \sum_j \sum_k N_{jkm} \times A_{jkm} \times EF_{ijk} \tag{10.1}$$

where N refers to number of vehicles, A is the activity of those vehicles per day within grid m, EF are respective emission factors, and subscripts i, j, k, and m refer to pollutant type, vehicle type, fuel type, and grid number, respectively. Data on different components have been collected from the Bangladesh Road Transport Authority (BRTA 2010a, b), and other reasonable sources (Urbanemissions 2010; Bond et al. 2004; Rouf et al. 2008; Reynolds and Kandlikar 2008). Some data, such as fuelwise distribution of vehicles and vehicle activities were collected using field surveys and cross-checked with an existing estimate (Khaliquzzaman 2006). We assume that all the vehicles will be converted to CNG by the end of 2012 and assume a vehicle growth rate of 6 % for years 2011–2012. This results in a reduction of direct $PM_{2.5}$ emissions from 10,400 to 3,200 ton/year (68 %) and SO_2 from 3 to

0.8 ton/year (74 %). The majority of the reductions in $PM_{2.5}$ emissions arises from the conversion of a rather smaller number of diesel vehicles.

The spatially disaggregated emissions inventory is fed into an air quality model specifically developed for Dhaka city (Rahman 2010). The air quality model considers secondary formation of $PM_{2.5}$ from primary pollutants of NO_x and SO_x. Model results indicate that there will be a reduction in ambient concentration of $PM_{2.5}$ due to emissions changes in $PM_{2.5}$ and SO_x resulting from the policy.

In order to determine the avoided premature mortality resulting from the improved air quality due to CNG conversion, we utilize concentration-response functions (Pope et al. 2002), along with the gridwise changes in ambient concentration of $PM_{2.5}$ from the air quality model, existing mortality rate, and gridwise population:

$$Deaths\ avoided = \sum_m \sum_n \Delta c_m \times CR_n \times mortality\ rate_n \times population_m \qquad (10.2)$$

where m is the grid number, n is the cause of mortality (respiratory or cardiovascular), and Δc_m is the change in ambient concentration of $PM_{2.5}$ in grid m. CR functions for all cause mortality are generally used in modeling policy interventions (USEPA 2005, 2007; Kunzli et al. 2000). But inasmuch as the causes of deaths vary significantly between the developed and developing countries (Cropper and Simon 1996), using an all-cause CR function from a developed country would have been misleading. Data on population and cause-specific mortality rates were collected from WHO (2009) and BBS (2009b). The calculation results in around 9,000 avoided premature deaths in 2012 in Dhaka City Corporation and 11,100 in Greater Dhaka.

The most common approach to determine monetary benefits due to these avoided premature deaths is to use a value of statistical life (VSL), defined as the amount people are willing to pay (accept) to reduce (increase) the mortality risks (probability of death) they face. Although the VSL approach has its critics,[4] it is still widely used (USEPA 2005, 2007). Health benefits are calculated as

$$Health\ benefits = Premature\ deaths\ avoided \times VSL \qquad (10.3)$$

VSL is a widely researched area with many hundreds of studies published, although estimates for developing countries are not as frequent. The published estimates also vary widely (see Viscusi and Aldy 2003 for a review). The willingness to pay to reduce health risks was around USD 1 million for China (Krupnick 2006), similar to those in developed countries when estimated using the same techniques (contingent valuation) and corrected for purchasing power parity (PPP). This represents USD 190,000 for Bangladesh, equivalent to BDT (Bangladesh Currency) 13 million. Using a literature survey and income elasticity of VSL of 0.55 (Viscusi and Aldy 2003), VSL in Bangladesh becomes USD 53,000. We use a median of USD

[4] Leksell and Rabl (2001) recommend using loss of life expectancy (LLE) for valuing premature mortality due to long-term exposure to $PM_{2.5}$.

120,000 as VSL in this study. The total benefit of the 11,100 avoided premature deaths in year 2012 is USD 1.33 billion. This represents a benefit of 1.3 % of the GDP of the country in 2012 (assuming a real GDP growth rate of 6 %).

10.4 Global Climate Change Benefits

The climate change benefits model follows a slightly different approach than the impact-pathway model of Fig. 10.3. Although relevant emissions are required from the emissions inventory model as before, the pollutants are different, and the spatial disaggregation is no longer required because of the global impact of GHG emissions. Among motor vehicle emissions, CO_2 and CH_4 are established GHG, contributing directly to global warming (UNFCCC 2010). However, recent studies (Reynolds and Kandlikar 2008) show aerosols such as sulphates (SO_2), black carbon (BC), and organic carbon (OC) can also have an important influence on the earth's radiation balance and thus global climate. Black and organic carbons are components of $PM_{2.5}$ of which black carbon has a potentially large impact on warming (Bond et al. 2004). On the other hand, SO_2 (precursor to sulphates) and organic carbon have cooling effects on the climate through facilitating the formation of aerosols (Reynolds and Kandlikar 2008). Although NO_x emissions can also have an impact on global warming through secondary effects (formation of nitrates and shortening lives of CH_4, both of which have a cooling effect, or formation of ozone, which has a warming effect), we assume that changes in NO_x emissions from fuel switching are negligible (following Reynolds and Kandlikar 2008). We therefore concentrate on five global emissions (CO_2, CH_4, SO_2, black carbon, organic carbon) before and after conversion of the vehicles.

Exhaust emissions factors for SO_2 come from the literature (Urbanemissions 2010), although we modify their CO_2 emissions factors to reflect fuel economy for vehicles in Dhaka. We assume a 5 % fuel economy penalty (but a net carbon benefit) for emissions of CNG vehicles converted from petroleum (as per Reynolds and Kandlikar 2008). Although a 25 % loss in fuel economy and a net carbon penalty for diesel to CNG conversion is present in policy analysis (Reynolds and Kandlikar 2008), some reports suggest that CNG vehicles have net carbon benefits over diesel vehicles, even considering larger methane emissions (Nijboer 2010). We assumed that there is no direct carbon emissions benefit or penalty (represents around 15 % loss in fuel economy) and consider CH_4 emissions separately for diesel to CNG conversions. Our CH_4 emissions factors are from the literature as well (Reynolds and Kandlikar 2008). CH_4 leakage can be a particularly strong source of GHG, because almost all the motor vehicles in Bangladesh are retrofitted, with high possibility of leakage. We therefore include leakage emissions as well.

Black carbon and organic carbon are emitted as part of particulate matter. The fractions of OC and BC as part of PM_1 and PM_1 as part of PM_{10} can depend on vehicle and environmental characteristics such as vehicle type, combustion

Table 10.2 Total changes in emissions, global warming factors, and benefits in 2009 attributable to the policy

GHGs and particulates	Changes in emission 1,000 ton/year	Global warming potential	CO_2-equivalent changes in emissions (1,000 ton/year)
CO_2	−97.86	1	−97.86
CH_4	26.70	23	614.04
BC	−2.98	455	−1,356.23
OC	−0.90	−35	57.81
SO_2	−2.28	−100	227.94
Total			−580.8

technology, fuel type, and operating conditions. Emissions factors are again from the published literature (e.g., Bond et al. 2004; Reynolds and Kandlikar 2008).

The emissions inventory model shows that there is a net reduction in CO_2 emissions but an increase in CH_4 emissions, because previously there were no (or negligible) CH_4 leakage emissions from the vehicles (Table 10.2). SO_2, BC, and OC all decrease, by 74, 69, and 54 %, respectively, over preconversion emissions. Inasmuch as the per unit impact on earth's radiative balance of these different global pollutants is not the same, we use global warming potentials of each of these pollutants to normalize them to an equivalent scale. The normalization allows us to use a common metric, CO_2 equivalent emissions, which can be added or subtracted (depending on net warming or cooling effect) to generate net warming-weighted emissions of the different pollutants. Although global warming potentials for CO_2, CH_4, and NO_x are well established in the literature, the factors for BC, OC, and SO_2 are still not well established. We use 100-year global warming factors of BC, OC, and SO_2 (from Reynolds and Kandlikar 2008). The global warming factors for OC and SO_2 are negative, because an increase in these emissions results in net cooling of the atmosphere.

Table 10.2 also presents the CO_2 equivalent changes in emissions, considering the warming or cooling impacts. Therefore, although SO_2 emissions decrease, considering the cooling impact of SO_2 there is net warming as a result of the reduction in emissions, and the CO_2 equivalent changes for SO_2 are positive. There is a net warming impact due to increases in CH_4 emissions and decreases in SO_2 and OC emissions, and there is a net cooling impact due to reductions in CO_2 and BC emissions. Note that, considering the warming impact of only CO_2 and CH_4 emissions, the common GHG gases, CNG conversion can have a large net warming impact. However, once we include the effects of the aerosols and its precursors, there is a net cooling and beneficial effect resulting from the conversion.

There are two approaches to determine the monetary benefits associated with saving a ton of carbon, as calculated above. Carbon is now traded in forums such as the EU-Emissions Trading Scheme (EU-ETS), therefore we can use the price of carbon in that market. However, the EU-ETS prices work under a given carbon cap, and do not represent the social damage cost. We therefore use the social cost of carbon instead. The social costs of carbon in the literature vary by three orders of magnitude, from USD 1 to USD 1,500 per ton (Yohe et al. 2007). Peer-reviewed literature on the social costs of carbon finds that the mean social cost of carbon is USD 43 per ton, with a standard deviation of USD 83 per ton. We use a social cost of carbon of USD 45 in 2012 for our calculations. We note that the UK government uses a carbon cost of GBP 25 (around USD 43, in year 2007 prices; Price et al. 2007), and therefore believe our carbon cost is reasonable. This results in a net carbon saving worth USD 26 million in year 2012.

10.5 Foreign Currency and Energy Security Benefits

Using an indigenous energy source such as natural gas instead of imported petroleum does not automatically impart economic benefits because natural gas could be exported to earn foreign currency. This is especially true in context of the CNG prices in the country, which is often kept lower than the price on the international market (although prices have been increased recently). However, the foreign currency balance and savings in foreign currency are important in a small and developing country such as Bangladesh. If all the vehicles in Dhaka city are converted to CNG by year 2012, the scenario we have modeled here, it would save 1,300 million liters of petroleum and diesel a year. This would save the country USD 620 million a year in foreign currency in year 2012 alone. This is around 8.7 % of the annual trade deficit in Bangladesh (year 2008 data). If the prices of crude oil continue to increase, a likely scenario, these benefits will be even larger.

We do not model the energy security benefits because of a lack of a consistent methodology to determine such benefits. Although the oil dependence costs of the US road transport sector was modeled earlier (Parry et al. 2007), their methodology is not applicable to Bangladesh, because of a different geopolitical relationship with the Middle East, the major oil-producing region.

10.6 Uncertainties

Any benefits analysis such as ours has some uncertainties associated with it. Each step in our model can have significant uncertainties, depending upon the performance of the underlying modeling techniques and data quality. The uncertainties also increase from left to right of the impact-pathway chain (Fig. 10.3). In addition, our estimated air pollution and climate benefits possibly have a larger uncertainty than similar estimates in the developed countries, primarily because of the lack of reliable data for the emissions inventory model and the air quality model. Especially, robust estimates of emissions factors of in-use and retrofitted CNG vehicles in developing countries are not available, greatly affecting our estimates. Estimates of VSL also have large uncertainties. We present in Table 10.3 our qualitative evaluation of the uncertainty in the individual components of our model. We note that, even if the VSL is an order of magnitude lower, the air pollution benefits would still be larger than the climate change benefits for our estimate. At the same time, given

Table 10.3 Qualitative evaluation of the uncertainties in this study

Model component	Uncertainty	Remarks
Vehicle data	Small	Accurate scrappage information is not available.
Vehicle emission factors	Large	No testing on in-use vehicles in Bangladesh, especially black carbon, organic carbon, and CH_4 emission factors have large uncertainties; uncertainty in black carbon and CH_4 emissions has a large impact.
Vehicle activity data	Medium to large	No survey of in-use vehicles for travel activities. CNG vehicles possibly run longer because of lower running costs. The climate and health benefits will then be smaller than calculated here. Climate benefits could even be negative.
Fuel breakdown	Medium to large	Uncertainty in diesel to CNG conversion has a large impact on health and climate results due to diesel's higher PM and BC emissions.
Air quality model	Medium	Nonlinearity in emissions to concentration not included in the air quality model.
CR function	Medium	CR could be lower than what we used, because people could be less susceptible to pollutants, and could be higher if Laden et al. (2006) is the true CR.
Value of statistical life	Large	Our central estimate for VSL appears reasonable, but there are uncertainties in original estimates and on VSL's income elasticity; uncertainty in VSL is especially important when comparing health and climate benefits.
Global warming factors	Small to medium	Uncertainty for CH_4 is small, global warming potentials for black carbon, organic carbon, and SO_2 are still not well established.
Carbon price	Medium	Although market price for carbon is now available from EU-ETS, the market price is a function of the cap. Damage cost estimates are larger than the market price, but have more uncertainty.

the uncertainties, it is also possible that there are no net benefits to climate-changing emissions or even net penalties (e.g., if CH_4 emissions of retrofitted vehicles are only double our estimates, there will be no climate benefits). Net penalties in climate-changing emissions may also be possible if CNG conversion had led to larger vehicle ownership or larger vehicle travel, because CNG is much cheaper than competing petroleum fuels. Further research on these uncertainties is required to understand the full impact.

10.7 Conclusions

Our exercise clearly indicates a large air quality benefit of 1.3 % of national GDP to the residents of Dhaka as a result of CNG conversion of vehicles. As a comparison, total air pollution costs in China were estimated to be 3 % of China's GDP (World Bank and State Environmental Protection Administration 2007).

The rather high accrual of benefit from this policy in Dhaka is a result of different factors:

(a) Traffic is a major source of air pollution in Dhaka, and any reduction in emissions results in an almost proportional improvement in air quality.
(b) Dhaka is a densely populated city, which means any improvement in the air quality directly benefits a large number of people.
(c) Bangladesh is a poor country with a small GDP, therefore benefits to the GDP ratio get inflated.

The global warming impact of the CNG conversion is not straightforward. Considering the established greenhouse gas emissions (CO_2, CH_4), the conversion policy aggravates the global warming problem. However, if we consider the impacts of aerosols and their precursors, the CNG conversion results in net cooling of the atmosphere. This is primarily because of the lower emissions of BC, which has the largest impact on warming per unit of emissions among the pollutants considered. Conversion of diesel vehicles is responsible for most of the benefits, which is not surprising inasmuch as higher PM emissions resulting from these vehicles also contain higher BC. Reducing CH_4 leakage and improving CH_4 combustion in the retrofitted vehicles may enhance the climate benefits even further.

The monetary benefits of avoided damages due to global warming attributable to the policy, however, is far smaller (smaller by two orders of magnitude) than the monetary benefits of reduced local air pollution in Dhaka. This is especially true because we did not consider the health impacts other than mortality. Therefore, the carbon credit generation and associated financial benefits from such CNG conversion projects or policies under the clean development mechanism may not be large.[5] This means that the conversion of petroleum vehicles to CNG can be justified simply on the basis of local air pollution benefits alone. For large metropolitan areas in the developing countries with poor air quality, this general conclusion is likely to hold.

Our results must be put in the context of the uncertainties. Future investigations should focus on these uncertainties to generate a more robust estimate of benefits of CNG conversion. Despite the uncertainties, our exercise reveals that the CNG conversion policy in Bangladesh had large positive benefits in air quality improvement and foreign currency savings, and possible benefits through GHG reductions as well. This indicates that careful policies can be formulated, focusing on an individual country's strengths, to meet the energy and environmental challenges for a sustainable future.

Acknowledgment This work was undertaken as a research project of the Centre for Advanced Studies and Research (CASR) of Bangladesh University of Engineering and Technology (BUET). Any opinions, findings, conclusions, or recommendations expressed in this material are those of the authors and do not necessarily reflect the views of CASR or BUET. Thanks to Ms. Tanzila Khan for research support.

[5] However, the "cash" nature of the CDM funding/carbon credits may help initiate such projects, and then generate those nonmonetary health benefits.

References

Bangladesh Bureau of Statistics (2009a) BBS statistical pocket book of Bangladesh 2008. Retrieved July, 2009 from http://www.bbs.gov.bd

Bangladesh Bureau of Statistics (2009b) BBS Bangladesh data book. Retrieved July, 2009 from http://www.bbs.gov.bd

Bangladesh Road Transport Authority (2010a) BRTA statistics: registered vehicles in Dhaka. Retrieved April, 2010 from http://www.brta.gov.bd/statistics.php

Bangladesh Road Transport Authority (2010b) BRTA statistics: registered vehicles in Bangladesh. Retrieved December, 2010 from http://www.brta.gov.bd/statistics.php

Bond TC, Streets DG, Yarber KF, Nelson SM, Woo J, Klimont Z (2004) A technology-based global inventory of black and organic carbon emissions from combustion. J Geophys Res 109:1–43

Cropper ML, Simon NB (1996) Valuing the health effects of air pollution. DEC notes, No. 7, April. World Bank, Washington, DC

European Commission (2003) Externalities costs: research results on socio-environmental damages due to electricity and transport. Brussels

ExternE (2005) In: Bickel P, Friedrich R (eds) Externalities of energy, methodology 2005 update. European Commission, Luxembourg

Gurjar BR, Butler TM, Lawrence MG, Lelieveld J (2008) Evaluation of emissions and air quality in megacities. Atmos Environ 42(7):1593–1606

International Association for Natural Gas Vehicles (2010) Natural gas vehicle statistics. Retrieved February, 2011 from http://www.iangv.org/tools-resources/statistics.html

IRIN (2009) Bangladesh: air pollution choking Dhaka. Retrieved July, 2009 from http://www.irinnews.org/Report.aspx?ReportId=83772

Khaliquzzaman M (2006) Emission inventory for motor vehicle fleet. In: Internal paper. The World Bank, Dhaka

Krupnick A (2006) Mortality risk valuation for air quality policy. Retrieved December, 2007 from www.irr-neram.ca/CQ5Vancouver/Krupnick.ppt

Kunzli N, Kaiser R, Medina S, Studnicka M, Chanel O, Filliger P, Herry M, Horak F Jr, Puybonnieux-Texier V, Quenel P, Schneider J, Seethaler R, Vergnaud J-C, Sommer H (2000) Public-health impact of outdoor and traffic-related air pollution: a European assessment. Lancet 356:795–801

Laden F, Schwartz J, Speizer FE, Dockery DW (2006) Reduction in fine particulate air pollution and mortality: extended follow-up of the Harvard Six Cities study. Am J Respir Critic Care Med 173:667–672

Leksell I, Rabl A (2001) Air pollution and mortality: quantification and valuation of years of life lost. Risk Anal 21(5):843–857

Nijboer M (2010) The contribution of natural gas vehicles to sustainable transport. International Energy Agency, Paris

Parry IWH, Walls M, Harrington W (2007) Automobile externalities and policies. RFF discussion paper 06-26. Resources for the future, Washington, DC

Petrobangla (2008) Annual report 2007, Bangladesh oil, gas and mineral corporation. Dhaka

Pope CA III, Burnett RT, Thun MJ, Calle EE, Krewski D, Ito K, Thurston GD (2002) Lung cancer, cardiopulmonary mortality, and long-term exposure to fine particulate air pollution. J Am Med Assoc 287:1132–1141

Price R, Thornton S, Nelson S (2007) The social cost of carbon and the shadow price of carbon: what they are, and how to use them in economic appraisal in the UK. DEFRA. Retrieved April, 2010, from http://www.decc.gov.uk/en/content/cms/what_we_do/lc_uk/valuation/shadow_price/shadow_price.aspx

Rahman SM (2010) Air quality assessment and health effects of air pollution in Dhaka city through impact-pathway model. M.Sc. (Engg) dissertation, Bangladesh University of Engineering and Technology, Dhaka

Reynolds CCO, Kandlikar M (2008) Climate impacts of air quality policy: switching to a natural gas-fueled public transportation system in New Delhi. Environ Sci Technol 42(16):5860–5865

Rouf MA, Hasib HH, Nasiruddin M (2008) Studies on smoke opacity of diesel vehicles in Dhaka city. Bangladesh J Sci Ind Res 43(4):559–570

Rupantorito Prakritik Gas Company Ltd, RPGCL (2009) CNG Activities of RPGCL. Retrieved December, 2009 from http://www.rpgcl.org.bd/op_act.php?ltype=op_cng

United Nations Framework Convention on Climate Change (2010) Global warming potentials. Retrieved April, 2010 from http://unfccc.int/ghg_data/items/3825.php. Accessed Apr 2010

Urbanemissions (2010) Emission factors. Retrieved April, 2010 from http://www.urbanemissions.info/simair/vapis.html

USEPA (2004) Air quality criteria for particulate matter. National Center for Environmental Assessment, Office of Research and Development, U.S. Environmental Protection Agency, Research Triangle Park

USEPA (2005) Regulatory impact analysis for the final clean air interstate rule. Office of Air and Radiation, U.S. Environmental Protection Agency, Research Triangle Park

USEPA (2007) Draft regulatory impact analysis: control of emissions of air pollution from locomotive engines and marine compression-ignition engines less than 30 litres per cylinder, assessment and standard division. Office of Transportation and Air Quality, U.S. Environmental Protection Agency, Research Triangle Park

USEPA (2010) Clean alternative fuels: compressed natural gas. Retrieved June, 2010 from http://www.afdc.energy.gov/afdc/pdfs/epa_cng.pdf. Accessed June 2010

Viscusi WK, Aldy JE (2003) The value of statistical life: a critical review of market estimates throughout the world. J Risk Uncertain 27:5–76

World Bank and State Environmental Protection Administration (2007) Cost of air pollution in China: economic estimates of physical damages. Retrieved December, 2007 from www.worldbank.org/eapenvironment

World Health Organization (2009) WHO disease and injury country estimates: burden of disease. Retrieved July, 2009 from http://www.who.int/healthinfo/global_burden_disease/estimates_country/en/index.html. Accessed July 2009

Yohe GW, Lasco RD, Ahmad QK, Arnell NW, Cohen SJ, Hope C, Janetos AC, Perez RT (2007) Perspectives on climate change and sustainability. In: Parry ML, Canziani OF, Palutikof JP, van der Linden PJ, Hanson CE (eds) Climate change 2007: impacts, adaptation and vulnerability. Contribution of working group II to the fourth assessment report of the Intergovernmental Panel on Climate Change. Cambridge University Press, Cambridge, pp 811–841

Chapter 11
After Copenhagen, Revisiting Both the Scientific and Political Framings of the Climate Change Regime

Amy Dahan and Stefan Aykut

Abstract The chapter discusses the political results of the Copenhagen Conference and the evolutions in the international climate arena including geopolitical shifts, new issues on the agenda, and a changing cartography of the main actors. As recent attacks on the climate regime concern both its political governance and the peculiar relationship between science and politics that developed through its main institutions (IPCC and the Conference of the Parties), the first part retraces the construction of the climate arena and the second part analyzes the framing of the problem among climate science, expertise, and politics. Drawing on this historical sketch, we suggest the years 2000 were characterized by a convergence of top-down approaches in climate expertise and policies, structuring action and discourse around quantified reduction targets, temperature and concentration thresholds, and carbon budgets. The bottom-up character of the voluntary reduction commitments in the Copenhagen Accord—confirmed at Cancun and Durban—is a serious setback to this approach. We conclude by discussing several contributions coming from social scientists to the post-Copenhagen debate. These well-known intellectual figures shift the focus from the links between science and politics toward the relationships between science and societies.

Keywords Climate regime • Climate governance • Conference of Copenhagen • IPCC • Post-Copenhagen debate • Climate framing

A. Dahan (✉)
Directrice de Recherche émérite au CNRS, Centre Alexandre Koyré (EHESS-CNRS), 27, Rue Damesme, Paris, 13è, France
e-mail: dahan@damesme.cnrs.fr

S. Aykut
Postdoctoral Student, LATTS Lab, IFRIS, Paris-Est University
e-mail: s.aykut@gmail.com

J.B. Saulnier and M.D. Varella (eds.), *Global Change, Energy Issues and Regulation Policies*, Integrated Science & Technology Program 2, DOI 10.1007/978-94-007-6661-7_11,

List of Acronyms

CFC	Chlorofluorocarbon
COP	Conference of the Parties (COP 15: the 15th Conference)
GMO	Genetically modified organism
IPCC	Intergovernmental Panel on Climate Change
NGO	Nongovernmental organization
SBSTA	Subsidiary Body for Scientific and Technological Advice, which meets at least twice a year to provide advice to the Conference of the Parties (COP) on matters of science, technology, and methodology
UN	United Nations
UNEP	United Nations Environmental Program
WBGU	German Advisory Council on Climate Change (Wissenschaftliche Beirat der Bundesregierung Globale Umweltveränderungen)
WTO	World Trade Organization
WMO	World Meteorological Organization
UNFCCC	United Nations Framework Convention on Climate Change
US	United States

11.1 Introduction

For 20 years now, the so-called "climate regime" went on, following a precise scientific framing, in a political process. The Intergovernmental Panel on Climate Change (IPCC) has been a key actor in this process. Now, a few months after the Copenhagen conference, the failure is confirmed, and the regression seems to deepen. The present crisis of the "climate regime" testifies to the end of a cycle initiated more than 20 years ago. The crisis affects at once the political process, the scientific framing, and the relationship between the two. It is therefore important to take seriously attacks against the IPCC, because the visibility of critical voices and the skeptical offensives are signs of this crisis (Dahan et al. 2010).

The first part of this chapter briefly revisits this framing of climate change, its tacit assumptions (the interactions between science and politics, the role of expertise, the coproduction process) and recaps the different phases between Rio (1992) and Copenhagen (2009): the room for maneuver left to developing countries, the turning point of 2002–2003, and the rise of the stakes of adaptation.

The second part focuses on the analysis of Copenhagen's failure. Countries of the basic group (Brazil, South Africa, India, and China) and the United States rejected the top-down approach, which has prevailed since the Kyoto protocol, and is associated with the quest for an ambitious treaty that would be ratified by all countries and fix reduction targets for each one. As the Copenhagen Agreement stated, a bottom-up approach prevailed, where climate policies will be separately determined by each country according to its possibilities and its will. So, the crisis is deep and not contingent; it testifies to the new geopolitical order and violently contests the framing of climate change.

The third part of the chapter is devoted to the period begun by the failure of 2010. The crisis launched a critical offensive, even a contrarian movement, focused on the framing of the climate change problem, even on its very existence and on the role of the IPCC. However, a whole set of reflexive analyses provided by social scientists (sociologists, geographers, sciences studies scholars, etc.) seems particularly interesting to review because it shifts the focus from the links between science and politics, toward the relationships between sciences and societies and allows us to conceive climate change as a multiscalar challenge (local, national, and cosmopolitan).

Our position (in our research team), as observers of the climate regime (Dahan et al. 2009) for a number of years now, is above all reflexive. Our objectives are to analyze political dynamics, the role played by science and scientific and technical expertise; to study how questions of governance, democracy, and equity are posed (nongovernmental organizations, ONG, civil society); and, finally, to study how solutions for the future and climate politics are elaborated and constructed between interests, both pragmatic and utopian. It is the race, between climate degradation, and the implementation of effective policy measures for the fight against global warming, on the other hand, that is, for us, the determining tension according to which the lessons and outcome of the Copenhagen Conference, as well as the future of international negotiations, may be judged.

11.2 A Look at the History of the Climate Regime: From Toronto to Copenhagen

The construction of the climate problem on the international political stage is a process begun in Toronto in 1988/1989, following the success of the Montreal Protocol (1987), established for the sake of fighting ozone depletion. The success of the Montreal Protocol gave rise to a model for constructing international negotiations on climate change. Supported by the IPCC's establishment in 1988, the Rio Earth Summit in 1992, and the implementation of the United Nations Framework Convention on Climate Change in 1994, then by annual climate arenas (the Conferences of the Parties, COPs) and the IPCC reports (1990, 1995, 2001, 2007), the climate regime brought together a growing number of actors and partners, gave rise to new research practices, and witnessed the confrontation of diverse political stakes and economic interests.

Several notions arising from distinct disciplinary and epistemic universes intersect within the terminology of this concept of "regime": notions coming from the domain of international relations, from "science studies," which privileges the study of the role of science and expertise, and finally, from the discursive regime (Foucault's work 2001, 2004).

For 20 years, the climate regime was principally formed around three elements, which we summarize briefly.

11.2.1 A Political Process and a Climate Expertise That Are Separate but Closely Linked

The IPCC was created in 1988 by two organisms linked to the United Nations: the World Meteorological Organization (WMO) and the United Nations Environmental Program (UNEP). From that date onward, scientists and politicians proceeded jointly. From the onset, the science of climate change formed the cornerstone upon which discussions and negotiations were constructed. The IPCC is constantly referred to in the climate arenas. Each of the IPCC's reports was behind important moments of political decision. The 1990 report prepared the way for Rio and the creation of the Convention; the 1995 report is directly linked to the elaboration or the Kyoto Protocol; the 2001 report fostered the Adaptation turn; only the 2007 report, associated with the Nobel Peace Prize, has not accomplished as much. Admittedly the risks revealed by the methodology of global numerical modeling and of extremely sophisticated expertise, did give rise to many suspicions on the part of developing countries throughout the 1990s (not to mention the oil-producing countries). The IPPC was accused of extreme dramatization of the climate threat, and of wanting to impose, in an authoritarian manner, a perspective that was judged as too physicist and globalizing. Nevertheless, the efforts at education and reflexivity deployed by the IPCC, the attention paid to the concerns of the South (land use, forests, extreme events), and the alliance formed with NGOs that accompanied the rise of the theme of adaptation, appeared to overcome these tensions. With the political failure of Copenhagen, it is these critiques that are coming back with force.

From the beginning, the IPCC has been involved in the political process; being an intergovernmental body, it is subject to political influence. Moreover, the summaries of its reports, written for decision makers, have to be agreed on and adopted word-for-word by scientific redactors and political government representatives (Encinas de Munagorri 2009; Dahan 2008). However, in its discourse, the IPCC always hastened to affirm a linear and purified vision of its relation to the political process. This stance is expressed in the IPCC's famous credo: "policy-relevant, but not policy-prescriptive." This discrepancy between a "science-speaks-truth-to-power" rhetoric and a far more complex and hybrid practice, makes the IPCC vulnerable and today has placed it on the defensive.

In the science studies literature, the notion of coproduction refers foremost to the idea of a joint evolution of the scientific and political order and of the mechanisms that accompany or authorize it. The notion of the boundary is thus an important element (Jasanoff 1987): indeed, to say what is "scientific" defines, from the outset, the domain of the "political," and vice versa. On a first level, the climate regime is coproduced in the sense that the scientific framing of the question through general circulation models, and its political treatment in the UN arenas, through treaties intended to be signed by almost the entire international community, mutually reinforce each other. The political definition of the atmosphere as a "global common good" and the scientific definition of the global as a pertinent scale of analysis, combined with "downscaling" as an adequate method for returning to smaller units of analysis, echo each other. Nevertheless, this joint evolution cannot be reduced to a parallel simultaneous evolution. More or less complex interactions intervene. Indeed, we have shown how the IPCC-SBSTA (Subsidiary Body for Scientific and Technological Advice) couple crystallized this coproduction, where the two a priori antagonistic tracks of political conflict and scientific consensus have mixed in a complex game of competition and cooperation (Dahan 2008). It is possible to take this farther. In this same literature, several works have shown, with concrete examples, how "hybrid" or boundary objects are coproduced in a back-and-forth between science and politics: this is particularly the case with the 2 °C threshold (Jasanoff et al. 1995).

In the climate arena, several numbers and units of measure circulate, and are reified by actors. The number that traditionally dominated negotiations is 2 °C, often coupled with 450 ppm CO_2-eq (see also Chap. 1). This number is tied into the framing of the climate question, which defines it as a problem of collective action, and thereby also defines the atmosphere as a "global common good" (Nordhaus 1994). The number of 2° thus corresponds to a threshold that is not to be surpassed, by analogy to pollution or use-thresholds in the more traditional cases of managing a common good.

11.2.2 A "Sharing the Burden" Strategy, with CO_2 Emission Reduction Numbers and Stabilization Objectives

The strategy of "sharing the burden" was inscribed in the Kyoto Protocol with a horizon of 2010 and with a view to developing countries. It aims to distribute reduction emissions at the global level and within a given timeframe. The strategy has remained operative both in the quest to prolong the Kyoto Protocol in the post-2012 period, and in the search for a new treaty that would include the United States and the large emerging economic powers. In other words, the explicit framework of the negotiation process evolved more and more clearly toward the quest for an international treaty that fixes objectives for all countries, as well as a stage-by-stage implementation calendar, with emissions ceilings evolving with time. This was intended to happen based on a general formula, which is considered to reflect at once historical responsibilities, present capacities, and equity conditions. Now, this framework appears to the United States today, and to its democratic administration, to be an illusory ambition (Aldy and Stavins 2008). It is also rejected, due to diverse reasons such as sovereignty and an absolute right to development, by the large emerging economies.

In the 1990s, the ambition of reduction numbers and objectives was the principal object of Kyoto Protocol negotiations between industrialized countries. These numbers pertain to politics. With the accentuated evolution toward a top-down and coproduced climate regime, these numbers and objectives tend to become elements of the framing itself, and are linked to scientific expertise.

11.2.3 A Clear Distinction Between Industrialized and Developing Countries

This distinction is made between countries, but also between the subjects that concern them: reduction and attenuation versus adaptation and financing. This distinction was recognized starting with the crucial phase of the process' formalization, the elaboration of the Kyoto Protocol (1994–1997), which excluded the so-called non-Annex 1[1] countries from any reduction commitment and thereby recognized a differentiated responsibility of industrial countries on the question of global warming. The distinction was never questioned prior to the Bali COP 13 Conference (2007), even though it was the reason for the US refusal to ratify the Protocol.

From 2002 to 2007, developing countries, united under group G77+China, managed, through an exceptional activism, to establish adaptation as an important theme in climate negotiations. The climate problem seemed to encompass all of the problems of development and even of sustainable development, a concept whose potential for renewal was greatly dimmed by the background noise of the communication of all sorts of public and private institutions. This meant that all other environmental questions (biodiversity, water and soil management, nutrition, and fishing), or the question of north–south equity, which comprise the ensemble of questions around sustainable development, were progressively reconfigured by the climate change regime, subjected to the rhythm of its progression and the geopolitical dynamic that was developing within it. At COP 13 in Bali, this evolution reached its apogee with the adoption of a mandate around four constitutive elements that had to be jointly negotiated in the Copenhagen COP 15 two years later. These elements are:

- Actions to reduce emissions
- Solutions for adapting to the impacts of climate change
- Technological transfer
- Financial mechanisms

11.3 The Failure of Copenhagen

For many NGOs (Radanne et al. 2010), for concerned parties of the climate regime, for a portion of the global civil society, the future of planetary problems (climate, environment, development, north–south equity), were veritably on the agenda for

[1] Non-Annex 1 countries are essentially developing countries.

Copenhagen. The global fabrication of the crucial character of this deadline, fed by the prospect of a meeting of 120 heads of state, is, moreover, astonishing in itself. This exaggerated hope—doubtless a fiction—misfired.

The ascent, over the course of the first decade of the twenty-first century, of large emergent economies, above all China, who in 2007 became the biggest CO_2 emitter in the world, made the separation between industrialized countries and the undifferentiated block of developing countries altogether unacceptable, in particular for the United States. Furthermore, the economic and financial crisis accentuated the economic competition between the emerging powers and western countries. Moreover, the bloc of emerging powers is far removed from any environmental discourse, put forth by NGOs in particular, on the planetary ecological crisis. In Copenhagen, the climate problem bluntly appeared, for the first time, not so much as an environmental problem but as a problem of the decarbonization of productive capitalist economies, bringing into play in that transformation enormous competing interests and energy stakes. An age of realpolitik seemed to impose itself. With it, the questions of adaptation, along with its not-so-clear links to questions of development, were relegated to secondary importance.

The 2° goal retained at Copenhagen did indeed arise from a veritable scientifico-political coproduction, with several central actors, including the IPCC, the European Union, different expert committees such as the German Advisory Council on Climate Change (WBGU) in Germany, and more recently the G8 of l'Aquila. Moreover, focusing on a single number raised certain problems, because the number constitutes a "black box," which hides power struggles and scientific controversies from its construction.[2] Therefore, it is not evident to which temporal horizon the number refers, and this leaves the door open to a multitude of reduction scenarios, including the "overshooting" scenario (in which emissions surpass the fixed threshold). Even more important, in its 1995 report, the IPCC affirmed that stabilization at 550 ppm would correspond to a probable warming of two degrees. Today, this threshold is considered to be at 450 ppm,[3] possibly less, which is an altogether different level of ambition!

11.3.1 Thresholds, Scenarios and Budgets: The Ascent of the Top-Down Approach to Climate Change and the Problem of Governance

The 2° threshold is inscribed in a more general framework of climate governance and its scientific handling. Without going into detail on these points, we want to underline the increasingly prominent evolution toward top-down approaches in the

[2] On numbers transformed into black boxes, see Latour (1999).

[3] A stabilization at 450 ppm CO_2-eq corresponds to "better estimation" of the warming of 2 °C, and a probability of 50 % for exceeding 2 °C (IPCC 2007).

framing of climate change. This evolution passes first by a highly globalized consideration of CO_2 concentrations in the atmosphere, more so than a consideration of emissions, and culminates in what is called the carbon "budget" approach. The approach adopted in the IPCC's fifth report confirms the choice to privilege concentrations. Working on CO_2 concentrations allows for a change in methodology and for a consideration of "stocks" instead of "flux." The advantage consists of being able to determine an accumulated quantity of gas that is admissible in the atmosphere, of calculating the quantity that was emitted since the beginning of industrialization, and of concluding on the "budget" that remains. Subsequently, one descends from the global budget to a budget attributable to each country.

So, a growing and worrying hiatus was thus manifested in Copenhagen between an alarmist scientific expertise, constructed around key numbers, thresholds, and carbon budgets, which presupposes an effective global governance and corresponds to a top-down approach to the climate problem on a planetary scale; and on the other hand, a bottom-up approach which prevailed in the Accord, imposed by the United States and China, who advocate for national policies and against restrictive objectives. Scientific expertise is thereby all the more weakened.

11.3.2 The Myth of a New Treaty (KYOTO 2) Shattered

The Copenhagen accord can be deceiving, but an analysis shows that the failure was not contingent. It corresponds precisely to the fundamental wishes of the powers that today dominate the global geopolitical scene, and it represents the limits beyond which they do not seem willing to go. We could have a premonition about this at the time of the crisis that marked the end of the Bali conference: the emergent countries' refusal of any restrictive objective for 2050 because it was up to developing countries to show the way and to take up ambitious reductions at home for 2020: the United States's refusal of any goal for 2020 or 2050 that would constrain them, without placing equal demands on the big emergent economies. And Europe, which was full of the best intentions, was incapable of influencing a change in this opposition. It is necessary to take into account these structural limits in order to define effective directions for action, and in order not to recreate illusory objectives for Cancun.

A paradox reveals itself at the very heart of the mediation of the Copenhagen conference. There is the staging of a planetary voluntarism, which idealizes heads of state as the world's saviors, capable if they so wish, of orchestrating necessary energetic, economic, even political changes. This vision is, however, poorly adapted to the fact that states are limited in their decisions, by their exterior (international competition, other international regimes such as the WTO, World Trade Organization), as well as in their interior (tensions inside the federal states of the United States and Germany, the importance of lobbies, the role of public opinion, the limited room to maneuver in democratic regimes).

Instead of waiting, in vain, for a new Kyoto 2-type treaty, is it not more effective to acknowledge geopolitical realities, the room to maneuver that states actually have, and the concrete advances made at other scales of governance, without overestimating the capacity of UN summits to regulate the ensemble of the problem? At Copenhagen, we observed regions, Länder, states (including American and Canadian states), megacities, that are making important efforts to reduce their emissions, and village alliances, which also develop numerous initiatives. Indeed, emissions reduction goes hand in hand with other urban concerns: developing public transit, fighting smog and pollution, reducing household electric bills, and so on (see Chaps. 4 and 13). There is a noticeable and significant presence of scales of governance smaller than that of the states that are already acting. We may wish for greater and better action, but this evolution is nevertheless palpable.

Our critique thus has to do with the staging of a global voluntarism, which regularly becomes something much bigger than what is warranted by the facts, and the demobilizing effect of this regime of messianic hope. It is important to begin once again to first concentrate efforts on domestic solutions, and then European policy. Those NGOs highly mobilized around the climate question, should they not rethink, if even partially, their role, in order to recover a critical attitude toward the questions within the national framework? If it is necessary to push the powers that be at all scales to go further, it is also necessary to prepare public opinion for transitions yet to come. The massive presence of civil society at summits is important, but it is not sufficient. The "global civil society" does not exercise direct pressure on a global power. In their final press conference, we saw important heads of state, Barack Obama and Nicolas Sarkozy, address above all the public opinion of their own nations.

What today blocks the adoption of more ambitious climate policies is not so much the leadership (the heads of state) as it is the absence of favorable social forces in various countries, or the existence of public opinion in the position to constrain this leadership. The multiple debates on ecological fiscality in numerous countries, and the timidity of its implementation, point to the road we have yet to travel.

11.4 The Critical, Even Skeptical, Offensive of 2010

Two dimensions of the climate regime seem to be under particularly strong attack today: the relation between science and politics and the UN governance framework for negotiations:

- The science–politics relation in the climate regime is determined by a singular expert organization—the IPCC, a major actor in the regime—and by a novel relationship between politics and expertise. Numerous lines of disagreement seem superimposed around questions of the validity of the consensus, the legitimacy of the experts, and the neutrality of their opinions. The climate problem thereby constitutes a challenge for the analysis of science–society relations.

- The system of climate governance, constructed since the 1990s around a strategy of distributing reduction objectives, has focused all efforts on the global level. In its center, the UN system, slow and heavy, is prone to various deadlocks: those imposed by states opposed to ambitious reduction policies and those tied to the least-developed countries' desires to bring all of their problems into the negotiations. The climate governance system seems less and less adequate to the new geopolitical order at the dawn of the twenty-first century.

We focus here on the offensive against the science–politics framing. It has developed on the terrain of Copenhagen's failure, even if certain elements already existed (e.g., the so-called Climate Gate scandal). We don't want to remain silent on conflicts of interest, and on the targeted, interested, and systematic deconstruction of climate science by diverse interests and lobbies in the public arena (Oreskes and Conway 2010). However, in this chapter, we underline several analyses coming from the social sciences (from sociologists, geographers, anthropologists) that strongly criticize the framing of the climate problem without directly contesting the anthropogenic greenhouse effect and being openly skeptical. We treat two groups, which have in common one figure, the geographer Mike Hulme (Hulme 2009).[4]

11.4.1 The Hartwell Paper

This contribution to the post-Copenhagen debate merits attention. Published by two institutes of the prestigious London School of Economics and Oxford University and financed by the Japan Iron and Steel Federation, the Hartwell Report (Prins et al. 2010) brings together authors well versed in the climate debate, who have regularly taken a critical position vis-à-vis the political process as well as the IPCC (Prins and Rayner 2007). The report takes as its point of departure a twofold assessment of the failure. Its first target is the "Kyoto approach," which came to a halt in Copenhagen. For the authors, this path was condemned from the start, because the framing in terms of the "pollution problem," with a fixed number of objectives for reduction and a constrictive treaty, is not the solution, but an obstacle to solving the problem. Second, for them and we don't share at all this analysis, Climate Gate, that scandal of pirated emails from the University of East Anglia last November 2009, seems to the authors to have significantly eroded trust in climate science in general, and in the IPCC in particular on a permanent basis.

In contrast to the pollution paradigm, climate change is qualified as a "wicked problem," designated as belonging to a category of problems, which, due to their complexity and extent, cannot be entirely described in scientific terms. According

[4]Founding director of the Tyndall Center for Climate Change Research, principal IPCC author for the third report, and now professor at the University of East Anglia. In 2009 he published a book (Hulme 2009) in which he acknowledges the gravity of the climate problem while taking a pessimistic stance toward our capacity to "resolve" it.

to the authors, in the research that governs the domain of climate change, the answers depend on the questions asked, and each new piece of knowledge raises new questions. Thus, according to their argument, because more knowledge does not lead to greater certitude, the IPCC's failure is systematic. Its very construction, they say, is bound to fail, because it is founded on the model of public education (the "deficit model"), and thus on the belief that science can guide politics. Inasmuch as the scientists do not succeed in having everyone agree, the political process does not advance.

The authors thus propose to change paths radically, by abandoning any hope of a global and restrictive treaty based on reduction objectives, and any form of carbon markets. The problem, they say, should be dealt with by "indirect" measures. This formula consists of making climate policy for other reasons, as a collateral product of the solution to pollution problems in developing countries, of energy independence problems, and of approaches based on green growth and energy efficiency, and finally—and this constitutes the last turn of this at points astonishing paper—of access to cheap energy for all.

The Hartwell paper questions science–politics relations in terms close to those developed by science and technology studies over the past 30 years (Jasanoff et al. 1995), beginning with the critique of the "deficit model" and the "linear model," of which IPCC partisans were prisoners, as well as the impossibility of closure in technoscientific controversies. However, the "wicked problem" category used by the authors is not convincing, because their statement that it is impossible to establish causal chains completely is not peculiar to climate science. The impossibility of leaving it to science and technology to close debates definitively is, according to Beck, a general characteristic of our time; it is even constant through time, according to Bruno Latour (Beck 1992). Worse, the report seems itself to be a prisoner of the linear model when it postulates that because there will be no scientific consensus, any direct political solution is compromised (Grundmann 2006).[5] Science and technology studies teach us that mechanisms for debate closure are always complex, never purely "scientific," and that it is always beneficial to look at them in detail. The example of the two-degrees target, neither purely scientific, nor purely political, contradicts one of the fundamental arguments of the Hartwell Paper, according to which, "a distinctive characteristic of the climate change debate has been of scientists claiming with the authority of their position that their results dictated particular policies" (Prins et al. 2010, p. 18). This unfounded statement is put forth without any concrete examples. The cursory condemnation of the IPCC does not leave any place for accurately evaluating its role in the climate arena, the manner in which it pushed the process forward, and contributed, through its forms of governance and action, to creating trust between states. Today, the IPCC reports serve as a foundation to a highly heterogeneous community of actors, and no state denies the

[5] This assertion is all the more surprising inasmuch as Reiner Grundmann had himself shown how the ozone problem was effectively fought without pre-existing scientific certainty (Grundmann 2006).

reality of climate change in negotiations any longer. What is more, the Hartwell Paper remains silent on conflicts of interest, and on the targeted, interested and systematic deconstruction of climate science by diverse interests and lobbies in the public arena. Finally, although it turns away from any global solution, the local policies recommended by the Hartwell paper seem particularly timorous and bereft of ambition.

11.4.2 The Role of Social Sciences in the Climate Problem

Let us turn to the special issue of the review *Theory, Culture, Society* (2010) devoted to the climate change problem. It brings together very diverse authors coming from different disciplines, and who are not known as skeptics. However, all challenge the scientific and political framing of the problem over the past 20 years, which led to very few results in terms of emission reduction. Following their assessment of climate risk (because the spectrum of positions is rather broad in this respect), they propose different measures and orientations. But what unites them is the novel and ambitious place they want to give to the social sciences in the climate problem, and which they think poorly acknowledged by the IPCC. We look at three of these authors: Mike Hulme, Sheila Jasanoff, and Ulrich Beck, whose reputation and work extends far beyond this issue of the magazine.

Climate change, they claim, contributed to dissolving three dichotomies of modernity: nature/culture, local/global, and present/future. But few lessons were learned from this, and there is a consistent attempt to re-create the boundaries between nature and culture, or to think in a deterministic way, relationships between the local and the global. For example, climate change is a hybrid phenomenon between nature and culture par excellence; however, extreme events (Katrina in 2005, Pakistan floods in 2010, or others) command attention only insofar as they are of anthropogenic origin, or in the proportion that they are.

What is common to all of these perspectives?

- The climate question is not a problem of pollution; CO_2 is not analogous to CFCs (chlorofluorocarbons) in the ozone hole. It is a problem that pervades all human activity (a "wicked problem," as the signatories of the Hartwell Problem called it).
- The Kyoto strategy was doomed to failure, and therefore had feeble results.
- They generally argue against catastrophist visions, emergency discourses, statements according to which only 100 months remain to save the planet, from which stems a strong critique of the manner in which the Copenhagen meeting was constructed as a crucial and pivotal moment, by scientists, NGOs, and the media.

Those most critical of IPCC science and scientists say that the framing is too scientific, that it releases people from responsibility, and that the IPCC remains attached to a traditional hierarchy of sciences with the natural sciences at its summit where only economics, sufficiently formalized and dignified, can participate in the expert enterprise and in defining climate politics. Thus it is clear that these critiques are strongly marked by tensions between diverse epistemic communities.

Such critiques are particularly characteristic of geographers, notably Mike Hulme, author of an interesting and stimulating book, *Why We Disagree about Climate Change* (Hulme 2009), who is also a signatory to the Hartwell Paper. According to Hulme (2010), the IPCC manifests a climate determinism and reductionism, by considering that the climate is the essential factor for societies and that it determines their possibilities for development or survival. It is from this critique that stems an opposition to the vision of Group II of IPCC (which is in charge of vulnerability of socioeconomic and natural systems to climate change, its consequences, and the adaptation options) about adaptation strategies. Although acknowledging the validity of the IPCC's climate predictions for 2050, Hulme thinks that the social and political picture of a future in which these predictions would mix is very risky. This is due to the fact, he argues, that there is an irreducible level of uncertainty and the future is always open.

In general, geographers (and Hulme is a perfect illustration) severely judge articulations from the local to the global, as they operate in reports and in the IPCC's conceptions. The prevailing globalization of climate, which has reached its peak (the concept of average temperature, average ocean levels, etc.), is, in their opinion, de-culturized, and tends to efface our anthropological experience of climate and weather, which remains subjective and local.

Jasanoff and Beck, belonging to other epistemic communities (science studies, sociology), notably engage in an examination of what climate change provokes or, rather, ought to provoke, as reflections in the social sciences, as research needs in those disciplines. Jasanoff (2010) insists on the need to construct bridges between large-scale, abstract scientific representations of environmental knowledge and the smaller scales of social signification. Using her concept of "civic epistemology," Jasanoff emphasizes that sustainable representations of the environment don't arise solely out of scientific activity, but are undergirded by cultural and normative comprehensions of the world. There is no neutral interpretative field, and the same numbers and facts acquire different meanings in different nations (think of GMOs, genetically modified organisms, nuclear energy, etc.). In her opinion, it is impossible that a global consensus could emerge solely out of expert consensus. Thus, although the IPCC was constructed to produce knowledge for the sake of a global politics, it cannot respond to particular national traditions of political legitimacy. That would be asking too much of it.

In his article, *Climate for Change, or How to Create a Green Modernity*, Beck (2010) meets Jasanoff on several points: for example, his first argument is that discourse on climate politics is an expert elitist discourse, in which the people, societies, workers, and citizens do not see themselves, and in which their interests are neglected. For him, it is a question of new growth or of green modernity that is on the agenda, or rather, an economic transition toward other modes of production and consumption (a new phase of capitalism or a new stage of civilization). Hence, the central question for a sociologist like Beck is to know from where could arise support for profound ecological changes that presuppose radical changes in behavior, consumption, life-style, and growth, in these particularly uncertain times?

According to Beck, it is necessary to rethink radically the category of the environment, which is too narrow when it excludes human action and the social, and too broad, and even, as he says, suicidal, when it encompasses them, because then it no longer thinks politics at the level of metachange and loses sight of modern society itself. By trivializing the terms of "climate politics," we devitalize them, he claims, because we ignore that climate politics are precisely not about climate, but that they rather have to do with the transformation of all of the fundamental categories and institutions of nation–state modernity.

For Beck, social inequalities and climate change are two faces of the same problem: what he calls risk societies are, by virtue of the historical dynamics of their national systems and of international legislative systems, prisoners of a repertoire of behaviors that are completely surpassed by the gravity of the global ecological crisis. Like Jasanoff, he has a tendency to reverse accepted statements: for him, a global public discourse will not emerge from a consensus on political decisions (based itself on scientific consensus) but, rather, from dissentions (and controversies) about the consequences of decisions. Thus, the uncertainties (or insecurities) created could open the path to a transnational reflexivity, a global cooperation, and coordinated responses. Indeed, a cosmopolitan vision of ecology to him seems essential, in particular for developing countries. Beck insists on the danger of a politics (and a thought) of limits, which would be anti-immigration, antiglobalization, antimodern, antigrowth, and so on, because it would reinforce the conditions of an international caste system in which the poor of developing countries would be consigned to poverty (notably energy poverty) for perpetuity.

Thus, in this panel of authors, in spite of several divergences, climate change seems to offer new opportunities for reflecting on and conferring a determinant role to the interpretative social sciences.

11.5 Conclusion

In the post-Copenhagen debate, the panoply of propositions reflects not only the different framings of the problem, but above all the diverse assessments of climate risk. The hope that everyone will agree around a scientific basis seems an illusion. By way of a conclusion, we emphasize four points that might serve to indicate orientation for action:

1. Climate change is not a global problem. It is a multiscalar problem that can and must be treated on all pertinent scales. The staging of a global voluntarism tends to mask the setbacks and advances made on all other levels of governance; it delays actions that are compatible with local agendas and beneficial secondary effects. We thus ask for the principle of subsidiarity in the climate arena, to support and broaden local and national initiatives, and to lighten the international negotiations agenda.
2. Abandoning a planetary voluntarism in the climate arenas does not entail effacing those arenas. Climate arenas remain exceptional meeting points, where political states representatives, multiple actors from think tanks, academia, business, and civil society, can launch and disseminate ideas and technologies, compare solutions, and debate questions, that do not (or no longer) have other forums (since Johannesburg) that capture their scope.

These questions are: (green) development, equity, responsibility, and global solidarity. The COPs are meetings in which inventories can be taken of the road covered and that which remains to be covered. In the record left by these 15 years of COPs the Off was ultimately more fruitful than the On. Abolishing these arenas would therefore be a regression and a waste. The exceptional mobilization of civil society in Copenhagen shows the importance of these sites, and the UN erred in fearing civil society and excluding it from the last days of negotiation; the presence of civil society is an important argument for keeping the UN framework.

3. The absence, at Copenhagen, of any mention of the Kyoto Protocol is not accidental. It is a manifestation of the new dominant geopolitical order. Europe neither defended the Kyoto Protocol, which had given it an incontestable political capital in these arenas, notably vis-à-vis developing countries, nor proposed an alternative solution. Its lack of unity and the weakness of its modes of governance condemned it to a position of marginality and poor visibility. The fact that NGOs and developing countries formed an alliance to claim themselves loudly as representatives and followers of the Kyoto Protocol is paradoxical when one recalls the criticisms they once made regarding the Protocol and its ineffective results. In order to regain its leadership, Europe must implement, within its borders, policy measures including technological solutions and regulations that work and lead to effective reductions.
4. We are experiencing a convergence of several crises: the environmental crisis, the energy crisis, the climate crisis, and the financial and economic, and doubtless, social crisis. The authors of the Hartwell Paper (whose criticism of the IPCC we do not share), or the Nobelist economist Elinor Ostrom are not wrong in suggesting that certain climate policies can be adopted "for other reasons." Even if this assessment calls for greater coordination between the political fields and separate international arenas such as the OMC and the Climate Convention, or even between public health policy and CO_2 emission reduction politics. This point of view tends to become more general in this moment. US President Obama just gave a press conference on energy issues in which he recommends these types of positions and actions.

Placing the climate question in the mainstream of the unsolved problems might serve to trivialize the problem, to rid it of its absolute urgency. But it is necessary to note that we are far from being able to convince public opinion or governments that this is a matter of absolute peril (see studies of American public opinion or the concern of emerging economies). Hence the debate on measures and policy must, as a result, acquire greater autonomy in relation to scientific assessment, and it must be led in a public and open manner.

The central question is to know from where might come support for ecological changes as profound as those demanded by the climate question, which presupposes such radical changes in behavior, consumption, life-style, and growth. Indeed, climate policies are not precisely about climate, but concern the transformation of all the fundamental categories and institutions of nation–state modernity.

References

Aldy J, Stavins RN (2008) Designing the post-Kyoto Climate Regime: lessons from the Harvard project on international climate agreements. Report. Belfer Center for Science and International Affairs. Harvard Kennedy School, Cambridge

Aykut S, Dahan A (2011) Le régime climatique avant et après Copenhague: sciences, politiques et l'objectif des deux degrés. Nat Sci Soc 19:144–157

Beck U (1992) Risk society: towards a new modernity. Sage Publications, London, (original German edition 1986)

Beck U (2010) Climate for change, or how to create a green modernity? Theory Cult Soc 27(2–3):254–266

Dahan A (2008) Climate expertise: between scientific credibility and geopolitical imperatives. Interdiscip Sci Rev 33(1):71–81

Dahan A (2009) Entre Poznan et Copenhague: le régime climatique au milieu du gué. Nat Sci Soc 17(3):271–282

Dahan (dir.) A (2007) Les Modèles du Futur. Ed La Découverte, Paris

Dahan A, Aykut S, Guillemot H, Korczak A (2009) Les arènes climatiques: forums du futurs ou foires aux palabres? La Conférence de Poznan. Rapport de Recherche, Koyré Climate Series, n°1, Feb 2009

Dahan A, Aykut S, Buffet C, Viard-Cretat A (2010) Les leçons politiques de Copenhague: Faut-il repenser le régime climatique? Rapport de Recherche, Koyré climate series, n°2, Feb 2010

Dahan A, Buffet C, Viard-Crétat A (2011) Le compromis de Cancun: vertu du pragmatisme ou masque de l'immobilisme? Rapport de Recherche, Koyré climate series, n°3 (en ligne. www.koyre.cnrs.fr/IMG/pdf/climate.pdf)

Encinas de Munagorri R (2009) Expertise et gouvernance du changement climatique Droit et Société, vol 51, Maison des Sciences de l'Homme(L.G.D.J.), Paris

Foucault M (2001) Dits et Ecrits II. Gallimard, coll "Le Quarto", Paris

Foucault M (2004) Naissance de la Biopolitique. Gallimard, Le Seuil, coll "Hautes Etudes", Paris

Grundmann R (2006) Ozone and climate. Scientific consensus and leadership. Sci Technol Hum Values 31(1/Jan):73–101

Grundmann R (2007) Climate change and knowledge politics. Environ Polit 16(3/June):414–32

Hulme M (2009) Why we disagree about climate change. Cambridge University Press, Cambridge

Hulme M (2010) Cosmopolitan climates: hybridity, foresight and meaning. Theory Cult Soc 27(3–4):267–276

IPCC (2007) Climate Change 2007. Contribution of Working Group I. to the Fourth Assessment Report of the Intergovernmental Panel on Climate Change. Cambridge University Press, Cambridge

Jasanoff S (1987) Contested boundaries in policy-relevant science. Soc Stud Sci 17(2/May): 195–230

Jasanoff S (2010) A new climate for society. Theory Cult Soc 27(3–4):233–253

Jasanoff S, Markle G, Petersen J, Pinch T (eds) (1995) Handbook of science and technology studies. Sage, London

Latour B (1993) We have never been modern. Harvard University Press, Cambridge

Latour B (1999) Pandora's hope: essays on the reality of science studies. Harvard University Press, Cambridge

Nordhaus WD (1994) Managing the global commons. The economics of climate change. MIT Press, Cambridge

Oreskes N, Conway E (2010) Merchants of doubts. Bloomsbury Press, New York

Ostrom E (2009) A polycentric approach for coping with climate change. World Bank policy research working paper 5095. The World Bank, Washington, DC

Prins G, Rayner S (2007) Time to ditch Kyoto. Nature 449:973–975

Prins G, Galiana I, Green et al. (2010) The Hartwell paper: a new direction for climate policy after the crash of 2009. Joint research paper of the Institute for Science, Innovation and Society and the MacKinder programme for the study of long-wave events. Institute for Science, Innovation and Society, Oxford

Radanne P, Diaz E, Goetz E (2010) Les enseignements de la Conférence de Copenhague sur le climat. Publication Association 4D, Paris

Review Theory, Culture and Society (2010) Special issue on climate change, 27:3–4

Sarewitz D, Pielke R Jr (2000) Breaking the global-warming gridlock. *The Atlantic Monthly*: 55–64. In: Pielke R Jr (2007). The honest broker. Making sense of science in policy and politics. Cambridge University Press, Cambridge

Shackley S, Wynne B (1996) Representing uncertainty in global climate change science and policy: boundary-ordering devices and authority. Sci Technol Hum Values 21(3/Summer):275–302

Chapter 12
The Drafting of the Future International Climate Regime: From the Copenhagen Accord to the Cancún Agreements

Sandrine Maljean-Dubois and Vanessa Richard

Abstract Discussed at the international level since the 1980s, climate change is from now on at the top of the international political and diplomatic agenda. The urgency to act has been shown over the last years in many aspects. Climate change policies fit into the scheme of what political scientists call "multilevel governance," which emphasizes the role of international negotiations but also the multiplicity of public and private stakeholders—NGOs, businesses, unions—either with a global, regional, domestic, or local basis, and the diversity of on-going processes at different levels from global to local and from local to global. Because stakes are global, the international climate change regime nevertheless plays a pivotal and decisive role. The international climate change regime is, however, shaped slowly and step by step. The Kyoto Protocol (1997) has 193 parties but the United States, the primary GHG emitter in 1997 and second nowadays, did not ratify it. Major developing emitters are not bound by any GHG emissions reduction commitment under the Protocol. The first commitment period will expire at the end of 2012. For effectiveness reasons, the "post-2012" system must include the United States and major developing emitters, and drastically reinforce the reduction targets. One cannot help but notice that the post-2012 regime is still in the process of, and far from, being drafted. The adoption of the Copenhagen Accord (2009) did not stop the negotiation process, which is still going on. The Cancún conference (2010), although much less the focus of media attention, led to the adoption of a "Copenhagen Accord-Plus," revived a process that had almost come to a standstill, and made the content of the Copenhagen Accord integrate the heart of the UNFCCC. These evolutions give rise to many issues concerning the level of ambition, the nature and content of differentiation between parties, and the legal architecture of the whole regime.

S. Maljean-Dubois (✉) • V. Richard
Aix-Marseille University/CNRS, CERIC (UMR 7318), Espace René Cassin,
3 avenue Robert Schuman, 13628 Aix-en-Provence, Cedex 1, France
e-mail: maljean.dubois@gmail.com; vanrichard@yahoo.com

J.B. Saulnier and M.D. Varella (eds.), *Global Change, Energy Issues and Regulation Policies*, Integrated Science & Technology Program 2, DOI 10.1007/978-94-007-6661-7_12,

Keywords Climate change • International law • Kyoto Protocol • NAMAs • Implementation • Differentiation • International regime • UNFCCC • Common but differentiated responsibilities • IPCC

List of Acronyms

AWG-LCA	Ad Hoc Working Group on Long-term Cooperative Action under the Convention
BICS	Brazil, India, China, South Africa
CBDR	Common but differentiated responsibilities
CDM	Clean development mechanism
COP	Conference of the Parties
COP/MOP	Conference of the Parties to the Convention serving as the Meeting of the Parties to the Protocol
EU	European Union
GHG	Greenhouse gases
GNP	Gross National Product
ICJ	International Court of Justice
ILC	International Law Commission
IPCC	Intergovernmental Panel on Climate Change
KP	Kyoto Protocol
MOP	Meeting of the Parties
MRV	Measurement, reporting, and verification
NAMAs	Nationally appropriate mitigation actions
NGO	Nongovernmental organization
OECD	Organisation for Economic Co-operation and Development
REDD	Reducing emissions from deforestation and forest degradation
SBI	Subsidiary Body on Implementation
UNDP	United Nations Development Programme
UNEP	United Nations Environment Programme
UNFCCC	United Nations Framework Convention on Climate Change
USD	United States dollar
WMO	World Meteorological Organization

12.1 Introduction

From the mid-1980s climate change has given rise to heated scientific debates on its anthropogenic origins, scale, or consequences. The acuteness of these controversies led to the creation of an expert's mechanism without precedent as regards its dimensions and functioning terms. In the uncertain context of a "*controversial universe*", (Godard 1993) the Intergovernmental Panel on Climate Change (IPCC) was jointly created in 1988 by the World Meteorological Organization (WMO) and the United Nations Environment Programme (UNEP) with a clear decision-support mission. Nearly 20 years later, the IPCC was jointly awarded the 2007 Peace Nobel Prize with Al Gore, whose book, *An Inconvenient Truth*, inspired a documentary that was awarded the 2007 Academy Award for Best Documentary Feature, "For their efforts to build up and disseminate greater knowledge about man-made climate change, and to lay the foundations for the measures that are needed to counteract such change."

Beyond its political significance, this Nobel Peace Prize acknowledges the IPCC's contribution. Between 1990 and 2007, the IPCC made the diagnosis of climate change at the global scale through four assessment reports, and little by little reduced the initial uncertainty margin. In 1990, the first IPCC report was quite measured. The second IPCC report states that "The balance of evidence suggests a discernible human influence on global climate" (IPCC 1995), and the third IPCC report establishes that "most of the warming observed over the last 50 years is attributable to human activities" (IPCC 2001). The fourth and last IPCC report, based on the work of more than 2,500 scientists from 130 countries, is most alarmist. It bears out that the "warming of the climate system is unequivocal" and that "most of the observed increase in global average temperatures since the mid-twentieth century is very likely due to the observed increase in anthropogenic GHG [greenhouse gases] concentrations" (IPCC 2007a). The impact of human activities goes beyond the increase in global average temperatures; it also results in rising global average sea level, perturbations in wind patterns, extra tropical cyclone path change, increased drought and heat wave risks, and increased frequency of heavy rainfall, among others. The IPCC report adds that it is "very likely" that these changes will be more important in the twenty-first century than those observed during the twentieth century. Moreover, "Anthropogenic warming and sea level rise would continue for centuries due to the time scales associated with climate processes and feedbacks, even if GHG concentrations were to be stabilized" (IPCC 2007a).

Placed on the agenda since the 1980s, climate change is from now on at the top of the international political and diplomatic agenda. The urgency to act has been shown over the last years in many aspects. According to the findings of the last IPCC assessment report, any delay in the reduction of GHG emissions will significantly reduce the chances to achieve the stabilization of GHG emissions at lower levels and increase the risk of more serious climate change impacts (IPCC 2001). From an economic angle, the *Stern Review on the Economics of Climate Change* (Stern 2006) puts the cost[1] of inaction as higher than the cost of action in the mid-term, an assessment that is relatively little disputed (its calculation is questioned but the cost of the lack of action could even be underestimated). More broadly, there is a social urgency as shown by recent reports from the United Nations Development Programme (UNDP 2008) and the World Bank (World Bank 2010).

---

[1]Costs are estimated as GDP parts.

Climate change policies fit into the scheme of what political scientists call "multilevel governance," which emphasizes the role of international negotiations but also the multiplicity of public and private stakeholders—NGOs, businesses, unions—either with a global, regional, domestic or local basis, and the diversity of on-going processes at different levels from global to local and from local to global (Compagnon 2010). Because stakes are global the international climate change regime nevertheless plays a pivotal and decisive role. First, a structuring role: it is expected to ensure the horizontal consistency of different policies implemented at the international level (trade, development, investment, finance, etc.) and vertical consistency by allowing the interlocking of different policy levels, again from local to global and global to local. Second, the international climate change regime must also have the revitalizing role of a driving force that makes different positions progress in negotiation processes and allows the setup of an international consensus and the promotion of more ambitious climate policies.

The international climate change regime is, however, shaped slowly and step by step (Maljean-Dubois and Wemaëre 2010). In a context of scientific uncertainty and acute controversies on the impact of human activities on climate change, a first step was the adoption of the United Nations Framework Convention on Climate Change (UNFCCC) in May 1992, and its opening up to signature during the 1992 Rio Conference. The UNFCCC entered into force in 1994 and has 194 contracting parties. It lays down the core principles of the international climate change regime but only contains general obligations. As a framework convention, it was to be complemented by (an) additional instrument(s) aimed at specifying the GHG emissions reduction commitments. One year after its entry in force, the parties adopted the Berlin Mandate which took note of the need to define more ambitious and binding commitments for industrialized countries, in the form of a protocol or any other legal instrument (for instance, an amendment to the UNFCCC). After less than 3 years of negotiation, the Kyoto Protocol was adopted on the 11th of December 1997. It only entered into force on the 16th of February 2005. Today, the Kyoto Protocol has 193 parties but the United States, the primary GHG emitter in 1997 and second nowadays, did not ratify it. Major developing emitters are not bound by any GHG emissions reduction commitment under the Protocol. The contracting parties agreed on quantified reduction commitments that only bind industrialized countries and cumulatively correspond to a reduction effort of 5.2 %[2] of their GHG emissions compared with the baseline of their 1990 GHG emission levels, and must be achieved by the end of the first commitment period, expiring at the end of 2012.

For effectiveness reasons, the "post-2012" system must include the United States and major developing emitters, and drastically reinforce the reduction targets. One cannot help but notice that the post-2012 regime is still in the process of, and far from, being drafted. Progress is all the slower than the benefits of some significant emissions reduction are both variable in space (some regions of the Earth could

[2]In practice the collective target is lower because the United States does not participate in the Kyoto Protocol.

even benefit from the changes and in any case different regions will be differently affected), and diffuse in time (the benefits are mostly expected for future generations when difficult measures must be adopted today). Climate change nonetheless already gave birth to original legal and institutional constructions, either at the international level or at the regional level. The hopes pinned on the Copenhagen conference in December 2009, its media coverage, the presence of 125 heads of state and government, of more than 40,000 delegates or accredited participants, show the importance given nowadays to the stakes of the fight against climate change. The adoption of the Copenhagen Accord did not stop the negotiation process, which is still going on. The Cancún conference (December 2010), although much less the focus of media attention, led to the adoption of a "Copenhagen Accord-Plus" and revived a process that had almost come to a standstill, and made the content of the Accord integrate the heart of the UNFCCC. But is the regime taking shape ambitious? Which differentiation does it make between developed and developing states? Finally, what could its legal architecture look like?

12.2 Ambition Level

The post-2012 climate regime must be ambitious enough to meet the scientists' recommendations. That involves commitment to reduce GHG emissions drastically in the mid-term (2020) in order to maintain emission trajectories on the right trend in the long term (2050) and far beyond the commitments agreed upon for the first commitment period of the Kyoto Protocol.

To keep a fifty–fifty chance to limit the temperature increase at the level of 2 °C, GHG emissions concentration must be stabilized around 450 ppm of CO2-eq,[3] which implies both that the peak in the emissions must occur before 2020–2025,[4] and massive emissions reduction. According to the IPCC Working Group III (Mitigation of Climate Change), so as to reach stabilisation at 450 ppm developed countries should reduce their emissions by 25–45 % below 1990 levels by 2020, and developing countries "need to deviate below their projected baseline emissions" (IPCC 2007b) compared with business-as-usual scenarios.

From this point of view, the Accord reached in Copenhagen refers from its first paragraph to the objective of global warming stabilisation at 2 °C in order to achieve the UNFCCC ultimate objectives, and explicitly bases this target on scientific knowledge.[5] The +2 °C objective is also cited as a point of reference with regard to

[3] Note that in order to stabilize global warming at 2 °C with a 70 % probability, emissions should be stabilized at 400 ppm, when we have already reached 387 ppm.

[4] The IPCC insists on the urgency to act insofar as one considers cumulative emissions and not the levels of concentration at a given time.

[5] "(...) we shall, recognizing the scientific view that the increase in global temperature should be below 2 degrees Celsius, on the basis of equity and in the context of sustainable development, enhance our long-term cooperative action to combat climate change" Copenhagen Accord, para. 1.

the achievement of drastic global GHG emissions reduction (Para. 2). And yet the Accord does not set any quantified target by 2050, either to achieve a reduction by 50 % of global emissions or to achieve a reduction by 80 % of developed countries' emissions. There is no base-year, no specific date for the peaking of global emissions. The second paragraph only calls for the peaking to occur "as soon as possible" but recognizes that it will take longer in developing countries.

After all, nothing guarantees that the temperature increase will be held below or at the 2 °C threshold. This 2 °C ambition (long-term objective) is indeed inconsistent with existing agreed (short-term) commitments, the voluntary and bottom-up approach adopted, and the weakness of control mechanisms that makes commitments and actions impossible to compare. On the 15th of January 2011, 140 countries, among them about 40 industrialized countries, had effectively claimed they supported the Accord. Some 87 countries, representing 82 % of global GHG emissions, had committed to reduce their emissions or submitted mitigation actions as stipulated in the Copenhagen Accord.

The +2 °C Global Warming Threshold

This value has become a reference threshold used in a number of reports. Many scientists and many governments do agree that the world needs to keep the average global temperature rise at a +2 °C level (compared with preindustrial times, which means +1.4 °C from now on). Moreover, a recent update on previous IPCC studies explains the serious "reasons for concern" (Smith et al. 2008) even if this threshold of +2 °C temperature increase is not crossed.

One must take into account that because of the different uncertainties involved in the models, it is necessary to use probability tools (Schneider et al. 2007) both as regards the 2 °C value and the values of stabilized concentrations of CO_2 attached to this +2 °C threshold.

As a quick reference point, one can say, for instance, that to keep a 50 % chance to limit the temperature increase at this level of +2 °C, CO_2 emissions concentration must be stabilized around 450 ppm. According to the level of uncertainty, models' results lead to say that the probability is in fact between 25 and 80 %.

See also Chap. 1.

Developed countries' pledges amount to a 12–18 % emissions reduction by the year 2020 compared to 1990 emissions levels, depending on the terms (use of offsets, accounting of emissions and sinks resulting from land-use, land-use change, and forestry) and the (high or low)[6] threshold each country decided to set

[6]The low threshold refers to situations where a minimum commitment level was proposed, whereas a high threshold reflects stronger commitments but on certain conditions, for instance, on condition that other parties make comparable efforts.

(UNEP 2010; Casella et al. 2010). The addition of industrialized countries and some developing countries' pledges show we are rather moving forward a 3– to 3.5 °C increase by 2100.[7] On the basis of such emissions reduction commitments, there is a 70 % probability that we will go past 2 °C. There is thus something cynical about mentioning in the Accord the 1.5 °C called for by the Maldives: indeed, the Accord makes provision for "an assessment of the implementation of this Accord to be completed by 2015, including in light of the Convention's ultimate objective. This would include consideration of strengthening the long-term goal referencing various matters presented by the science, including in relation to temperature rises of 1.5 degrees Celsius" (Para. 12). The Cancún Agreements do not go further, but the objective of 2 °C and the mention of the 1.5 °C threshold are repeated in the "Shared Vision" and are from now on written in a COP decision (Decision 1/CP.16). The Cancún Agreements also make the Copenhagen Accord pledges, from both developing and developed countries, to enter the heart of the UNFCCC (see *infra*). They do not, however, lead to any rise in the ambition level.

12.3 The Differentiation of Obligations

Considering these emissions trends, for effectiveness reasons the post-2012 international climate regime must be global and bring together all of the major GHG emitters including the United States and major developing emitters, while respecting the equity principle set down by the UNFCCC through the common but differentiated responsibilities (CBDR) principle.

The Bali Action Plan which was adopted in 2007 (Decision 1/CP.13) and has been structuring the negotiations since then, specifies that the "differences in (...) national circumstances" have to be taken into account when defining mitigation commitments (Paras. 1-b)-i and 1-b)-v). Already enshrined in the 1992 Framework Convention, this differentiation influences and structures the Kyoto Protocol. It is precisely why the United States, which considers it to be discriminatory and inequitable because all parties are not granted the same treatment, did not ratify the Protocol (Shelton 2007). Differentiation, between whom, and to what extent, is a major issue in the negotiations. The CBDR principle the UNFCCC and the Kyoto Protocol both lay down is not very helpful in the end inasmuch as each Party understands it differently.

12.3.1 Differentiation Between Whom?

During the negotiations that preceded the Copenhagen conference, developing countries rejected any differentiation between them (except perhaps with regard

[7] See the Climate Action Tracker website.

to more vulnerable categories such as the "small island developing states" and the "least developed countries") and remained strongly attached to the binary divide between two groups, Annex I and non-Annex I countries. One must remember that Annex I is the first annex to the United Nations Framework Convention on Climate Change and lists the states who are subjected to "specific commitments" (industrialized countries and countries "that are undergoing the process of transition to a market economy"). Such a division was made on the basis of a voluntary choice to belong to one category or the other and by and large reflects the OECD/non-OECD watershed. As Marie-Pierre Lanfranchi notes, the border "is besides not always drawn clearly as shown by the case of Turkey,[8] or that of central and eastern European countries: the latter are listed as 'Annex I Parties' while Asian or Latin American economies in transition are not. In addition, the cases of Kazakhstan and Belarus remind us that by definition the border can be easily crossed."[9] (Lanfranchi 2010). From this point of view, the Bali conference has broken a taboo down. The Action Plan is the first text to use "developed" and "developing" countries instead of "Annex I" and "non-Annex I" countries, thus breaking with the logic of the UNFCCC and the Kyoto Protocol and creating an opportunity for structuring the negotiations from new foundations. In practice, the negotiations that come under the two working groups are relatively compartmentalized. The possible commitments of developing countries are being negotiated within the framework of the Convention (in the Ad Hoc Working Group on Long Term Cooperative Action: AWG-LCA). "*Further Commitments for Annex I Parties under the Kyoto Protocol*" are being negotiated within the framework of the Protocol (in the Ad Hoc Working Group on Further Commitments under the Kyoto Protocol: AWG-KP).

In the end, the Copenhagen Accord keeps the distinction between Annex I and non-Annex I countries but also refers to developed/developing countries categories. In any case, developing countries with economies in transition are not treated differently than other developing countries. Among the latter, special attention is simply paid to the situation of the most vulnerable ones (least-developed countries, small island developing states, Africa; Para. 3) who have priority in the access to funding (Para. 8). The refusal of countries with economies in transition to form an intermediary category or to join Annex I countries is used as a means of aligning themselves with all developing countries with regard to mitigation, to reject any GHG emissions reduction commitment beyond Article 4.1 of the UNFCCC. These distinctions are confirmed in the Cancún Agreements.

[8] Turkey was granted the right not to be listed in Annex II (countries with financial commitments) while being listed as an Annex I country.

[9] Translated by the authors. Kazakhstan wanted to be listed in Annex I but encountered a radical opposition from the G77 and China, who feared this could create a precedent: Doc. FCCC/CP/2001/13/Add.4, p. 42. As for the Belarus, the amendment that allows the Belarus to be listed in Annex B of the Kyoto Protocol was adopted by the Meeting of the Parties (Decision 10/CMP.2) but is not yet in force.

12.3.2 The Differentiation Between the Actions of Annex I and Non-Annex I Countries

The main divides between parties during the Barcelona and Bangkok meetings that preceded the Copenhagen conference crystallized on differentiation. The key issue was that of the creation (or not) in the final text adopted in Copenhagen of a *chapeau* above Paragraph 1-b)-(i) on nationally appropriate emissions reduction commitments and actions of developed countries and Paragraph 1-b)-(ii) on nationally appropriate mitigation actions (NAMAs) of developing countries.

Far from being anecdotal, this issue has catalyzed political differences because it amounted to "ponder on what is common and what is differentiated between developed countries and developing countries" (Guérin and Wemaëre 2009).

Whereas some (among which the United States) were convinced it was necessary to build a "bridge" between those two paragraphs, others (developing countries) wanted them to be separated by a "firewall." In particular, negotiators disagreed on the interpretation of the CBDR principle. Where the United States insisted on the common responsibility, developing countries emphasized it was differentiated. The principle is after all so broad, and almost an oxymoron, that it gives birth to opposite interpretations. According to developing countries, the Bali Action Plan separates the commitments and actions of each category impenetrably. In the others' opinion, both paragraphs use the same terms that have to be granted a single definition (in particular as regards the expression "measurable, reportable, and verifiable" commitments and/or actions). Having in mind the case of China notably, they wanted the obligations of both groups to be as similar as possible regarding their legal nature, and the measurement, reporting, and verification mechanisms.

In the end, the Accord refers to the common but differentiated responsibilities from Paragraph 1: "We emphasise our strong political will to urgently combat climate change in accordance with the principle of common but differentiated responsibilities and respective capabilities." It mentions equity on several occasions. Paragraph 2 grants developing countries a more flexible schedule, in particular with regard to the peaking of emissions, so as to take into account the fact that "social and economic development and poverty eradication are the[ir] first and overriding priorities."[10] India indeed considered that setting a date for the peaking of developing countries' emissions was unacceptable during the last night of the negotiations.

There is a clear distinction between developed countries' actions (emissions reduction commitments) and developing countries' actions (mitigation measures; Paras. 4 and 5). One can see it in the structure of the Accord—distinct paragraphs are devoted to each type of action—as well as in its content. Developed countries (here called "Annex I

---

[10]"(...) We should cooperate in achieving the peaking of global and national emissions as soon as possible, recognizing that the time frame for peaking will be longer in developing countries and bearing in mind that social and economic development and poverty eradication are the first and overriding priorities of developing countries and that a low-emission development strategy is indispensable to sustainable development."

parties") can notify their quantified economy-wide emissions reduction commitments to the Secretariat to make them enter the Appendix I of the Accord, with the base year of their choice. It makes it more difficult, but not impossible, to compare each Party's commitments in terms of effort level. Once again, the Accord is the product of a bottom-up approach where each one chooses its own target in the mid-term.

The Accord makes an explicit, albeit faint, reference to the Kyoto Protocol: "Annex I Parties that are Party to the Kyoto Protocol will thereby further strengthen the emissions reductions initiated by the Kyoto Protocol" (Para. 4). It is thus ambiguous as regards the continuation of the Kyoto Protocol, an issue that is still open. In practice the Copenhagen Accord pledges made by Annex I countries that are parties to the Protocol will serve as a basis for the AWG-KP negotiations. Registered pledges "will be measured, reported and verified in accordance with existing and any further guidelines" that should "ensure that accounting of such targets and finance is rigorous, robust and transparent" (Para. 4). If the European Union did not succeed in making its positions triumph, the United States showed a formidable political efficiency. It did not yield an inch, secured the bottom-up approach, and the international verification of developing countries' actions (see infra), all this in the context of serious uncertainties in Copenhagen as to whether the United States were able to adopt a domestic climate legislation, which is all the more in jeopardy now that President Obama does not have a majority in the Senate anymore.

As for developing countries ("non-Annex I Parties"), they will implement mitigation measures. They do not have to notify reduction targets, but their NAMAs will be registered by the Secretariat in the table of the Appendix II of the Accord. The deadline fixed, January 31, 2010, was the same for developed and developing countries, and was very/too close for a lot of them. Many states indeed submitted their NAMAs later.

The Appendix II where developing countries can register their mitigation actions is even less detailed than Appendix I, inasmuch as it does not even refer to base years. As it is a matter of actions, not commitments, that are economy-wide, and because of the lack of indication with regard to their submission, developing countries have a wide scope and, at the same time, do not have any marker, either quantitatively or qualitatively, to define the perimeter of their actions (activities, programs, sectoral approaches, etc.) and the effort level each action pursues. Consequently, Appendix II suffers from high heterogeneity in terms of perception of the actions that developing countries plan.

Developing countries' mitigation actions are also the product of a bottom-up approach. In addition the Accord states that NAMAs are submitted in a manner consistent with Article 4.7 of the UNFCCC, which provides that "The extent to which developing country Parties will effectively implement their commitments under the Convention will depend on the effective implementation by developed country Parties of their commitments under the Convention related to financial resources and transfer of technology and will take fully into account that economic and social development and poverty eradication are the first and overriding priorities of the developing country Parties." One can understand this reference as conditioning the commitment to implement the pledged NAMAs to the financial support of developed countries, in accordance with the Convention, and most of the developing countries understand it as such in the implementation of the Copenhagen Accord. In any case, China understands it this way, as shown by its letter to the UNFCCC Executive Secretary dated January 28, 2010, a letter in which it notifies the same carbon intensity target it had announced in Copenhagen: "China will endeavor to lower its carbon dioxide emissions per unit of GDP by 40–45 % by 2020 compared to the 2005 level, increase the share of non-fossil fuels in primary energy consumption to around 15 % by 2020 and increase forest coverage by 40 million hectares and forest stock volume by 1.3 billion cubic meters by 2020 from the 2005 levels. Please note that the above-mentioned autonomous domestic mitigation actions are voluntary in nature and will be implemented in accordance with the principles and provisions of the UNFCCC, in particular Article 4, paragraph 7."[11] In the absence of criteria or guidelines on the way NAMAs should be entered in Copenhagen Accord's Appendix II, it is quite impossible to compare actions and to assess their level of ambition and if they can contribute effectively to the achievement of a global GHG emissions reduction target by 2050.

Nationally Appropriate Mitigation Actions (NAMAs)
NAMAs refer to a set of policies and actions that countries undertake as part of their commitment to reduce greenhouse gases emissions. First mentioned in the Bali Action Plan (2007), they are core elements of the Copenhagen Accord (2009) and the Cancún Agreements (2010). Reflecting a bottom-up approach, they are based on these ideas:

- Different countries mean different nationally appropriate actions on the basis of equity and in accordance with the common but differentiated responsibilities principle and taking into account their respective capabilities.
- Whether developing countries will effectively implement national mitigation actions depends on whether developed countries effectively comply with their commitments to provide financial support and technology transfers.
- Recognition that developing countries' priorities are economic and social development and poverty eradication.

By definition, the nature and scope of NAMAs can significantly vary from one country to another. Indonesia, for example, might focus on integrating climate change policy with other aspects of economic development, such as progressive reduction in oil subsidies, poverty reduction through promotion of alternative income to reduce illegal logging, and exploit more fully the country's renewable resources, especially geothermal energy. Maldives announced that it aims to achieve carbon neutrality by 2020. Chile announced that it will implement NAMAs in order to achieve a 20 % reduction below the business-as-usual emissions growth trajectory in 2020, as projected from the year 2007. Energy efficiency, renewable energy, and land use, land-use change, and forestry (LULUCF) measures will be the main focus of Chile's NAMAs, and so on (NAMAs 2011); see also Chap. 13.

[11] Letter from NDRC for the attention of Yvo de Boer, January 28, 2010.

Although there is a clear distinction between developed countries and developing countries' obligations, the deal between the United States and developing major emitters led the latter to make the significant concessions on verification and transparency to which the United States were attached. In this respect, a distinction remains between developed and developing countries but it is softened. Regarding developed countries, their commitments shall be measured, reported, and verified under existing or further guidelines and shall in particular allow a rigorous, robust, and transparent accounting of targets. As for developing countries, all their actions and national inventories shall be submitted through national communications every 2 years. Concerning the verification of actions, the Accord provides for verification through consultations and analysis at the international level under "clearly defined guidelines that will ensure that national sovereignty is respected" (Para. 5). Reference to international guidelines is actually a concession even though it is immediately followed (and locked?) by the mention of national sovereignty. In addition, the terms chosen are interesting: "consultations" and "analysis" do not amount to control, inspection, monitoring, verification, and the like. Balanced by the reference to national sovereignty, these terms do not entail real constraints even if they slightly open a (tiny) door.

The Accord specifies that when these actions receive support, they enter the registry and then are subjected to international control: "These supported nationally appropriate mitigation actions will be subject to international measurement, reporting and verification in accordance with guidelines adopted by the Conference of the Parties."[12] The control of NAMAs was made possible only on the condition that these actions are financially supported by developed countries. The snag of a partial control for supported actions only is that it makes it difficult to get a whole picture of the emissions of all developing countries and will not allow the knowledge of whether global emissions remain in the right direction towards the achievement of the 2 °C temperature increase limitation target.[13] In this respect the Accord shows, if need be, the surge of power of developing major emitters who gain the recognition of their actions at little expense.

From this angle, the Cancún Agreements confirm the content of the Copenhagen Accord.[14] Moreover, Decision 1/CP.16 establishes new processes within the Subsidiary Body on Implementation (SBI) to ensure the comparison between the parties' mitigation efforts, called "international assessments" for developed countries, and "international consultations and analysis" (a phrase from the Copenhagen Accord) for developing countries. As for the latter, the decision specifies that the process is "non-intrusive, non-punitive, and respectful of national sovereignty"; focuses on unsupported actions; does not consider the "appropriateness" of domestic

[12] See also Decision 5/CP.15.

[13] Which is the Accord's target (Para. 1).

[14] Decision 1/CP.16, Outcome of the Work of the Ad Hoc Working Group on Long-term Cooperative Action under the Convention.

policies; includes an analysis by technical experts; and results in a summary report. The decision sets work programs to flesh out these elements but no deadline.

The Copenhagen Accord only mentioned carbon markets implicitly, recognizing their role without taking any decision on the matter. It called for the creation of incentives in favor of developing countries as they adopt a low-carbon development. It only mentioned carbon markets in a mitigation perspective and not as a source of financing for developing countries' actions. In a section on "Various approaches, including opportunities for using markets, to enhance the cost-effectiveness of, and to promote, mitigation actions, bearing in mind different circumstances of developed and developing countries," the Cancún Agreements confirm they favor market-based mechanisms. Market-based mechanisms created by the Kyoto Protocol are intended to continue in order to ensure developed countries achieve their commitments. A decision on the grounds of principle confirms that the two existing project-based mechanisms—the clean development mechanism (CDM) and the joint implementation (JI)—will be maintained if there is a second commitment period.

Emissions Trading (ET)

Article 17 of the Kyoto Protocol allows countries that have spared emission units—that is to say whose emissions do not reach their quantified emission units assigned amount—to sell the remaining emission units to countries whose emissions exceed their assigned amount. Thus, a new commodity was created in the form of emission reductions or removals. Because carbon dioxide is the principal greenhouse gas, people speak simply of trading in carbon. Carbon is now tracked and traded as is any other commodity. This is known as the "carbon market".

See also Chap. 13.

Both mechanisms in parallel have received decisions aimed at facilitating their enhancement and development.

The need remains to clarify the mandate of concerned institutions during the gap between the commitment periods, and to decide what to do with the Russian "hot air" as well. Finally, negotiations on "new" market-based mechanisms must be pursued. One can in particular think of sectoral agreements to ensure the creation of some systems of emission-assigned amounts that would be exchangeable within a same sector at the international level, or else granting carbon units to NAMAs in particular in the framework of the "Reducing emissions from deforestation and forest degradation initiative" (REDD). Sketched out by the Copenhagen Accord although no decision was made specifically for that purpose, the "REDD-plus" mechanism was created by the Cancún Agreements. However, as they were unable to resolve related financial issues, in particular on the role of market-based finance, the parties asked the AWG-LCA to explore financing options to make recommendations at COP 17. The market-based mechanism option is not ruled out for the moment even though certain forestry countries strongly oppose it.

12.3.3 Support, Assistance, and Funding

In Copenhagen developing countries, and in particular African countries, called for a very significant increase in financial assistance (1–5 % of the GNP of developed countries devoted to official development assistance), much beyond what developed countries offered or could offer.[15] The Copenhagen Accord enshrines the collective commitment by developed countries to provide at the appropriate level new and additional, predictable and adequate resources.[16] The fast-start budget aimed at preparing developing countries for their entire participation to the further international climate regime amounts up to USD 30 billion for the 2010–2012 period, expenses being equitably shared between mitigation and adaptation (Copenhagen Accord, Para. 5).

The Copenhagen Accord takes note of the will of developed countries to jointly mobilize USD 100 billion a year as of 2020. But will developing countries really get "new and additional, predictable and adequate funding as well as improved access" to funding? Until now, the developed countries who actually honored their official development assistance commitments are scarce. In the field of climate change, mitigation-related bilateral aid has been insufficient. It has remained stable and technology transfers have only been made in the energy efficiency field even if technology transfer cooperation has globally increased (Lanfranchi 2010). Unless additional resources are found by new or alternative ways (several propositions are being discussed), the last 40 years' experience calls for caution. In the end, although promises are ambitious and were a core element in the final deal and conditioned the acceptance of African countries, nothing guarantees new and additional resources or the creation of innovative new resources.

The Accord calls for the creation of a Copenhagen Green Climate Fund through which a significant part of the funding should pass. The Fund was created during the Cancún conference and its governance and functioning terms were specified. It will be governed by a management board of 24 members (half from developed countries and half from developing countries). Until the selection of a trustee to manage the international financial assets, the World Bank ensures its first 3 years of operational implementation (Delbosc and Jeulin 2011). Finally, the Cancún Agreements create a Technology Mechanism (Decision 1/CP. 16, from Para. 102) as suggested by the Copenhagen Accord to speed up the development and transfer of technology in support of mitigation and adaptation actions (Para. 11).

---

[15] Developed countries do not even achieve the objective of 0.7 % of their GNP although it was set long ago by the United Nations General Assembly and many UN conferences.

[16] "Scaled up, new and additional, predictable and adequate funding as well as improved access shall be provided to developing countries (…)": Para. 8.

12.4 The Legal Architecture of the Regime

The Copenhagen outcome is a series of decisions adopted by the Conference of the Parties (to the Convention: COP) and the Conference of the Parties to the Convention serving as the Meeting of the Parties to the Protocol (COP/MOP). The COP simply took note of the Copenhagen Accord that had been negotiated and adopted outside the conventional framework, without endorsing it in the absence of a consensus (Decision 2/CP.15). The Cancún Agreements, consisting in a "balanced set of decisions," brought the content of the Copenhagen Accord back in the framework of the Convention. They were adopted by consensus minus one vote, that of Bolivia, which deserves to be emphasized as it is theoretically enough to consider that consensus was not reached.

12.4.1 On the Legal Nature of the Copenhagen Accord

Although the Copenhagen outcome had been negotiated for many weeks in both AWGs, and although the COP/MOP was meeting, the Accord was negotiated on the fringes by five countries (the United States and the BICS: Brazil, India, China, South Africa), agreed upon by 28 parties (among them the European Union) and then submitted in a somewhat cavalier manner to the COP/MOP for adoption, sparking the legitimate indignation of many states.

But for all that, the Accord has connections with the conventional framework. As for its form, it looks very much like a COP decision. As for its content, it refers on many occasions to the UNFCCC and even to the Kyoto Protocol, showing that some of the countries who participated in its negotiation wanted to see it enter the conventional framework, which was impossible in Copenhagen. After long hours of discussion and confronted by the opposition from several countries (among which were Tuvalu, Bolivia, Venezuela, and Pakistan) the COP finally refused to confirm or approve the Accord and only "took note" of it. As regards the COP/MOP, none of its decisions refers to the Accord. Because the COP took note of it, it however appears in an appendix to Decision 2/CP.15. The UNFCCC Secretariat hosts a special webpage about the Accord on the website of the Convention. It has nevertheless specified that the Accord does not have a legal nature under the auspices of the Framework Convention (DBO/drl, 2010).

In the original version distributed, no country name was mentioned at the beginning of the Accord. There was a blank space after "The Heads of State, Heads of Government, Ministers, and other heads of the following delegations present at the United Nations Climate Change Conference 2009 in Copenhagen: ... " The Accord was steeped in vagueness at that time and in no way could it have binding legal effects. The situation has changed since then because in the report of the Copenhagen COP, 114 states and the European Union, representing 80 % of global GHG emissions, are listed as having agreed on the Copenhagen Accord.

Twenty-six additional states even later expressed their intention to be listed as countries who agree on the Accord.

The Copenhagen outcome is thus in itself a very fragile instrument with a low, not to say with no, binding character. But the fact that the Accord does not bind the parties does not mean they will not comply with it. By receiving the support of 140 parties, it undoubtedly gained some legitimacy.

In any case one can still wonder about the consequences of the notification of the parties' commitments (Annex I/developed countries) and actions (non-Annex I/ developing countries) to the UNFCCC Secretariat. Although it is difficult to grant them a legal qualification, notifications represent a political pledge that could create expectations and must be implemented in good faith. One can also point out that each notification can be seen as a unilateral act of states, an autonormative act like a "promise" which expresses a will that tends to create legal effects and thus is … binding. Such a notification indeed meets the conditions to constitute a unilateral act of states. International law is not strict in this matter. The said act must be attributable to a state acting within the limits of its jurisdiction and made publicly enough. In the *Nuclear Tests* cases, the International Court of Justice admitted that unilateral acts could be an autonomous source of law: "It is well recognized that declarations made by way of unilateral acts, concerning legal or factual situations, may have the effect of creating legal obligations. Declarations of this kind may be, and often are, very specific. When it is the intention of the State making the declaration that it should become bound according to its terms, that intention confers on the declaration the character of a legal undertaking, the State being thenceforth legally required to follow a course of conduct consistent with the declaration. An undertaking of this kind, if given publicly, and with an intent to be bound, even though not made within the context of international negotiations, is binding" (ICJ 1974). The International Law Commission is in line with this judgment when it lays down the principle according to which "Declarations publicly made and manifesting the will to be bound may have the effect of creating legal obligations. When the conditions for this are met, the binding character of such declarations is based on good faith; States concerned may then take them into consideration and rely on them; such States are entitled that such declaration be respected" (ILC 2006). The essential "will to be bound" remains of course difficult to prove as the International Court reminded in the *Burkina Faso v. Mali Frontier Dispute* case (ICJ 1986). One must indeed remember that in the *Nuclear Tests* cases the ICJ performed a bold application of the principle of good faith because the parties had agreed on the fact that France's declarations were political declarations which did not legally bind France (Daillier et al. 2009). Let us add that even if the binding character of notifications is established, commitments are still reversible. "Repentance" is allowed and states can backpedal on certain conditions, in particular through recourse to dispute settlement mechanisms. A unilateral commitment cannot be arbitrarily retracted (ILC 2006).

Naturally, the assessment of the binding (or not) character of a given notification must be made on a case-by-case basis. In the opinion of the ILC, "To determine the legal effects of such declarations, it is necessary to take account of their content, of all the factual circumstances in which they were made, and of the reactions to which they gave rise" (ILC 2006). As stated by the ICJ, where there is a doubt "A restrictive interpretation is called for" (ICJ 1974). And one cannot but notice that the parties' notifications are extremely diverse in that respect. Among the commitments of developed countries, many are weakened by the fact they are conditional (e.g., Canada's −17 % commitment was set at this level to align with the provisional emissions reduction objective of the United States, thus making this target depend on the final US energy and climate legislation); some states commit without condition to reach a target (e.g., Australia: −5 % GHG emissions or the EU: −20 %) but put conditions for reaching more ambitious objectives (Australia: −15 to −25 %, the European Union: −30 %). States seldom commit without condition. The only example is that of Kazakhstan, who commits to reduce its emissions by 15 %.

As for developing countries, they do not formally notify "commitments" but "actions." Therefore, do they commit? Some of them only make pledges with commitments they already had under the UNFCCC, like Afghanistan who commits to create a GHG inventory and to submit its national communication.

In any case, the Cancún Agreements reintegrate the Copenhagen Accord pledges in the heart of the UNFCCC. The COP decision "Takes note of quantified economy-wide emission reduction targets to be implemented by Parties included in Annex I to the Convention as communicated by them", and "of nationally appropriate mitigation actions to be implemented by non- Annex I Parties as communicated by them" and listed in two documents (FCCC/SB/2011/INF.1/Rev.1 and FCCC/AWGLCA/2011/INF.1).

12.5 Conclusion

The mandate of both working groups—AWG-LCA in the framework of the UNFCCC and AWG-KP in that of the Kyoto Protocol—had been extended until the Cancún meeting in late November 2010 and again extended until the Durban meeting in December 2011. The future of the Protocol or more broadly the issue of the legal form of a future agreement was not solved, either in Copenhagen or (at least officially) in Cancún. The Cancún Agreements leave the Protocol's continuation option open. However, because the Protocol was very explicitly "dropped" by Russia, Japan, and Canada in Cancún, its fate seems to be sealed. In any case, neither the Copenhagen Accord nor the Cancún Agreements have instilled any ambition in terms of emissions reduction. The consensus develops out of the lowest common denominator and even though confidence seems to have returned after the Cancún conference there is no real international dynamic nonetheless. With regard to the form of an outcome, everything remains possible: a series of COP decisions and/or a binding agreement, as "nothing in this decision shall prejudge prospects for, or the content of, a legally-binding outcome in the future."[17] With regard to the

[17] Preamble, Decision 1/CP.16.

outcome's content, however, the Copenhagen Accord and the Cancún Agreements outline the future climate regime. Its center of gravity slides from the Protocol towards the Convention. The top-down approach gives way to a bottom-up approach and constraints to incentives. Along the same logic transparency, guaranteed by the MRV[18] system, is replacing observance. It is thus a quite different regime than the existing one. Will it be able to take up the challenge of climate change with which humanity is confronted?

References

Casella H, Delbosc A, de Perthuis C (2010) Cancun: year one of the post-Copenhagen era, CDC climate research, climate report no. 24. Last visited 14 Feb 2011. http://www.cdcclimat.com/Climate-Report-no24-Cancun-Year,512.html?lang=en

Compagnon D (2010) Les défis politiques du changement climatique: de l'approche des régimes internationaux à la gouvernance transcalaire globale. In: Cournil C, Colard-Fabregoule C (eds) Changements climatiques et défis du droit. Bruylant, Bruxelles

Daillier P, Pellet A, Forteau M (2009) Droit international public. LGDJ, Paris

Delbosc A, Jeulin M (2011) What should we take away from cancun? CDC climate research, climate brief no. 3. Last visited 15, 2011. http://www.cdcclimat.com/What-should-we-take-away-from.html

Godard O (1993) Stratégies industrielles et conventions d'environnement: de l'univers stabilisé aux univers controversés. INSEE-Méthodes Environnement, Économie 39–40:145–174, Paris

Guérin E, Wemaëre M (2009) Négociations climat: compte-rendu de la conférence de Barcelone (2–6 Nov 2009). IDDRI, Idées pour le débat no. 06

ICJ (1974) *Nuclear tests*, Australia & New Zealand v. France, 20 December 1974, para. 43 (Australia v. France) and para 46 (New Zealand v. France)

ICJ (1986) Frontier dispute, Burkina Faso v. Mali, 22 December 1986, para. 39

ILC (2006) Unilateral acts of states, guiding principles applicable to unilateral declarations of states capable of creating legal obligations, ILC 58th session, 2006, Principle 1

IPCC (1995) Second assessment report. http://www.ipcc.ch/

IPCC (2001) Third assessment report. Climate change 2001: synthesis report – summary for policymakers. http://www.ipcc.ch/

IPCC (2007a) Fourth assessment report. Climate change 2007: synthesis report – summary for policymakers. http://www.ipcc.ch/

IPCC (2007b) Fourth assessment report. Contribution of working group III to the fourth assessment report of the Intergovernmental Panel on Climate Change – technical summary. http://www.ipcc.ch/

Lanfranchi M-P (2010) Le statut des pays en développement dans le régime climat: le principe de la dualité des normes revisité ?. In: Le droit international face aux enjeux environnementaux, Proceedings of the 2008 annual conference of the French Society for International Law. Pedone, Aix-en-Provence/Paris

Maljean-Dubois S, Wemaëre M (2010) La diplomatie climatique. Pedone, Paris

NAMAs (2011) Wikipedia article. Last visited 11 Dec 2011. http://www.wikipedia.org/wiki/Nationally_Appropriate_Mitigation_Action

Schneider SH, Semenov S, Patwardhan A, Burton I, Magadza CHD, Oppenheimer M, Pittock AB, Rahman A, Smith JB, Suarez A, Yamin F (2007) Assessing key vulnerabilities and the risk

[18] Measurement, reporting, and verification.

from climate change. In: Parry ML, Canziani OF, Palutikof JP, van der Linden PJ, Hanson C (eds) Climate change 2007: impacts, adaptation and vulnerability. Contribution of working group II to the fourth assessment report of the Intergovernmental Panel on Climate Change. Cambridge University Press, Cambridge

Shelton D (2007) Equity. In: Bodansky D, Brunnée J, Hey E (eds) The oxford handbook of international environmental law. Oxford University Press, Oxford

Smith JB, Schneider SH, Openheimer M, Yohe GW, Hare W, Mastrandrea MD, Patwardhan A, Burton I, Corfee-Morlot J, Magadza CHD, Füssel HM, Pittock AB, Rahman A, Suarez A, van Ypersele JP (2008) Assessing dangerous climate change through an update of the Intergovernmental Panel on Climat Change (IPCC) "reasons for concern". PNAS Early Edition. Last visited 15 Dec 2011. http://www.pnas.org/cgi/doi/10.1073/pnas.0812355106

Stern N (2006) Stern review on the economics of climate change. Treasury, London. http://www.hm-treasury.gov.uk/sternreview_index.htm

UNDP (2008) Human development report 2007/8, fighting climate change: human solidarity in a divided world. UNDP, New York

UNEP (2010) The emissions gap report. Are the Copenhagen Accord pledges sufficient to limit global warming to 2 °C or 1.5 °C? A preliminary assessment. http://www.unep.org/publications/ebooks/emissionsgapreport

World Bank (2010) World development report 2010: development and climate change. World Bank, Washington, DC

UNFCCC Documents and Decisions

"Report of the Conference of the Parties on its seventh session, held at Marrakesh from 29 October to 10 November 2001, Addendum", Doc. FCCC/CP/2001/13/Add.4.

Decision 10/CMP.2, "Proposal from Belarus to amend Annex B to the Kyoto Protocol", 17 November 2006.

Decision 1/CP.13, "Bali Action Plan", 15 December 2007.

Decision 5/CP.15, "Work of the Consultative Group of Experts on National Communications from Parties not Included in Annex I to the Convention", 19 December 2009.

Decision 2/CP.15, "Copenhagen Accord", 19 December 2009.

Decision 1/CP.16, "Outcome of the Work of the Ad Hoc Working Group on Long-term Cooperative Action under the Convention", 10 December 2010.

UNFCCC Secretariat, "Notification to Parties, Clarification relating to the 18 January 2010", 25 January 2010, Doc. DBO/Drl.

"Compilation of information on nationally appropriate mitigation actions to be implemented by Parties not included in Annex I to the Convention", Note by the secretariat, FCCC/AWGLCA/2011/INF.1, 18 March 2011.

"Compilation of economy-wide emission reduction targets to be implemented by Parties included in Annex I to the Convention, Revised note by the secretariat", FCCC/SB/2011/INF.1/Rev.1, 7 June 2011.

Chapter 13
Governing Urban Infrastructure in Developing Cities: The Role of Carbon Finance

Jun Li

Abstract One of the key issues in the post-Kyoto climate regime is to reach consensus on how to finance actions needed in fast-growing economies that will enable altering their business-as-usual emission pathways. Specifically, cities in developing countries will play a significant role in climate mitigation and adaption given their contribution to greenhouse gas emissions and inherent vulnerability to induced global change (e.g., sea-level rise, increased water scarcity and drought, forced migration, etc.). Changing their pathways and increasing climate resilience in these countries requires significant incremental investments in urban infrastructures today. However, financial and institutional capacities are much lower as compared to the developed world. International financial and technology transfer are bound to bridge the gap under a well-designed institutional framework. This chapter discusses different climate finance mechanisms, possible improvement and instrumentalization in light of enhancing urban infrastructure governance in developing cities, in conjunction with their policy relevance, implementability, and economic and environmental effectiveness. Also, we posit institutional implications for using carbon finance to facilitate the development of climate-resilient urban infrastructure in fast-growing cities in developing countries.

Keywords Greenhouse gas • Cities • Urban infrastructure • Climate finance • NAMA

List of Acronyms

BAP	Bali Action Plan
BAU	Business as usual
BEE	Building energy efficiency

J. Li (✉)
CIRED, 45 bis, avenue de la Belle Gabrielle, 94736 Nogent-sur-Marne, Cedex, France
e-mail: jun.li@centre-cired.fr

J.B. Saulnier and M.D. Varella (eds.), *Global Change, Energy Issues and Regulation Policies*, Integrated Science & Technology Program 2, DOI 10.1007/978-94-007-6661-7_13,

BIC	Brazil, India, China
CCAP	Centre of Clean Air Policy
CDM	Clean development mechanism
CD4CDM	UNEP's Capacity Development for CDM Project
JI	Joint implementation
CER	Certified emission reduction
COP	Conferences of the Parties
CPAs	CDM Programme activities
DC	Developing countries
EU	European Union
FDI	Foreign direct investment
gCO_2/p-km	Gram CO_2/passenger km
gCO_2/v-km	Gram CO_2/vehicle km
GHG	Greenhouse gases
Gteq C	Gigaton of carbon equivalent (= 1,000 Mteq C)
Mteq CO_2	Gigaton of CO_2 equivalent (1 Gteq CO_2 = 1 Gteq C *44/12)
Mteq C	Megaton of carbon equivalent
Mteq CO_2	Megaton of CO_2 equivalent (1 Mteq CO_2 = 1 Mteq C *44/12)
IEA	International Energy Agency
IPCC	Intergovernmental Panel on Climate Change
MOP	Meeting of the Parties to the Kyoto Protocol
MRV	Monitoring, reporting, and verifying/measurable, reportable, and verifiable
NAMA	Nationally Appropriate Mitigation Actions
NDRC	National Development and Reform Commission, a macroeconomic management agency of People's Republic of China.
NLT	No-lose target
OECD	Organisation for Economic Co-operation and Development
PAM	Policies and measures
PCDM	Programmatic CDM
PoA	Programs of activities
SA	Sectoral approach
SCM	Sectoral crediting mechanism.
SD	Sustainable development
SD-PAMs	Sustainable development policies and measures
SNLT	Sectoral no-lose target
teq CO_2	Ton of CO_2 equivalent
UN	United Nations
UNEP	United Nations Environment Program
UNFCCC	United Nations Framework Convention on Climate Change

13.1 Introduction

Cities are on the forefront of combating global warming. Around two thirds of the world's energy (7,900 Mtoe in 2006) are consumed in cities contributing roughly 80 % to global greenhouse gas (GHG)[1] emissions (OECD 2009; Allix 2009; World Bank 2009), and it is projected that cities would account for 73 % of world energy consumption in 2030 in the reference scenario of the International Energy Agency (IEA 2009). The prospects for the development trajectory of cities will play a determining role in addressing the climate change challenge. The urbanization is expected to continue throughout the world, particularly in developing countries, global population will grow to nine billion in 2050, of which more than 70 % will be in urban areas (UN 2007). This implies that more than 60 billion m^2 new residential buildings are to be built in the cities (equivalent to fourfold current EU-15 housing floor area). In the meantime, most of the increased urban population will come from developing countries over the next decades. For instance, China's urban population had an over threefold increase during 1970–2000 and would double again during 2000–2030! This implies a dramatic rise in energy demand and resultant GHG emissions in these newly urbanized areas in the absence of implementing relevant public policies effectively over the next decades.

Although it is widely recognized that cities can play a proactive role in shaping the low-carbon economy perspective, few climate mitigation and adaptation policies have been reflected in local strategies due to the dearth of financing and weak governance as well as a lack of institutional capacity in developing cities. This chapter attempts to investigate the means of utilizing financial instruments as well as a combinatory portfolio for reconciling the GHG mitigation targets and sustainable socioeconomic development programs in the developing cities. Our work builds on a cross-comparative analysis of recent literature analyzing different carbon finance mechanisms in light of facilitating climate-friendly projects and resilient infrastructure development in developing countries. It endeavors to contribute to bridging the gap of understanding the major financing issues of climate mitigation in cities and suggests necessary improvements in these instruments in light of furthering the climate agreement negotiations and contributing to coherent governance of international climate finance.

> The *Kyoto Protocol* is an international agreement linked to the United Nations Framework Convention on Climate Change. The major feature of the Kyoto Protocol is that it sets binding targets for 37 industrialized countries and the European community for reducing greenhouse gas (GHG) emissions. These amounted to an average of 5 % against 1990 levels over the five-year period 2008–2012.

(continued)

[1]Because carbon dioxide (CO_2) from energy use constitutes the major part of greenhouse gases emissions in urban area, we use GHG (greenhouse gas) and carbon/CO_2 interchangeably throughout the text for simplicity.

(continued)

The major distinction between the Protocol and the Convention is that although the Convention encouraged industrialized countries to stabilize GHG emissions, the Protocol commits them to do so. Recognizing that developed countries are principally responsible for the current high levels of GHG emissions in the atmosphere as a result of more than 150 years of industrial activity, the Protocol places a heavier burden on developed nations under the principle of "common but differentiated responsibilities."

The Kyoto Protocol was adopted in Kyoto, Japan, on December 11, 1997 and entered into force on February 16, 2005.

13.2 Disentangle the Major Difficulties of the Climate Finance Tools

The major climate finance tools.

Under the Treaty, countries must meet their targets primarily through national measures. However, the Kyoto Protocol offers them an additional means of meeting their targets by way of three market-based mechanisms.

Emissions trading (ET), as set out in Article 17 of the Kyoto Protocol, allows countries that have emission units to spare—emissions permitted them but not "used"—to sell this excess capacity to countries that are over their targets. Thus, a new commodity was created in the form of emission reductions or removals. Because carbon dioxide is the principal greenhouse gas, people speak simply of trading in carbon. Carbon is now tracked and traded like any other commodity. This is known as the "carbon market".

The *clean development mechanism* (CDM) allows a country with an emission-reduction or emission-limitation commitment under the Kyoto Protocol to implement an emission-reduction project in developing countries. Such projects can earn saleable certified emission reduction (CER) credits, each equivalent to 1 t of CO_2, which can be counted towards meeting Kyoto targets. A CDM project activity might involve, for example, a rural electrification project using solar panels or the installation of more energy-efficient boilers. The mechanism stimulates sustainable development and emission reductions, while giving industrialized countries some flexibility in how they meet their emission reduction or limitation targets.

The mechanism known as *joint implementation* (JI), allows a country with an emission reduction or limitation commitment under the Kyoto Protocol (Annex B Party, the list is very similar to annex 1 one; see Sect. 13.3.2) to earn emission

(continued)

(continued)

reduction units (ERUs) from an emission-reduction or emission removal project in another Annex B Party, each equivalent to 1 t of CO_2, which can be counted towards meeting its Kyoto target. Joint implementation offers Parties a flexible and cost-efficient means of fulfilling a part of their Kyoto commitments, and the host Party benefits from foreign investment and technology transfer.

13.2.1 Inadequate Support for Climate-Friendly Infrastructure in Developing Cities

So far, most proposals of engaging developing countries in GHG mitigation are targeted to the political or economic entities such as national governments for economywide commitments, or sector-based (energy- and carbon-intensive sectors among others) mitigation approaches. Intense efforts in project-based instruments, primarily the Kyoto's Clean Development Mechanism or sectoral approach have largely focused on industry and energy sectors in climate change negotiation. However, cities, which contribute nearly 80 % of global emissions, have not yet been recognized as a legitimate entity to implement different GHG mitigation policies and measures with relevant technical and financial assistance schemes. In the existing climate negotiations framework, involvement of cities in carbon finance is still confined to project-based CDM and/or JI which is not adaptive for scaling up the actions in the major urban emissions sectors such as transport and buildings sectors, and thus moving urban development towards sustainable trajectory UNEP (2007).

Likewise, few previous studies on negotiation addressed the eligibility and feasibility of making cities candidates as carbon finance players and the subsequent implications for institutions in urban infrastructure (e.g., transport, buildings) management. The link between carbon finance instruments (project-based and sector-based) and mitigation programs in cities is still wanting to allow cities to finance energy efficiency in buildings and transport so as to harness considerable GHG mitigation potentials in the urban infrastructures. It is thus necessary to bridge the gap between the carbon finance insufficiency and the means required to enhance the climate risk resilience of urban infrastructure in cities for the years to come.

Policies and strategies to reduce GHG emissions in urban transport and building sectors need to be included in the ongoing international climate negotiations, however, a coherent and integrated governance structure is still wanting in many developing cities. In fact, despite national decentralization programs and the inclusion of good governance principles in national policies and strategies, many governments do not aptly consider local governments as important partners in the articulation of action plans (UN-Habitat 2008). Cities in developing countries need to seize the momentum to shape the perspective of sustainable pathways to avoid locking urban infrastructure into irreversible carbon-intensive assets with negative economic and environmental impacts; note that retrofitting or renovating large-scale carbon-intensive infrastructure due to wrong decision-making made today will pose tremendous economic and social challenges in the fast-growing cities in the future.

13.2.2 Imperative of Changing Climate Finance Models

It is widely perceived that the actions of emission reductions in developed countries alone are insufficient to solve the climate puzzle and are often more expensive to make than in developing countries (DC). Contributions by developing countries are a critical means to achieve additional emission reductions beyond the climate targets at the global level. Developed countries are still responsible for the major part of the cumulative GHG emissions (Raupach et al. 2007) and are more advanced in climate-friendly technologies, albeit the emerging countries will be the major contributors of global GHG emissions increase in the next decades. Therefore, technology and knowledge transfer and financial assistance from developed countries to developing countries are indispensable to allow them to transform their urban infrastructures in compliance with the goals of climate-friendly urbanization and economic growth. Implementing GHG-friendly sectoral policies such as energy efficiency and renewable energy in developing cities will by no means be easy and costless; considerable investment and efforts will need to be mobilized to accelerate the uptake of climate-friendly technologies, thus financial assistance and technology transfer from developed countries to developing cities are essential to help the latter tackle GHG management.

The set of actions will consist of a palette of components such as regulatory changes, infrastructure investment, and tailored financial support and training, and is often referred to as Nationally Appropriate Mitigation Action (NAMAs, Neuhoff et al. 2009; see Sect. 13.4.3). As mentioned in Ward et al. (2008), external funding or technological support other than carbon financing (e.g., funding of sustainable development policies and measures: SD-PAMs) may also be used by these countries to assist them in achieving outcomes beyond what their self-funded programs may be able to achieve, whereby another type of sectoral approach may prove to be effective.

The prospective investment flow through the global carbon market is likely to play an important role in helping developing countries to fulfill the targets of climate mitigation in the post-Kyoto period.

Development funding through both domestic and international agencies needs to be mobilized at scale. International finance will be critical, as will the mobilization of domestic investment for implementing sustainable development policies in developing countries (Winkler et al. 2008). Then, the underlying question is how developing cities can benefit from the climate/nonclimate funding or financial assistance to enable them to build up the technical, financial, and institutional capacity to embark on a sustainable and low-carbon urban development trajectory.

13.3 Gap Between Existing Tools and Finance Needed for Mitigation in Developing Cities

13.3.1 Inadequate Financial Resources

The funding currently available under the UNFCCC (United Nations Framework Convention on Climate Change) Convention and Kyoto Protocol is inadequate and the exiting funding arrangements suffer from inappropriate governance; the resources that are now being allocated are out of phase with what is required in addressing mitigation and adaptation actions in developing countries (Gomez-Echeverri and Müller 2009). The finance issues were one of the thorniest subjects in the Copenhagen negotiation, where the parties still failed to reach a concrete agreement, although the vague target of 100 billion US$ by 2020 has been accepted by developed nations. Nevertheless, the political pledge made by developed countries during the Copenhagen summit per se is still far from being sufficient to enable developing countries to tackle the complex of climate change-related issues. The financial transfer target is vague and does not provide a concrete implementation scheme comprising technical assistance packages and a local capacity building plan in developing cities.

Even when CDM that accounts for the overwhelming share of the financial flow from developed to developing countries for GHG emission mitigation is considered, the projects registered by 2012 in China are expected to generate about 320 Mteq[2] CO_2 CERs, accounting for only a tiny part of the 2020 emission intensity reductions commitment. Moreover, as pointed out by Lütken (2010), China is likely to retain most of CERs for domestic emission reduction targets instead of selling to the global carbon market: as China accounts for nearly 60 % of CERs, this would make the developed countries, in particular the EU, fall short of capacity of outsourcing reduction significantly in the post-Kyoto regime. Compared with the trade balance and foreign direct investment (FDI), currently, the magnitude of carbon finance (primarily the CERs) is largely dwarfed by the latter. For example, the yearly FDI flow into China and India totaled around 120 billion US$ for 2008–2009, whereas the CERs generated in 2012 is at most 5 billion US$ (assuming a price of 11 US$/teq CO_2). Likewise, CDM CERs transactions account for a minuscule part in the trade. For instance, the total trade volume in China in 2008 was over 2,000 billion US$. Thus the current carbon finance model is unlikely to deliver adequate financial resources for supporting the longer term actions in developing and less-developed countries.

The recent literature surrounding the post-2012 climate regime indicates that the current individual project-based financing tools need to be scaled up by sectoral approach (SA) and other nonemissions reduction-driven policies. Different mechanisms of

[2] All GHG gases are concerned and the effects of non-CO_2 ones are converted in equivalent quantities of CO_2.

engaging developing countries in significant sectoral GHG mitigation have been widely discussed. Some studies discuss the possibility of scaling up CDM investment in the transport and building sector through alternative crediting or noncrediting instruments such as sectoral approach and SD-PAMs. For instance, both barriers and solutions are explored to harness CDM finance to address energy efficiency in buildings (Cheng et al. 2008). A reformed framework is argued to be necessary to scale up CDM in urban transport in DCs (Sanchez 2008). More specifically, a NAMA could encompass several approaches such as sectoral crediting, SD-PAMs, CDMs, or PCDMs (programatic CDMs; Payá 2008; Wemaere 2009).

13.3.2 Some Issues of the CDM

The CDM is one of the flexible technology transfer and environmentally friendly development financing tools within the UNFCCC framework, introduced by the Kyoto protocol 1997, whereby Annex 1 countries with emissions cap objectives participate in GHG emissions cut projects in Non-Annex 1 (no explicit emission quantity cap).

Annex 1 Countries

Parties include the industrialized countries that were members of the OECD (Organisation for Economic Co-operation and Development) in 1992, plus countries with economies in transition (the EIT Parties), including the Russian Federation, the Baltic States, and several Central and Eastern European States.

Non-Annex 1 Countries

Parties are mostly developing countries. Certain groups of developing countries are recognized by the Convention as being especially vulnerable to the adverse impacts of climate change, including countries with low-lying coastal areas and those prone to desertification and drought. Others (such as countries that rely heavily on income from fossil fuel production and commerce) feel more vulnerable to the potential economic impacts of climate change response measures. The Convention emphasizes activities that promise to answer the special needs and concerns of these vulnerable countries, such as investment, insurance, and technology transfer.

The global CDM market has been increasing rapidly and has attracted international financial institutions. Primary CDM reached 389 Mteq CO_2 with a value of 6.5 billion US$, and secondary CDM transaction volume and nearly four- and threefold that of the primary CDM market with 1,072 Mteq CO_2 for 26 billion US$

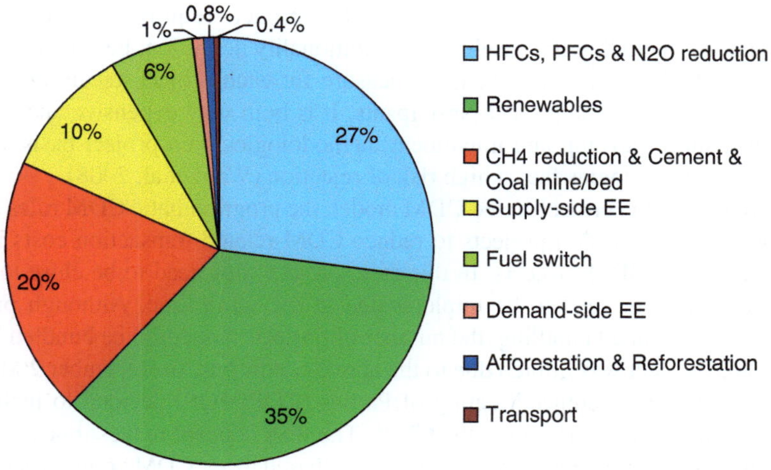

Fig. 13.1 Pipeline projects and CERs breakdown of CDM (*Source*: Fenhann 2009)

transaction (Capoor and Ambrosi 2009). However, the majority of CDM projects in the pipeline concentrate on renewable sectors (nearly half of total projects); expected CERs issuance in the CDM projects primarily deal with non-CO_2 GHGs in industries, power generation, and renewables (Fig. 13.1).

Furthermore, despite considerable transaction volumes and values in the global carbon market, there is only a handful carbon transaction projects (including CDM) dealing with mitigation in the buildings and transport sector in cities, which combined, however, account for more than 50 % of cities' carbon emissions. It is then necessary to identify the main cause of this gap and propose relevant solutions to expand the city-based carbon finance for emissions mitigation policies and measures in developing countries.

Major stumbling blocks and catalyzers to scaling up energy efficiency investment via CDM in buildings have been identified (Cheng et al. 2008), such as the fragmentation and complexity of construction projects, the small-scale and disperse emission points (spreading over hundreds of millions of housing) that make the registry and monitoring, reporting and verifying (MRV) costly and time consuming under the current CDM framework. One of the main challenges of integrating buildings into CDM is related to the baseline definition and the subsequent additionality demonstration inasmuch as CDM is project-based, also due to the fact that the combined technological intervention cannot be considered in the same methodology, making CDM less attractive for investing in building energy efficiency (BEE) because the energy performance of buildings often requires integrated technical and managerial measures (heating and cooling system optimization, efficient lighting, double or triple-glazing, enhanced ventilation, efficient appliances, etc.). More specifically, some "soft" measures (e.g., optimized architecture design for passive heating or cooling) are not quantifiable in terms of GHG mitigation, although essential for improvement in energy efficiency, thus not recognized and credited in the project provision (Cheng et al. 2008).

Furthermore, the current project-based approach requires that the CDM methodology, including the baseline and additionality definition, be established for each individual technical intervention measure for each project, resulting in high transaction costs for the project participants. It is both very expensive and technically difficult to develop and implement methodologies on a project basis, as the proposed methodologies face a high risk of rejection (Ward et al. 2008).

In contrast to the project-based CDM model, the programmatic CDM rules allow the "bundling" of several projects to reduce CDM-related transaction costs in the small individual CDM projects. In this case, project sites need to be close to each other, and projects need to be implemented at the same time. Although project developers welcomed bundling, the number of projects that could be bundled into a small-scale project was limited due to the above conditions. In December 2005, the Kyoto Protocol Conference/Meeting of Parties (COP/MOP) decided to include a "program of activities" (PoA) in the CDM. The PoA concept makes it possible for a large number of interrelated project activities (referred to as "CDM program activities or CPAs") in different locations, even in different countries, to be registered as a single CDM project and implemented in a coordinated fashion. Because a PoA is essentially a program to coordinate individual CDM activities, it is commonly referred as "programmatic CDM."

13.4 Raison d'être of Alternative Approaches to Financing Climate-Resilient Infrastructure

Four different sector-based or economywide carbon finance instruments are discussed in this section.

13.4.1 Sectoral Approach

Unlike a project-based carbon financing tool such as CDM, a sectoral approach allows broad coverage to maximize the efficiency of an emissions trading system; this approach mainly focuses on a limited number of sectors for inclusion in a future climate regime. It may be more practical for many countries to start on a sectoral basis than through a national approach, inasmuch as only a handful of key sectors (e.g., electricity production, land use change, and forestry) account for the majority of emissions in many countries (Bell and Drexhage 2005). This specificity then gives rise to the idea of a sectoral crediting mechanism (SCM), which is based on a sectoral approach and envisaged as expanding the coverage of the CDM from a project-by-project level to a sectorwide level.

The advantages of a sectoral-based approach include ease of administration, data availability, greater equity, increased technology transfer, and targeted emissions reduction, but there would also be some weakness such as doubt on cost effectiveness, limited extent (only a few sectors are targeted), and leakage, mainly to the uncovered sectors. Thus the boundary of sector definition is of particular importance (Schmidt et al. 2008). Indeed, implementing the sector-based approach would imply a negotiation over the baseline level of "unilateral" actions by developing countries and on additional support to implement policies and to facilitate the acquisition of technologies (Baron et al. 2009).

In addition, the sectoral no-lose target (SNLT) approach was proposed by the Centre of Clean Air Policy (CCAP) as a part of sectoral approach to GHG emissions reductions in the highest-emitting developing countries and considered as a key component of the post-2012 climate change mitigation framework. As indicated in the name, no penalties are incurred in case of failing to meet a target, but emissions reductions achieved beyond the target level earn emissions reduction credits (ERCs) that can be sold to industrialized nations (Schmidt et al. 2008). In other words, developing countries would not face compliance penalties if they did not meet their SNLTs. There could be two types of no-lose targets, a hard emissions cap or an intensity-based target.

The subsequent question arises about how to define a no-lose target for the cities in DCs. Participating developing countries establish initial no-lose emissions targets, based upon their national circumstances from sector-specific energy-intensity benchmarks that have been developed by independent experts. Industrialized nations then offer incentives for the developing countries to adopt more stringent emissions targets through a "Technology Finance and Assistance Package." which helps to overcome financial and other barriers to technology transfer and deployment. These sector-specific energy-intensity benchmarks could also serve as a means for establishing national economywide targets in developed countries in the post-2012 regime. The idea is illustrated in Fig. 13.2.

Indeed, in setting the crediting baselines beyond which carbon market financing would apply, it is necessary to understand what such external support is already available for low carbon investment in the sectors (e.g., transport) in question.

The feature that particularly distinguishes SNLTs from CDM-type policy instruments is that once SNLTs are agreed as part of a quantitative multilateral agreement, the concept of additionality no longer applies, nor do any of the CDM institutional constraints that go along with this. This is because in the quantitative elements of the post-2012 agreement the fixed and binding targets of industrialized countries would be set in the light of the scale of credits that would be expected to be generated by SNLTs from developing countries. Therefore SNLTs (and the credits that may stem from them) are explicitly accepted and factored into the elements of the overall agreement that set a quantitative emissions outcome (Ward et al. 2008). SNLT are expected to attract developing countries in sectors for which they seek significantly scaled-up private sector investment according to their sustainable development priorities, and where current carbon market policy tools, such as the CDM, are not considered adequate to the task.

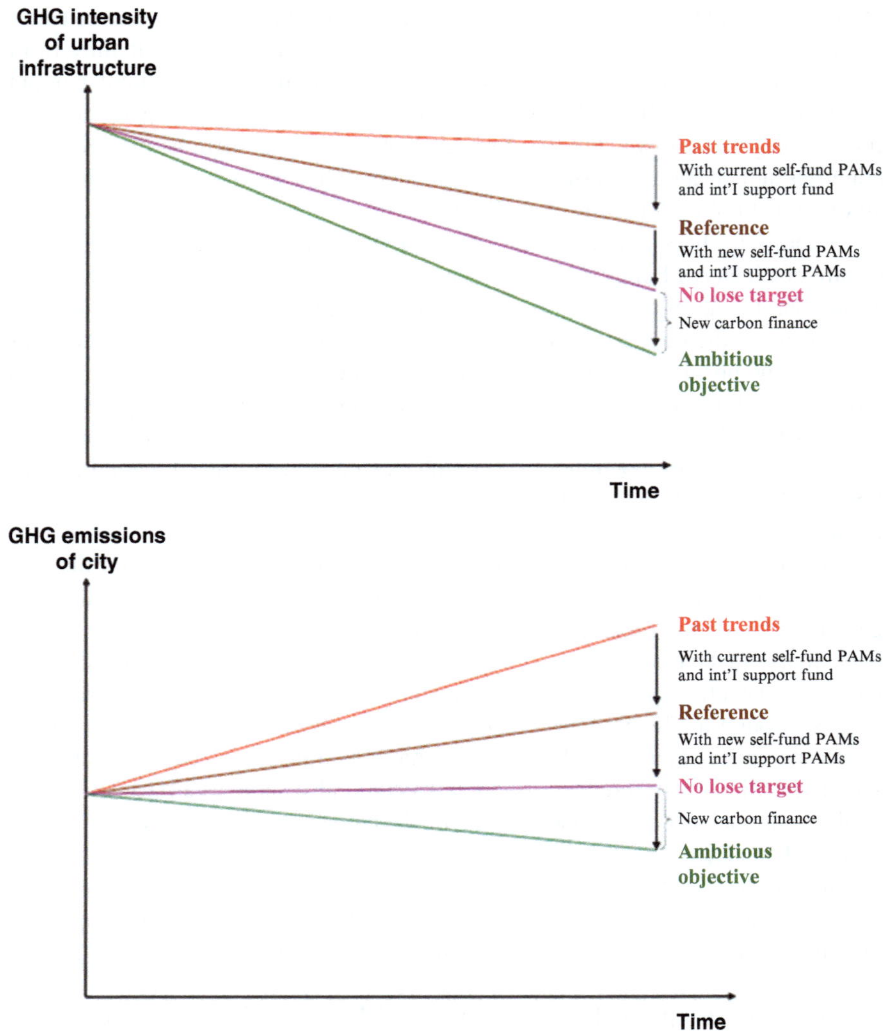

Fig. 13.2 Schematic representation of no-lose target mechanism in cities (*Source*: adapted from (Ward et al. 2008))

13.4.2 Sustainable Development Policies and Measures: SD-PAMs

The concept of sustainable development policies and measures was first introduced in a paper that examined the case of South Africa, where the energy development and housing are considered as important priorities within the national sustainability context (Winkler et al. 2002). The aim of SD-PAMs is to encourage the development of policies that contribute to developing countries' economic and social objectives, with the possibility of lowering GHG emissions at the same time

(Ellis et al. 2007). The question was investigated as to how parties to negotiation can develop a mechanism for formally recognizing and advancing the kinds of SD-PAMs (Baumert and Winkler 2005). Although there could be various definitions of SD-PAMs, it is generally acknowledged that SD-PAMs should be domestically driven and could include a large range of national and/or sectoral policies with a direct impact on GHG emissions such as improving energy efficiency or encouraging re-/afforestation activities. In other words, the SD-PAMs should be defined by individual countries because the priorities and circumstances differ significantly by country (Baumert and Winkler 2005). Typical SD-PAMs are characterized by large-scale PAMs, in contrast to individual projects such as CDM.

The strategic SD approach is to focus on PAMs that are firmly within the national sustainable development priorities of developing countries but which, through the inclusion in an international climate framework, recognize, promote, and support means of meeting these policy priorities on a lower-carbon trajectory. Incentives for SD-PAMs could come from both climate and nonclimate funding (Winkler et al. 2008). There are three different approaches to fitting SD-PAMs into the climate negotiation framework (Ellis et al. 2007).

Additionally, it is critical to know what types of activities can be proposed with regard to eligibility of PAMs. Countries pledge and review their unilateral actions, or commit to a list of eligible activities that are supported internationally. A key point is that it is necessary to know the business-as usual-emission (BAU) pathway to development goals. Some questions must be raised such as how to define and track impacts of PAMs. Is financing required for SD-PAMs support? (If so, it may be necessary to demonstrate the additionality of PAMs). Another open question is how strictly must emissions be MRVed (measured, reported, and verified) and whether greater uncertainties in emissions reductions can be allowed when no crediting incentive is provided. It is likely that technology transfer will be required for undertaking some PAMs. Last, CDM continues with provisions to prevent double-counting.

However, it must be pointed out that SA proposed in the current literature focuses on industrial and energy sectors (Bosi and Ellis 2005; Schmidt et al. 2008; Ward et al. 2008; Baron et al. 2009), and inherent barriers have to be removed to clear the way of scaling up CDM by reforming the current mechanism in developing countries (Ellis and Kamel 2007; UNEP 2008). More specifically, Houdashelt (2008) compared SA with SD-PAMs and concluded that the sector which is characterized by diffuse and small emissions points is more likely to be integrated into the SD-PAMs. In addition, SNLT has attracted much attention and appears to be embraced by major emerging economies, but concentrated in heavy industry and power sector (Galharret et al. 2008; UNEP 2009).

13.4.3 Nationally Appropriate Mitigation Action: NAMA

The concept of nationally appropriate mitigation actions (NAMA) was proposed in the Bali Action Plan (BAP) which included sectoral agreements as an option for achieving measurable, reportable, and verifiable actions to reduce emissions

(Ellis and Larsen 2008). NAMA is based on the consensus that all countries are not in the same situation, therefore differentiation might be one of the clues to reaching a successful agreement. BAP differentiates "quantified emissions limitation or reduction by developed countries" from NAMAs by developing countries (Payá 2008). BAP indicated that the NAMAs can be adopted by developing country parties in the context of sustainable development, supported and enabled by technology, financing, and capacity-building, in a measurable, reportable, and verifiable manner.

The NAMA should be designed as policies, legal requirements, and measures that integrate climate consideration within specific national sustainable development policies. A key point is that actions shall not include explicit emissions reduction objectives. However, emission reduction by improving cities' infrastructure management in buildings and transport should come as a consequence of implementing these sound development policies and measures[3] (UNFCCC 2008; Payá 2008).

However, there exists great uncertainty of the NAMA implementation; thus far, the major emerging countries such as China and India are fiercely against crediting NAMA actions in the domestic climate mitigation and adaptation programs.

13.5 Perspective of Financing Low-Carbon Development in Cities

13.5.1 Relevance of Applying Sectoral Non-lose Targets and SD-PAM in Cities

The discussions in previous sections suggest that a sector-based or citywide and bundling approach needs to be adopted for scaling up the financing tools in developing sustainable urban infrastructure in compliance with the climate policies. Notably, most literature indicates that the buildings and transport are most likely to be included in the SD-PAMs (Winkler et al. 2002; Baumert and Winkler 2005; Ellis et al. 2007; Bradley et al. 2007). A variety of technical and management options relating to energy-efficient buildings and transport and optimal land use planning may be included in the city-based PAMs or SNLT, irrespective of the nature of awardable or not of emissions credits. However, the resultant GHG mitigation from the taken efforts should be measurable and verifiable. Municipal monitoring and reporting agencies need to be established to fulfill the task of GHG inventory and registry based on the appropriate methodology available. For example, a motor emission intensity target (gCO_2/p-km or gCO_2/v-km) on the basis of rate of progress (percentage per year) or ultimate level could be introduced into the transport sector.

[3] By comparison, the MRVable nationally appropriate mitigation commitments or actions, including quantified emission limitation and reduction objectives, by all developed country parties, although ensuring the comparability of efforts among them, take into account differences in their national circumstances.

The citywide SNLT or SD-PAMs would have a very important role to play in shaping sustainable and low carbon urban development trajectory, however, the schemes and institutional aspects must be designed carefully to ensure the economic performance and environmental integrity in cities. The sustainable urban infrastructure development policies must be incorporated into a national development plan and public policy framework, for example, city PAMs or long-term building efficiency target with government commitment to improve the accountability and eligibility as well as lowering the excessive transaction costs.

The most important point is that carbon finance should not be devised simply as an incentive for local authority to withhold the energy efficiency policies in a target sector in the manner of rent seeking. Rather, with a clearly specified efficiency and emission target, local authority should leverage international carbon finance with pledged actions in the urban sectors by committing to improving efficiency or reducing carbon intensity by a certain percentage for the next years. These pledged actions will be monitored and evaluated by independent bodies before and after the credit is issued. Likewise, the performance in terms of carbon emissions reduction is subject to third party verification and certification. Carbon credits are ideally used to cover the incremental costs of implementation of new public policies with co-benefits of additional GHG and other local pollutant mitigation, consistent with the targets of the PAMs.

There is some debate on whether the actions taken in the name of the SD-PAM scheme should be awarded with reduction credits. Indeed, the actual implementation of sustainable buildings policies in DCs may not be necessarily climate-oriented but economic growth and environmental pollution alleviation-driven, nevertheless, these actions will make de facto and measurable contribution to GHG mitigation and should be recognized and supported by Annex I countries through financial assistance and technology transfer.

13.5.2 Relevance of Integrating Cities into NAMA

There is evidence that NAMA is a policy-based approach instead of a project-based mechanism like CDM. As developing countries consider NAMAs, a sectoral approach could be an attractive option for them to consider as a means to contribute to overall global emission reductions. In this regard, the cities' SD-PAMs may be factored into the NAMA framework. That means a NAMA incorporating urban (buildings and transport) development management PAMs might be envisioned provided that appropriate institutions will be put in place concomitantly. Energy-efficient buildings, renewable energy-integrated buildings, urban planning, high-efficient and low-carbon public transport, and integrated land use management in cities can be integrated into a comprehensive SP-PAMs package under the auspices of NAMA in DCs.

Translating NAMA into cities' development strategies will require the local government to define the right guidelines or directions in the context of sustainable urban development, supported and enabled by technology, financing, and capacity building in a measurable, reportable, and verifiable manner. Consequently, the key questions are what a city-based NAMA may look like. How should the NAMAs be supported within the international framework in terms of legal instruments and technology transfer? And what components should be included in a city's NAMAs? For example, a city-based NAMA may include various policies and measures for sustainable buildings and highly efficient public transit-oriented transport systems as well as a renewable energy deployment strategy. The ease of data collection and administration must be taken into consideration. The local government will need to prepare a roadmap of energy efficiency in buildings and transport planning such as comprehensive public transport infrastructure and fuel efficiency improvement in cities for the next decade.

13.5.3 Harnessing Climate Finance in Urban Infrastructure in Developing Cities

In the case where cities commit to devising a citywide SD-PAMs or SNLT scheme that falls into the category of the national NAMA scheme, local governments will need to prepare relevant sectoral targets consistent with the socioeconomic development goal and negotiate appropriate baselines with the COP and international partners and financiers. Relevant government bodies and institutions shall be involved in the PAMs based on science-based targets as well as a robust baseline definition. Supportive legal and institutional framework will also play an overarching role to allow the cities to implement the climate friendly urban development PAMs smoothly and effectively.

The processes of MRV will also require the involvement of both domestic and international institutions that will be accredited by UNFCCC and other qualified organizations. The quality of metrics of emissions reductions in the policies and measures carried out in the cities will have a significant impact on the outlook of climate finance mechanisms and the credibility of city governments with respect to commitments to sustainable urban development. Therefore cities must adhere to international rules in carrying out the programs under the framework of SD-PAMs, SNLT, and NAMA. Cities will learn to prepare the policy-based development program in an ascending order; the costs of policy implementations (transactions and learning costs) will decrease as a result of the learning-by-doing process. Institutional capacity building in developing cities could follow the continuum concept (Ward et al. 2008): the difference in scales of climate mitigation actions can be ordered in a continuum, project, program, sector-specific policies, cross-sectoral policies, and economywide actions.

Clearly, establishing a sectoral agreement will necessarily require international coordination and national management for any form of sectoral approach (Bosi and Ellis 2005). It would be even more complicated if the cities were playing a central role in baseline negotiations and carbon credits management. Regular reporting on the sustainable development and climate benefits of SD-PAMs could take place through national communications or a separate reporting mechanism. This will necessarily require harmonizing domestic GHG emissions inventory with international standards; qualified monitoring, verification, and reporting institutions will need to be established to make the quantification of SD-PAMs transparent and effective (Ward et al. 2008; Winkler et al. 2008).

The international community will need to agree on general guidelines for what constitutes an SD-PAM that is eligible to be pledged under the UNFCCC. The pledged SD-PAMs could then be recorded and tackled by the Convention Secretariat or other bodies (Baumert and Winkler 2005) and a broader assessing program could also incorporate the quantitative dimension of PAMs. The national or World Bank-related processes have been proposed (Ellis et al. 2007) to identify development priorities and could also help identify and prepare national SD-PAMs. Monitoring and evaluation processes could be similar to that of CDM. Implementation of SD-PAMs may be allowed under the UNFCCC framework. Local authority's technical and institutional capacity needs to be enhanced to identify key sectors and design and implement appropriate PAMs in cities' urban infrastructure consistent with NAMA and long-term socioeconomic development objectives.

A clear and consistent price signal is a fundamental factor in maintaining the effective functioning of global market. The IPCC's 2007 Assessment Report 4 (IPCC 2007) estimates that approximately 6.5–8.5 Gteq CO_2 can be reduced in the building and transport sector by 2030, representing 15–20 % of global energy-related carbon emissions under the BAU scenario of the IEA (IEA 2007). Therefore, a relatively high carbon price will give incentives for developing cities to commit to sustainable urbanization and economic development by adopting low GHG-intensive technology and practices. The practice of a mandatory price floor for CDM projects in some countries (e.g., in China, the CER transaction price in CDM projects should not be less than 9.5 US$/ton CO_2eq, as set by the NDRC) is a good example for maintaining the carbon price at a level consistent with climate policy targets. In this respect, the compulsory emission reduction targets in Annex 1 countries will need to be toughened to guarantee the balance between supply of and demand for carbon credits. The implementation of the future EU climate-energy package and further commitments in other developed countries will admittedly drive the demand for carbon emissions allowances, also, the probable policies of limiting CER export in China, India, and Brazil would drive prices to rise and give incentives for developed countries to go beyond the BIC countries to seek credits from the less-developed world such as Africa (this would definitely help channel investment flow into sustainable development projects in African countries), because the long-term commitment and targets will be fundamental to providing stakeholders' price signals and to a stabilized global carbon market. The futures of supply of and demand for carbon emissions allowances will necessarily rise steadily over the next decades accordingly if a meaningful climate regime can be established with relevant policies in cities in the forthcoming negotiation.

13.6 Conclusion

Cities in DCs need to change their pathway to development and governance to anticipate the future constraints of energy supplies and carbon emissions. A good governance and sound institutional infrastructure are of key importance in shaping the resilient trajectory vis-à-vis increasing probabilities of climate change-related environmental hazards and social unrest. An integrated approach must be adopted to designate cities as eligible entities in the ongoing international climate negotiations; this will help scale up carbon finance to accelerate changing pathways more effectively.

Various aspects relating to design/implementation and institutions of financial mechanisms for GHG mitigations in developing cities need to be taken into account with care to ensure sustainable development and environmental integrity. Our analysis suggests that the SD approach is preferred to CDM and industry-based sectoral approach to scaling up climate finance in cities. Notwithstanding, the latter can be well integrated into the cities' PAMs on commitment in urban infrastructure management. There must be negotiations on the respective efforts and commitment both in developed and developing countries in the forthcoming COP to give developing cities the necessary legitimacy and capability to participate in the SD-PAMs and NAMA packages' design and policy negotiation for a promising post-2012 climate regime.

References

Allix G (2009) Les atouts des villes dans la lutte contre la pollution (The cities' advantages in combating the pollution). Le Monde, 4 Apr 2009

Baron R, Buchner B, Ellis J (2009) Sectoral approaches and the carbon market. OECD/IEA information paper, Paris

Baumert K, Winkler H (2005) Chapter 2: SD-PAMs and international climate agreements. In: Bradley R, Baumert K, Pershing J (eds) Growing in the greenhouse: protecting the climate by putting development first. World Resources Institute, Washington DC

Bell W, Drexhage J (2005) Climate change and the international carbon market. International Institute for Sustainable Development, Manitoba, Canada

Bosi M, Ellis J (2005) Exploring options for "sectoral crediting mechanisms". IEA/OECD information paper, Paris

Bradley R, Baumert K, Childs B, Herzog T, Pershing J (2007) Slicing the pie: sector-based approaches to international climate agreements. World Resources Institute, Washington, DC

Capoor K, Ambrosi P (2009) State and trends of the carbon market. World Bank, Washington, DC

Cheng C et al (2008) The Kyoto protocol, clean development mechanism, the building and construction sector. UNEP SBCI, Paris

Ellis J, Kamel S (2007) Overcoming barriers to clean development mechanism projects. OECD, Paris

Ellis J, Larsen K (2008) Measurement, reporting and verification of mitigation actions and commitments. OECD/IEA information paper

Ellis J, Baron R, Buchner B (2007) SD-PAMs: what, where, when and how?. OECD/IEA information paper

Fenhann F (2009) UNEP Risø Centre, CDM pipeline, cd4cdm
Galharret S, Colombier M, Wemaere M, Guerin E (2008) Strawman study "No lose target/Sectoral crediting mechanism" for the power sector. Paper prepared for OECD
Gomez-Echeverri L, Müller B (2009) The financial mechanism of the UNFCCC: a brief history. European Capacity Building Initiative, Oxford
Houdashelt M (2007) Comparison of the sectoral approach and SD-PAMs. Powerpoint presentation at CCAP future actions dialogue, London
IEA (2007) World energy outlook 2007: China and India insights. OECD/IEA, Paris
IEA (2009) World energy outlook 2009. OECD/IEA, Paris
IPCC (2007) Summary for policymakers. In: Metz B, Davidson OR, Bosch PR, Dave R, Meyer LA (eds) Climate change 2007: mitigation. Contribution of Working Group III to the Fourth Assessment Report of the Intergovernmental Panel on Climate Change. Cambridge University Press, Cambridge/New York
Lütken S (2010) A grand Chinese climate scheme, Dtu climate paper series. Nature. Developing nations tackle climate. By J Tollefson Vol 459, July, pp 158–159
Neuhoff K, Fankauser S, Guerin E, Hourcade J, Jackson H, Rajan R, Ward J (2009) Structuring international financial support to support domestic climate change mitigation in developing countries. Climate strategies International Support for Domestic Action (ISDA) project, London
OECD (2009) Green cities: new approaches to confronting climate change. Debriefing paper at the Las Palmas workshop, June 2009
Payá H (2008) International climate policy and developing countries' national development policies: a necessary link for a successful international agreement. Powerpoint presentation in "climate law in developing countries post-2012: North and south perspectives", September, 2008
Raupach et al (2007) Global and regional drivers of accelerating CO_2 emissions. PNAS 104(24): 10288–10293
Sanchez S (2008) Reforming CDM and scaling-up. Finance for sustainable urban transport. In: Olsen KH, Fenhann J (eds) A reformed CDM – including new mechanisms for sustain able development, Perspectives series 2008. UNEP Risoe Center, Roskilde, pp 111–126
Schmidt J, Helme N, Lee J, Houdashelt M (2008) Sector-based approach to the post-2012 climate change policy architecture. Clim Policy 8:494–515
UN (2007) World urbanization prospect – the 2007 revision. UN, New York
UNEP (2007) Buildings and climate change: status, challenges and opportunities. Sustainable Buildings and Construction Initiative (SBCI), Paris
UNEP (2008). In: Olsen K, Fenhann J (ed) A reformed CDM-including new mechanisms for sustainable development. UNEP- *Risø* centre CD4CDM. Perspective series 2008
UNEP (2009) Industry sectoral approaches and climate action: from global to local level in a post-2012 climate framework. Rev Res Debates Pos
UNFCCC (2008) Investment and financial flows to address climate change: an update. United Nations, Bonn
UN-Habitat (2008) Cities and climate change adaptation. UN-habitat donors meeting, Seville, 15–16 Oct 2008
Ward M et al (2008) Sector no-lose targets: a new scaling-up mechanism for developing countries. In: Olsen K, Fenhann J (ed) A reformed CDM-including new mechanisms for sustainable development. UNEP- *Risø* centre
Wemaere M (2009) Key questions on "NAMAs" and MRV related issues. Presentation at informal high level conversations on climate change conference, Washington, DC
Winkler H, Spalding-Fecher R, Mwakasonda S, Davidson O (2002) Sustainable development policies and measures:starting from development to tackle climate change. In: Baumert K, Blanchard O, Llosa S, Perkaus JF (eds) Building on the Kyoto protocol: options for protecting the climate. World Resources Institute, Washington, DC, pp 61–87
Winkler H, Höhnen N, den Elzen M (2008) Methods for quantifying the benefits of sustainable development policies and measures (SD-PAMs). Clim Policy 8:119–134
World Bank (2009) Little Green Data Book 2009, Washington, DC.

Chapter 14
The Triple A Issue: Agriculture, Alimentation Needs, Agrofuels

Jean-Marie Loiseaux

Abstract The triple A issue is discussed with, as its starting point, the analysis of present agricultural output and its projection on to 2050, so as to place it in the context of the major evolutions required between now and 2050 with growing populations, energy needs, alimentation needs, and agricultural production, the unprecedented growth of carbon-free energy needs in a context where emerging countries enjoy strong economic growth, thus increasing the need for energy by a factor of two at the world level. The environmental impact of large-scale farming and its capacity to produce energy are reviewed quantitatively on a world scale. A rational appreciation of the uncertainties and of the risks concerning our future ability to satisfy human alimentation needs is put forward. From this analysis, we conclude that the concerns expressed by some are not entirely unfounded but that the problem of human alimentation today is not due to insufficient foodstuff output on a world scale. Moreover, it appears that there is the potential for significantly increased output from large-scale farming. The risk of competition between foodstuff production and that of agrofuels is not as imminent as has been said, for simple economical reasons and also because of the location of production. The impact of large-scale farming on greenhouse gas emissions or on the world energy supply is evaluated and appears small, or very small relative to overall needs and emissions. Still, the principal cause for concern about alimentation needs, could well be that states or world governance bodies abandon their obligation to ensure that populations' alimentation needs are covered decently, a mandatory ethical priority.

Keywords Agrofuels • Agriculture • Alimentation • Greenhouse gas • Prices

---

J.-M. Loiseaux (✉)
Laboratoire de Physique Subatomique et de Cosmologie, Université Joseph Fourier Grenoble 1, 53 rue des martyrs, Grenoble, Cedex 1, France 38026

PIE – CNRS, 1 rue Michel-Ange, Paris, France 75016
e-mail: jeanmarie.loiseaux@numericable.fr

J.B. Saulnier and M.D. Varella (eds.), *Global Change, Energy Issues and Regulation Policies*, Integrated Science & Technology Program 2, DOI 10.1007/978-94-007-6661-7_14,

List of Acronyms

AAA	Agriculture, alimentation needs, agrofuels
ADEME	Agence de l'environnement et de la maîtrise de l'énergie (France), French Environment and Energy Management Agency
A-F	Agrofuel
CNRS	Centre National de la Recherche Scientifique (France), National Center for Scientific Research
d.m.	Dry matter
DDGS	Distiller's dried grain soluble
DIREM	Direction des resources énergétiques et minérales (France), French Direction for the Energetic and Mineral Resources
EC-DDGS	Energy content of DDGS
EC-ethanol	Energy content of ethanol
EC-straw	Energy content of straw
FAO	Food and Agriculture Organization (United Nations)
G$	Gigadollar = 10^9 dollars
G20	The Group of Twenty Finance Ministers and Central Bank Governors
GDP	Gross domestic product
GHG	Greenhouse gas
GT3A	Groupe de Travail sur les AAA (CNRS), Working Group on AAA
GteC; 1 GteC = 10^9 teqC	Gigaton of Carbon equivalent
Gtoe	Gigaton of oil equivalent; 1 Gtoe = 10^9 toe = 1 billion toe
INRA	Institut National de la Recherche Agronomique (France), (French National Institute for Agronomic Research)
J	Joule (SI unit of Energy)
Mha	Million of hectare = 10^6 ha
MJ	MegaJoule = 10^9 J
Mtoe	Megaton of oil equivalent[1]; Mtoe = 10^6 toe = 1 million toe
SI	International System of Units
teqC	Ton of carbon equivalent
teqCO_2	Ton of CO_2 equivalent; 1teqCO2 = 0,273 teqC
toe	Ton of oil equivalent
toe-etha	Quantity of ethanol equivalent to 1 toe
UEC	Usable energy coefficient
UEC(byprod.)	UEC for by-products
UEC(etha)	UEC for ethanol

14.1 Introduction

After the initial enthusiasm aroused by (first-generation) agrofuels (A-F), which were to be a viable energy resource, doubts arose concerning the real asset they might represent concerning the mitigation of greenhouse gas (GHG) emissions and the amount of energy gain likely to be reached. The discussion turned to controversy by 2007 with the significant price increase of both agricultural products and petroleum as it brought to light the potential competition of agrofuels with food products. This controversy is still ongoing, the situation is confused, and a certain degree of defeatism has settled in, not only in public opinion but also among scientists, all the more so inasmuch as several countries that had clearly chosen to develop agrofuels in the frame of the mitigation of their GHG emissions and in view of acquiring a higher degree of autonomy with regard to their energy supply have revised their commitments.

Because the information on the subject is scattered and its use complex, it is not so easy to form an opinion, as we have verified for ourselves. In particular, the hypotheses retained in the evaluation of energy gains and the reductions of GHG emissions are mostly incompletely documented if not partial (in both senses of the term). At any rate, these evaluations are not convincing; they are done without an explicit methodology so that their logic is not clear. Similarly, general discussion on the "agrofuel versus foodstuff" issue is, more often than not, ideologically oriented. Finally, the issue of hunger in the world has an ethical dimension that dramatizes the stakes.

The alimentation for humanity question arises in a world that will have to change drastically between now and 2050 and a very constraining context cannot be ignored:

- Mandatory and severe reduction of worldwide GHG emissions: Emissions must be divided by a factor of two on a worldwide basis and, on average, by a factor of four in industrialized countries in order to mitigate global warming and avoid climate change or disturbance that could threaten the survival of a fraction of humanity.
- Strong world population growth, from 6.5 billion in 2010 to an expected 9 billion in 2050: Implying (also taking into account the fact that the lifestyle in emerging and developing countries will converge towards that of industrialized countries) the doubling of worldwide energy needs, that is, 20 Gtoe (billion tons oil equivalent) in 2050 and the quasi-doubling of humanity's alimentation needs.
- Drastic change in the sources of energy to meet the requirements just stated because although, in 2050, 20 Gtoe primary energy will have to be provided, only about 4 Gtoe of CO_2 emitting fossil fuels will be allowed. This leaves 16 Gtoe of non-CO_2 emitting energy that will have to be provided by 2050, eight times more than today.
- Trade globalization seems irreversible, implying strong dependence on transportation which, at least for air travel and the transport of goods, translates

into a strong and lasting reliance on fluid fuels (oil, gas, agrofuels, and any other substitution fuels). In addition, large economic entities are concerned with establishing their energy autonomy, aware as they are that oil and gas reserves are not infinite and these fuels could well be more difficult to acquire than in the past.

- After these considerations on the context in which the triple A issue is posed, it is fitting to review briefly why agriculture is such a sensitive subject. Agriculture covers very large areas, spreads large amounts of fertilizers and phytosanitary products that, although beneficial to the crops, are potential environmental pollutants and whose production consumes energy. The large areas occupied by agriculture do not generate, and may even jeopardize, biodiversity so that agriculture is often perceived as one of the causes of the reduction of biodiversity that conditions adaptability of life on earth.

Agriculture's primordial functions are also seen as being to feed the world, to do away with famine, and to produce good quality foods at affordable prices. This viewpoint is ethical in character; it is not an objective consideration that proceeds from the laws of nature.

Finally, agriculture is, by its very nature, a source of energy, through the massive use of products stemming from photosynthesis and, in all rationality, it is logical to evaluate its potential, notably in terms of fuel production.

The last item of this introduction is important and should be kept in mind, namely the economic weight of large-scale farming compared to that of the big changes mentioned above: reduction by a factor of two of worldwide GHG emissions and development of a new energy mix and the doubling of energy needs for alimentation by 2050. Indeed, the total value produced by large-scale farming is equivalent to a little less than 1 % of world gross domestic product (GDP) value and the primary energy consumed worldwide is equivalent to 7 % of world GDP (63,000 G$ in 2010). Thus, although agriculture seems to be a major issue for the rethinking of a new world, its economic impact is, in fact, particularly small.

In this context, and in order to best clarify the matter, the Programme Interdisciplinaire Energie (Interdisciplinary Energy Program) of the CNRS (France) has assigned to a dedicated working group (GT3A) the task of answering the following question:

> How pertinent is it to build up agrofuel production in terms of a partial solution to energy self-sufficiency, in terms of the reduction of GHG emissions, in terms of its impact on meeting humanity's alimentation needs?

We have chosen here, to discuss the subject in three parts:

In Part 1, we describe the production from large-scale farming on a worldwide scale, its past evolution, its specifics and contingencies, its uses, its relative importance, and its constraints up to 2050.

In Part 2, we address, quantitatively and again on a worldwide basis, the energy aspects of agricultural production and the main impacts of agriculture on the environment (GHG).

In Part 3, we use the data and discussions of Parts 1 and 2 to answer a limited number of questions which we detail below.

Question A: Can agriculture decently feed nine billion human beings in 2050?
Question B: Can agriculture (agrofood and/or agrofuel) play a significant role in the reduction by a factor of two of GHG emissions and in the supply of liquid fuels for transportation and nonstationary mechanization by 2050?
Question C: Would competition between alimentation and agrofuel production be threatening?
Question D: Hunger in the world; how extensive is it, what determinants, what remedies?
Question E: Is the development of agricultural production compatible with the conservation of biodiversity and the environment?

The conclusion comes back to the original question:

"How pertinent is it to build up agrofuel production in terms of a partial solution to energy self-sufficiency, in terms of the reduction of GHG emissions, in terms of its impact on meeting humanity's alimentation needs?" and it addresses the need for worldwide governance to ensure that the world is provided with food today and on into the 2050s.

14.2 Part 1: General Characteristics of World Agricultural Production

14.2.1 Large-Scale Farming Production—Some Quantitative Data

Our aim in this first part is to gather general data pertaining to worldwide agricultural production (Cereals International Council 2008; FAO 2009) and to give some indicators on their merchant values (dollar) and energy content (Mtoe) to be compared to fossil energies.

From Table 14.1 we can derive some observations as follows.

- The value of annual production from large-scale farming amounts to only 12 % of the estimated value of fossil energies and less than 1 % of world GDP.
- The energy content of crops from large-scale farming is on the order of 1 Gtoe, or 10 % that of fossil energies but the fact that these crops can be transformed into fuel explains the fear of possible competition with food production.
- The quantities produced worldwide in kg per capita and the annual cost of these mean amounts shown in the last column demonstrate the physical and economic importance of fossil energies as compared to food production.

Table 14.1 Annual worldwide agriculture production

Product	Mt/year	Cultivated area (Mha)	Mean prod./ha (ton)	Max prod./ mean prod./ha	Value of 1 year production (G$)	Energetic value of 1 year production (Mtoe)	Annual production (kg/capita) and value (dollar per capita)
Cereals							
Rice	600	153	3.92	2	**355 G$** (150 $/t)	800 Mtoe (3 t cereal per toe)	92
Wheat	688	222	3.1	3			106
Corn (maize)	780	145	5.38	2			120
Other	300	167	1.8	3			46
Total cereal	**2,368**	**687**	**3.45**	**2.5**			**364** (55$)
Soya oil	34	92	0.37		**52.5 G$** (500$/t)	90 Mtoe (1.1 t oil/toe)	5
Colza oil	16	26	0.62				2
Sunflower oil	10	21	0.48				2
Palm oil	45	12	3.75				7
Total oil	**105**	**151**	**0.69**	**1.5**			**16** (8$)
Cane sugar	120	20	6		**48.6 G$** (300$/t)	65 Mtoe (2.5 t sugar/toe)	18
Beet sugar	42	7	6				6
Total sugar	**162**	**27**	**6**	**1.5**			**25** (7.5$)
Cereal + oil + sugar	**2,635**	**865**			**456 G$**	**950 Mtoe**	**405 (70$)**
Fossil energies		**9 Gtoe at 420$/toe on average over the mix coal, oil, gas**			**3,780 G$**	**9,000 Mtoe**	**1,385** (580$)

14.2.2 Characteristics Specific to Agricultural Production

14.2.2.1 Fluctuations and Flexibility

Weather hazards during the growth of crops and at harvest time cause significant fluctuations in the amounts produced worldwide. On the contrary, the demand for agricultural products, whether for food or for feed, is quite inelastic. Thanks to world stocks, major food shortages can be avoided but the stocks do not seem to have much impact on market price fluctuations. Another factor in the lack of elasticity is that, in most cases and by nature, there is only one harvest per year. It is not rare to observe the occurrence of several years' underproduction following a few years' overproduction, because of crop rotation requirements or a change in mechanized equipment specific to certain farming practices. In short, agricultural production is an economic activity where the combination of fluctuating output, inertia of the response to inelastic demand, and the urgency of some demands, generates systemic instabilities.

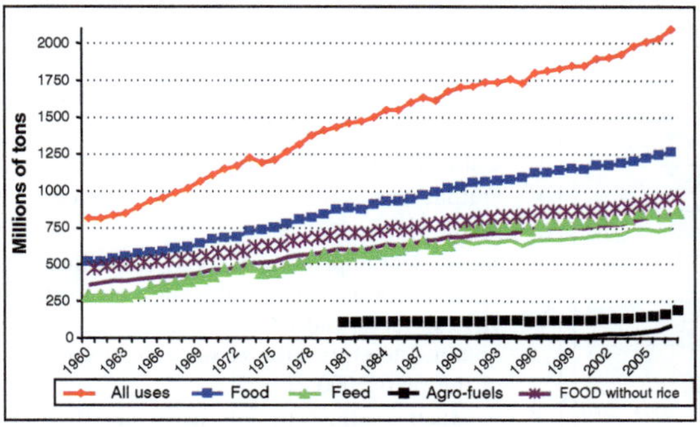

Comments

The distribution of uses has not changed significantly except for the slow takeoff of agro-fuel use, estimated in 2008 at 100 Mt or 4 % of total cereal production.

*First **hypothesis H1**: we will assume that the ratio of FOOD to FEED will not vary between now and 2050*

Fig. 14.1 Evolution of cereal consumption for humans (Food) and animals (Feed). Feed: 37 % of all cereal consumption or 45 % if rice is excluded (Dronne and Forslund 2009, with permission)

14.2.2.2 Uses of Agricultural Products

Cereals such as rice and wheat are used mainly as food for humans whereas corn and other cereals (besides rice and wheat) are mostly used as animal feed. Globally, approximately 58 % of cereals are dedicated to direct human consumption, 37 % serve as food for humans via animal feed and approximately 4 % of the total cereal production is used to produce agrofuels in the form of ethanol. Finally, about 15 % of sugar and oil production serve to produce agrofuels.

In Fig. 14.1 the evolution of cereal consumption for human and animal nourishment (feed) over almost a half-century is displayed. During this time period, the relative share of human and animal consumption varies little. The share of agrofuels is insignificant in spite of a slight takeoff over the last five years, with no impact to speak of on global foodstuff output.

14.2.2.3 Evolution of Agricultural Output Since 1978 and Projection to 2050

The observations in the box of Fig. 14.2 can be related to the survey of Table 14.1 where mean yields worldwide are often two to three times less than the best yields commonly found in countries that practice intensive farming. This tends to show that a significant margin remains for substantial world mean yield increase (Dronne and Forslund 2009).

The above points to the hypothesis that average yields per unit surface can still increase globally, at least for cereals. We assume a growth of mean yields on the order of 1.4 by 2050, thanks to improved seeds, better soil preparation, fertilizers, and the protection of crops and harvest from parasites.

We should also note the recent evolution in the food consumed by populations in emerging countries where the per capita consumption of meat is rapidly growing. In the short and mean term, the demand for meat foods should continue to grow.

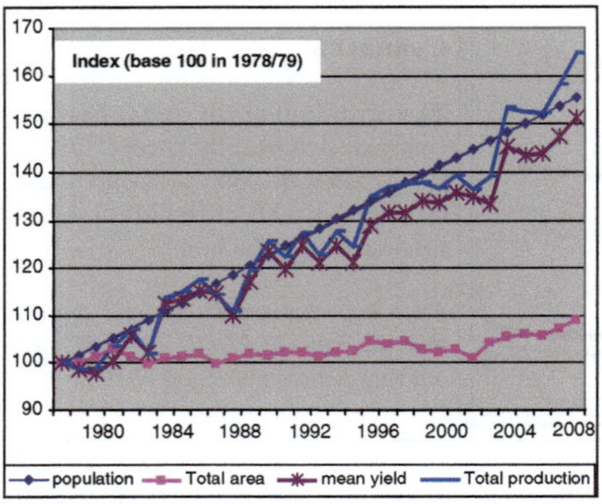

Comments

- ✓ Production increases by a factor 1.65 while population increases by a factor 1.56.
- ✓ Cultivated areas increase by 10% only.
- ✓ Output increases mainly because of rising mean yield,
- ✓ Mean yield improvement today is obtained mainly with increasing cultivation of good yield surfaces. Indeed, maximal outputs obtained tend to stabilize since the beginning of the years 2000,
- ✓ **Production adjusts to (solvent) demand!**

Fig. 14.2 Comparative evolution over 30 years of population size, area farmed, total cereal output, and output per unit area (Dronne and Forslund 2009, with permission)

Fig. 14.3 Wheat prices in \$1998/t since 1947 (Mazoyer 2008, with permission)

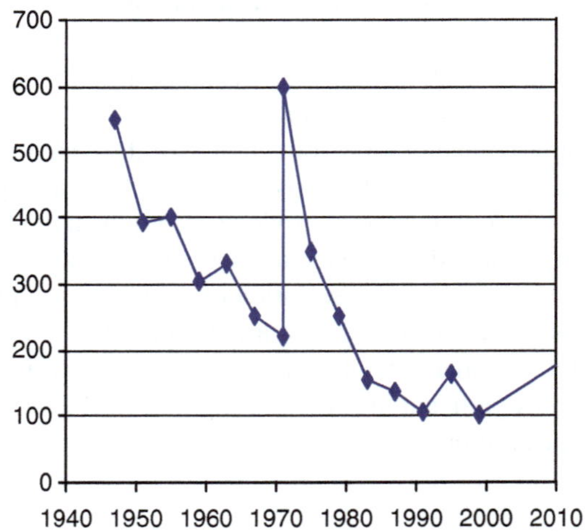

14.2.3 Agricultural Production: Economic Aspect

Let us comment now on the price evolution and fluctuations of agricultural production (Mazoyer 2008) (Fig. 14.3).

Primarily, we should recall that only 15 % of world agricultural production participates in international transactions, that domestic prices in large producing countries are more or less supported by subsidies. The international market is sometimes considered to be a surplus market.

The yellow curve in Fig. 14.4 shows significant variations in the price of wheat in current dollars with a rather strong correlation with the price of petroleum. But the fact that agriculture makes use of petroleum and gas for its mechanization and for the production of its input (fertilizers in particular) does not seem to account for

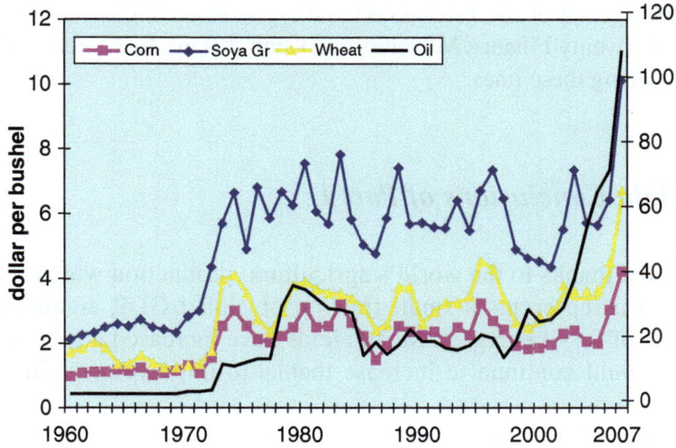

Fig. 14.4 Evolutions of prices in current $ from 1961 to 2007 (Dronne and Forslund 2009, with permission)

this strong coupling. Aside from that, if rapid wheat price fluctuations are linked to fluctuations in production, they are intensified by speculation.

14.2.3.1 Constant Dollar Cereal Price Decrease

The evolution of wheat prices in constant dollars, is interesting, as shown in the left graph of Fig. 14.4, covering the time period from 1950 to 2009 (Mazoyer 2008). The ton of wheat starts at $400 in 1950, it decreases to $130 (1998 $) in 2000; a price that is divided by 3 within 50 years and by 2 in the 30 years between 1980 and 2009.

In the large producing countries, this price decrease is accompanied by, or is the consequence of, a dramatic increase of the per-farm-worker productivity. Today, per-farm-worker productivity is evaluated at approximately 2,000 tons of wheat per year in well-mechanized farms, whereas per-farm-worker productivity in poorer countries without mechanization remains at approximately 1 ton per year. This productivity spread, 2,000 today, was 10 one century ago.

The consequence is that a farm worker's daily income in a poor country, excluding special assistance, is less than $1/day. Considering the price of wheat on the international market and with such low productivity in poor countries, this activity can no longer be profitable and there is a tendency to abandon it, or to replace it with nonfood producing crops. A vicious circle sets in then, inasmuch as the production of food-crops stagnates or even decreases whereby the countries where this happens rely increasingly on imported foods. In order to escape from this dangerous spiral, international aid should be established with the aim of reasonably mechanizing farming practices and improving productivity thanks to fertilizers, other inputs, and better-adapted seeds, helping these countries acquire sustainable food self-sufficiency.

That is on the agenda of the FAO (Food and Agriculture Organization) and G-20 (The Group of Twenty Finance Ministers and Central Bank Governors). They seem to want to act along these lines.

14.2.4 Main Conclusions of Part 1

Humanity is fed thanks to the world's agricultural production which, as such, is important, but it represents a small fraction of global GDP, significantly less than the share of fossil energies. Mean yields have increased a great deal since 1970. They should continue to increase thanks to improvements in the lower yield areas.

Uses of agricultural products

- The competition of direct human food production with that of animal feed or of agrofuels is not presently threatening.
- It could well be, however, that the proportions will have to evolve before 2050, as discussed in Part 3. Today, agriculture is rather in a potential overproduction phase.

Commercialization, trade prices, stocks, price instabilities, price decline

Economic conditions for agricultural products have changed considerably in the past 40–50 years. In constant dollars, prices have declined roughly by a factor of three. In large-scale farming countries, this decline has been compensated by a dramatic increase in per-farm-worker productivity, in particular thanks to mechanization. On the contrary, for farm workers in poor countries, this productivity increase did not happen and the price decline has pauperized that agriculture.

The market prices of agricultural products from large-scale farming fluctuate a lot for many reasons, most of which cannot be controlled in a free market. Taking into account the intrinsic characteristic of agricultural production and the present conditions of their commercialization, market prices cannot but fluctuate. Market regulation as well as the management of stocks by an international authority would be highly beneficial to both importing and exporting countries.

14.3 Part 2: Farming Energy and Environmental Impacts for Food, Feed, and Agrofuel Production

14.3.1 Energy Associated with Farming: A Re-Examination

In the framework of a renewed approach, we propose to examine the energy aspects of farming in both of its main destinations: foodstuff production for humans whether food or animal feed, and agrofuel production.

The energy context (limited liquid fossil fuel resources) and the necessary reduction of GHG emissions (to avoid climate disruption), as well as the need for increased agricultural output previously discussed, require clarity in the examination of this issue in a framework such that the data and calculations can be easily completed and the mistakes or approximations, inevitable in this sort of evaluation, be easily identified.

14.3.1.1 Overview of the Energy Issue: Data and Methods

Data:

Here, we consider three crops only: wheat, sugar beet, and colza which are relatively representative of Generation 1 agrofuel.

For foodstuff production, the crop comprises:

- For wheat: the grain and the straw
- For sugar beet: the sugar, the pulp, and the greens
- For colza: the colza seed and the straw

The quantities relative to by-products are given in dry matter (d.m.) units. For the products and by-products (straw, greens, pulp) the per ha mass and energy content are presented in the first four lines of Table 14.2.

Agrofuel production (A-F) uses only the grain of wheat, the sugar of sugar beet, and the oil of colza. However all the by-products have also to be taken into account.

The sugar from wheat grain, after being transformed into sugar + distiller's dried grain solubles (DDGS), as well as the sugar extracted from sugar beets, is submitted to an ethanol-producing fermentation that transforms them into ethanol + CO_2 according to the exothermic reaction:

$$C_6\,H_{12}\,O_6 \rightarrow 2C_2H_5OH + 2CO_2 + 1.2\text{ MJ per kg ethanol.}$$

For colza seed, the transformation consists in pressing the seeds to extract their oil with, as final products, the oil and the pressed residue which finds a use as dried oilseed cakes.

At this point, we observe that foodstuff production implies the simultaneous production of main products: wheat grain, sugar, colza seed, and of by-products (straw in the case of wheat or colza, pulp and greens in the case of beets). For the agrofuels considered here, the main products are ethanol or oil, the by-products are straw, CO_2, and DDGS for wheat, pulp, greens, CO_2, greens and residues for sugar beet, and straw and oil-seed cake for colza, as listed in Table 14.2.

The data concerning masses, in tons/ha are presented in columns 2–4–6 in Table 14.2; the energy content of the products is given in toe/ha in columns 3–5–7 of the same table.

Methodology

An evaluation of the energy balance of A-F production will thus have to take into consideration all the final products generated as well as all the energy input, first in the farming process, then in the transformation of the wheat grain, the sugar, or the colza seed into ethanol or oil.

Table 14.2

Type of production	Wheat masses (t/ha)	Wheat total EC (toe/ha)	Sugar beet product masses (t/ha)	Sugar beet EC (toe/ha)	Colza masses (t/ha)	Colza total EC (toe/ha)
Initial products	Grain [a] 9.0	3.06	Sugar [a] 11.7	4.56	Seed [a] 3.30	1.98
	Straw (d.m.) [b] 4.0	1.56	Pulp (d.m.) [b] 4.00	1.56	Straw (d.m.) [b] 3.0	1.17 [b]
			Greens (d.m.) [b] 3.00	0.99		
Σ initial values	**13.0**	**4.62**	**18.7**	**7.11**	**6.3**	**3.15**
Gross usable E gain (Foodstuffs)		**10.6**		**15.1**		**5.80**
Final outputs /ha in t or toe/ha (agrofuels)	Ethanol [a] 2.63	1.68	Ethanol [a] 5.11	3.27	Colza oil [a] 1.1	0.99
	Straw (d.m.) [b] 4.00	1.56	Pulp (d.m.) [b] 4.0	1.17	Straw (d.m) [b] 3.00	1.17
	CO_2 [c] 2.52	0.07	CO_2 [c] 4.89	0.14		
	DDGS (d.m.) [b] 3.85	1.16 [b]	Greens (d.m.) [b] 3.00	1.32	Oil seed cake [b] 2.2	0.99
			Residues [c] 1.60	0.64		
Σ final values	**13.0**	**4.47**	**18.6**	**6.54**	**6.30**	**3.15**
		Wheat		Sugar beet		Colza
N_2O emission in teqC/ha		0.25		0.11		0.24
Input E for farming in toe/ha		0.39		0.42		0.48
Input E for agrofuel processing in toe/ha		0.64		1.31		0.06
Gross usable E in toe/ha (A-F)		**3.58**		**5.46**		**2.5**
Gross usable E gain (A-F)		**3.48**		**3.15**		**2.5**
Net usable E (A-F) in toe/ha		**2.55**		**3.73**		**1.96**

This table gives the main numerical values for 1 ha of crop: the masses (ton/ha), the energy (E) content of main products and by-products in toe/ha, N_2O emissions in teqC. It contains also values such as gross (or net) usable energy in toe/ha, discussed in the Sect.12.2.1.2

We have chosen the case of intensive agricultural production and the corresponding data are taken from ADEME DIREM (2002). *This choice is made for economic reasons; namely it minimizes the per ton production cost and input per ton of products. Data for ethanol and byproducts are taken from (Ballerini 2006)*

The **bold characters** correspond to calculations explained in the text or in the first column

Notations:
E energy, *EC* = Energy content, *UEC* = Usable energy coefficient used for calculating gross usable energy gain, *d.m.* dry matter, *A-F* agrofuel production
[a]UEC = 1.0, [b]UEC = 0.7, [c]UEC = 0

Note and this is important, that the mass and energy conservation laws require that the initial and final states be clearly defined and that the entire output be considered (main products and by-products).

In Table 14.2, we present the mass and energy quantities per ha for each element of the crop and the mass and energy balance for the initial and final products of each crop. Mass and energy are conserved (disregarding losses that we neglect here). The energy associated with the CO_2 item in the table is the energy released during the exothermic reaction mentioned above that transforms sugars into ethanol. It is neglected hereafter.

14.3.1.2 Taking into Account the Energy Content of the Main Products and By-Products

The energy balances in the table are evaluated based on the caloric content of each of the products as expressed in toe/t or toe/ha. However, it seems obvious that the energy contained in the main products does not have the same use value as that of the by-products. For example, one MJ of straw does not have the same use value as one MJ of wheat grain or one MJ of ethanol. As a consequence, we propose a usable energy coefficient, UEC, as follows: UEC = 1.0 for the main products; UEC = 0.7 for all the recoverable by-products; UEC = 0.0 for CO_2. The value 0.7 for the by-products results from a rapid estimate and could be modulated according to the nature of each by-product.

Using these conventions, the following per ha quantities can be calculated for wheat farming in view of producing agrofuels (A-F).

In the case of wheat farming for A-F, then, we have:

Gross usable energy A-F (toe/ha) = UEC(ethanol)*(EC-ethanol) toe/ha + UEC(byprod.)*[EC-straw) + EC-DDGS)] toe/ha + UEC (CO_2)*(EC-CO_2) toe/ha = 1.0*1.68 + 0.7*(1.56 + 1.16) = 3.58 toe/ha

Gross usable Energy Gain(A-F) = [Gross usable energy (toe/ha)]/[Input E for farming + Input E for agrofuel processing] = [3.58 toe/ha]/[0.39 + 0.64] = 3.48

Net usable Energy A-F (toe/ha) = [Gross usable energy (toe/ha)] − [Input E for farming + Input E for agrofuel processing] = 3.58 − [0.39 + 0.64] = 2.55 toe/ha

Similar quantities have been calculated for the foodstuff production case.

14.3.1.3 Transition from the Production of Agrofuels (A-F) with By-Products to the Production of A-F *Without* By-Products or of By-Products *Without* A-F

Insofar as the by-products resulting from A-F production actually replace animal feed farming products, it seems legitimate to consider that for 1 ha of A-F crops, the production of ethanol itself, without by-products occupies only a fraction of this 1 ha surface and, likewise, the production of the by-products occupies only the other fraction of this same ha. With the convention that a fraction of the farming area is allocated to the sole production of A-F and the remainder of the area to the sole production of the by-products, the allotment can be done proportionally to the usable energy content of the A-F and the by-products. Then, for 1 ha of wheat, this leads to attributing 0.47 ha to the production of 1.68 toe of ethanol and 0.53 ha to the production of all the by-products. The energy input associated with farming and with the transformation of the main crop into A-F is allocated on the same basis, thus also for GHG emissions. This leads to the following values in ha/toe-etha: 0.47/1.68 = 0.28 ha/toe-etha. The same type of calculations gives 0.18 ha/toe-etha for sugar beet, 0.40 ha/toe-oil for colza oil.

Allotment of 1 ha wheat to the production of ethanol

This fraction is given by:

UEC(etha)*(EC-ethanol in toe/ha) divided by (UEC(etha)*(EC-ethanol) toe/ha + UEC (by-prod.)*[EC-straw) + EC-DDGS)] toe/ha = 1.0*1.68/(1*1.68 + 0.7*(1.56 + 1.16) = 0.47 ha

14.3.1.4 Analysis and Lessons of the Energy Balance

From Table 14.2, it appears that the gross usable energy gain is always large for foodstuff farming (wheat: 10; sugar beet: 15; colza: 5.8). As expected, agrofuel farming is characterized by much lower gross usable energy gain values (wheat: 3.5; beet: 3.2; colza: 4.6). Indeed, a large amount of energy (with UEC = 1) is spent in the transformation of wheat grain or sugar into ethanol + CO_2 + residues. Moreover, the usable value of the residues has been considered to be only 70 % of the full energy content.

The Worldwide Energy Impact of Foodstuff Farming

The energy consumed by worldwide foodstuff farming can be evaluated from the energy balance of 9 tons of wheat grain (Table 14.2), assuming that wheat farming is representative of the 2.6 Gton (Table 14.1) of foodstuff output (cereals, sugar, oil).

This worldwide energy consumption (foodstuff farming) is computed as follows:

Worldwide energy cost = (worldwide mass production (Table 14.1))*(Energy cost of farming/ha)/(mass/ha) = 2.6 Gton*(0.39 toe/ha)/(9 ton/ha) = 0.11 Gtoe

The worldwide energy cost for agrofoodstuff production is 0.11 Gtoe or 1.1 % of current total world annual energy consumption and 1.25 % of the 9 Gtoe of current fossil energy annual consumption (ADEME DIREM 2002). Similarly, the usable energy content of the crops including their by-products can be evaluated at approximately 1.2 Gtoe. The gross energy gain is more than 10, thanks to the solar energy.

The Energy Balance and Energy Yield of Agrofuel Farming

We recall (cf. Sect. 14.3.1.3) that, in order to take into account the associated by-products, we have assigned to ethanol or oil A-F farming the following areas per toe of ethanol or oil produced:

Wheat → 0.28 ha/toe-etha;
Sugar Beet → 0.18 ha/toe-etha;
Colza-oil → 0.40 ha/toe-oil.

This translates to the following output:

3.6 toe-etha/ha = 1/0.28 without by-products for the wheat sector, 5.6 toe-etha/ha without by-products for the beet sector, and 2.5 toe-oil/ha without by-products for the colza sector.

These relatively high values take into account the fact that the production of agrofuels implies that of by-products (DDGS, oilseed cake, straw, pulp, etc.) that can be used as animal feed, thus as substitutes to dedicated animal feed crops.

Evaluation of the Area Needed to Produce Agrofuel Yielding Gross 0.5 Gtoe

This value, 0.5 Gtoe of gross energy obtained from agrofuel, is often put forward. If we take as mean gross yield 4 toe/ha for all agrofuels, the surface needed is 0.5 Gtoe/4toe/ha = 125 million ha. The surface currently farmed worldwide is 865 Mha and should reach approximately (865 Mha*1.3) or 1,130 Mha in 2050. The surface dedicated to agrofuels would then be slightly larger than 10 % of total farmed surfaces. We can note that the 0.5 Gtoe covers twice the energy needed for farming at the world level and can be considered as a strategic production.

Improvement Potential for the Agrofuel Sector

The aim of the agrofuel sector is to maximize the energy provided by ethanol or oil production. Considering that in a farming–harvest–agrofuel transformation cycle, it is possible to use energies from sources other than petroleum or gas for all the stationary processes, the net energy yield in the form of liquid fuels can be significantly improved (by about 25 %).

Second Generation Agrofuels

This sector is in the development and optimization stages with ongoing basic research and process definition. However, we do not anticipate a radical change of the situation. Indeed, the cost of the energy input for the crop and for the transformation of the cellulose matter into liquid agrofuel will probably be as large as it is for wheat, for example. The contribution of second-generation agrofuels will be, rather, to enable the use of land that is not adequate for foodstuff farming, or to allow making use of the whole plant for A-F production. Finally, it remains to be seen how the production of these second-generation agrofuels can, in practice, be deployed in view of a significant output in economically viable conditions. However, the use of fertilizer will remain necessary.

14.3.2 Environmental Impact of Foodstuff and Agrofuel Farming

14.3.2.1 GHG Emissions Balances: Overview

The balances of CO_2 emissions are presented relative to one ha farmed land with the reference outputs defined above. They take into account the CO_2 emitted by the energies consumed, assuming that the primary energies used in the farming process as well as for the production of ethanol or oil are fossil energies (petroleum-like) implying CO_2 emissions. The N_2O emissions are also taken into account and the emission values used are those of the IPCC publications; other GHG emissions (methane and other gases) are not taken into account. The CO_2 emitted in the course of ethanol production is calculated in reference to 1 ha of wheat crop. The GHG emitted by the farming phase and by the transformation of the grain into ethanol + DDGS phase is allocated to ethanol proportionally to its energy content relative to the total energy content (ethanol and by-products). The quantity of GHG emissions avoided, then, is evaluated as the difference between the GHG emitted by the production and subsequent use of one toe ethanol and GHG that would have been emitted by the consumption of one toe petroleum-like fossil fuel. See the evaluation of GHG emissions avoided in the box at the end of Part 2.

14.3.2.2 Wheat Foodstuff Production: GHG Balance

For the wheat foodstuff production sector, the GHG emissions include those pertaining to fossil energies consumed by mechanized farming and the production of inputs (0.39 toe/ha) or 0.39 toe/ha*12/14 = 0.334 teqC/ha (this ratio 12/14 comes from the mass balance of the combustion equation of a typical C_nH_{2n} hydrocarbon). N_2O emissions are estimated (ADEME DIREM 2002) at 0.5 teqC/ha. This gives a total of 0.834 teqC/ha for the production of 9 tons of wheat and 4 tons of straw. As a result, the worldwide emissions, in teqC/ha, for the production of 2.6 billion tons of cereal, sugar, and oil are $0.834*2.6\cdot 10^9/9$ t = 0.24 GteqC representing 3 % of the 10 GteqC (Jancovici 2010) currently emitted annually worldwide.

14.3.2.3 Wheat-Based Agrofuel Production: GHG Balance

For the sector of agrofuel production from wheat, the per ha GHG emissions are larger because the contribution of the transformation of the crop (wheat grain, sugar, or colza seed) to ethanol or oil has to be added in. For standard agrofuels, we assume a mean output of 4 toe-etha or oil/ha and net GHG emissions avoided amounting to 0.5 teqC per toe ethanol or oil taking into account the use of by-products. (See the evaluation of GHG emissions avoided in the box at the end of Part 2.) For a gross agrofuel production of 0.5 Gtoe this would represent 0.25 GteqC of GHG emissions avoided, or around 2.5 % of the approximately 10 GteqC/year currently emitted worldwide.

We conclude, then, that the reduction of GHG emissions obtained through agrofuel production is not significant on a worldwide scale and cannot seriously be put forward as a justification for agrofuel production.

14.3.2.4 Potential Improvements

It is, however, possible to improve the GHG emissions balance, for example, by reducing N_2O emissions and using non-GHG emitting energy resources for the production of farming inputs or the transformation of wheat grain into ethanol, processes that are executed in stationary locations where carbon-free energy sources can be used (heat and CO_2 free electric energy produced from renewable energies or nuclear plants, or from coal-fired plants with CO_2 capture and storage).

14.3.2.5 What Would the Price of Petroleum Have to Be for Agrofuels to Become Economically Competitive?

A very simplified estimation can be obtained by considering the following situation. A manufacturer purchases 9 tons of wheat at a price P_{wheat}/t of \$100–\$200 per ton; he purchases the energy required for the transformation of this wheat into ethanol at a price 0.7 $P_{petroleum}$/toe. (He can use energies of lower prices than petroleum, such as coal, gas, or other sources). He sells his ethanol production (in toe) at the price of petroleum, $P_{petroleum}$/toe; he also sells the byproducts (DDGS) at 0.5 P_{wheat}/t. We assume that the margin between the products sold and the cost of the inputs (wheat, energy purchased for the transformation) is 1.3 times the total purchases in order to cover the investment amortization and the factory operation cost.

We start from the translation of our hypotheses as

1.3*(cost for buying wheat + cost for buying energy for the transformation of wheat into ethanol) = selling price of ethanol + selling price of by-products, we get:

$(P_{wheat}/t*9\ t+0.7*P_{petroleum}/toe*0.64toe)*1.3=1.68toe\text{-}eth*P_{petroleum}/toe$ $+3.85\ t\text{-}DDGS*0.5\ P_{wheat}/t.$

Solving this equation, gives

$P_{petroleum}/toe=8.9\ P_{wheat}/t$ or $\mathbf{P_{petroleum}/barrel=1.27\ P_{wheat}/t}$, with 1 toe = 7 barrels

According to this estimate, the price of petroleum must be at \$254 per barrel for wheat at \$200/t, \$190/barrel for wheat at \$150/t, and \$127/barrel for wheat at \$100/t. This gives, then, an estimate of the price of petroleum for which agrofuel production can start to become economically competitive, in the absence of a carbon tax on petroleum. Wheat and petroleum price volatility prevents further progress in this analysis. For the time being, the per barrel price of petroleum has constantly been less than the per ton price of wheat so that agrofuel production, where developed, has to be supported by a system of public subsidy. The competition between foodstuff and agrofuel farming will begin when the per barrel price of petroleum reaches approximately \$165 provided the price of wheat remains at \$130/t. Taking into account a carbon tax of 50\$ per ton of CO_2, lowers this estimation by about 20\$/barrel.

14.3.3 Conclusion of Part 2: Farming, Energy, Environmental Impacts for Food, Feed, and Agrofuel Production

The analysis of the energy aspects of farming and of its environmental impact can thus be summarized very simply:

- Foodstuff farming produces energy in the form of food or animal feed with an overall output of 1.2 Gtoe while consuming about 0.1 Gtoe.
- The gross energy produced by agrofuel farming amounts to roughly 4 toe/ha (without the by-products). The production of 0.5 Gtoe worldwide in 2050 would occupy about 10 % of the total surface farmed in 2050.
- The favorable balance for GHG emissions (excluding methane) obtained for agrofuel production is such that its impact on global GHG emissions is marginal as compared to the other global emissions, even if those are divided by 2 by 2050.

Because of their overall importance, however, particular attention must be placed on the reduction of methane emissions, in particular in the areas of rice farming, cattle raising, fossil energy handling, biomass decay, and so on.

The competition between foodstuff and agrofuel farming would become inescapable if the price of fossil energies were to rise significantly while agricultural prices remained stable. But, in Part 3, we demonstrate that other arguments show this occurrence is not as imminent as it may appear.

14.3.3.1 Remark on the Subject of Deforestation by Fire

The GHG emissions due to deforestation by fire should not be attributed to agriculture. There is no rational argument for that although it is frequently done. Admittedly, it is an ancestral practice, but it is no longer acceptable in a context where GHG emissions must be reduced. Burn deforestation must be detected and prevented without mercy and without delay. Deforestation with the sensible use of the harvested wood falling within the competence of governments is another story and a more complex one to decide.

Evaluation of GHG Emissions Avoided by A-F Production (ADEME DIREM 2002)

We present here the evaluation of GHG emissions that are avoided thanks to the production and use of ethanol as a replacement of fossil energies (wheat–ethanol sector). Note that, after ethanol combustion, all of the CO_2 taken from the atmosphere and allocated to ethanol has been re-emitted to the atmosphere.

We use the values of Table 14.2 and following conventions.

Em = Emissions,

Eetha or Eby-products = Energy content in ethanol or byproducts in toe,

1 toe-etha = quantity of ethanol equivalent to 1 toe,

1 teqC = 1 ton (CO_2) * 12/44,

UEC are usable energy coefficient.

The share of emission between ethanol and by-product is taken into account.

The balance of GHG emissions allocated to ethanol in teqC after the ethanol has been consumed, is, for one ha of wheat, in teqC/ha:

Em-etha (teqC/ha) for farming = [Einvested (toe/ha) * 12/14 (teqC/toe) + 0.25 teqC/ha (N_2O)] * (Eetha (toe/ha)/(Eetha + UEC*Eby-products), giving:

Em-etha (teqC/ha) for farming = (0.39*12/14 + 0.25) * 1.68/[1.68 + 0.7*(1.56 + 1.16)] = 0.274 teqC/ha.

Concerning the emission for grain transformation into ethanol, we get:

Em-etha (teqC/ha) for grain to ethanol transformation = (0.64*12/14) * 1.68/(1.68 + 0.7*1.16) = 0.370 teqC/ha**

**Here, the emission is allocated to ethanol and DDGS.

In conclusion

Em (teqC/ha) total = 0.274 teqC + 0.370 teqC = 0.644 teqC/ha or 0.383 teqC/toe ethanol taking into account the 1.68 toe-eth/ha energy output.

Because one toe-ethanol avoids the consumption of one fossil toe, the CO_2 balance for the production of one toe-ethanol as a substitute for one fossil toe is:

0.383 teqC/toe-eth – 12/14*1 teqC/toe = –0.474 teqC/toe-eth

The same evaluation for the beet–ethanol or colza–oil sectors gives balances close to –0.5 teqC/toe-eth or toe-oil.

The accuracy of the numbers above is to let the reader verify the calculation details. The real precision of this evaluation is on the order of 10–20 %.

14.4 Part 3: The Triple A Issue from Now to 2050, Agriculture, Alimentation Needs, Agrofuels

Questions and Answers about the Situation of Alimentation in the World from 2010 to 2050:

Question A: Can agriculture decently feed nine billion human beings in 2050?

Today, the state of agriculture is one of potential overproduction and the significant fluctuations observed are not due to insufficient mean production but to structural causes and a lack of collective market organization at the world level that would help cope with these fluctuations. Structural causes can be identified as: no stocks that can be readily called upon, large production fluctuations and long natural time delays for the adjustment of production to demand, inelasticity and urgency of the demand, and all these combined with a small exchange volume.

In Part 1, we have observed that, over long time periods, production adjusts to demand with an overall output growth of 1.5 %/year over 50 years. We also saw that the price of wheat has decreased by a factor of 3 since 1970 in constant dollars. Today, its "normal" price is approximately $150/ton.

From now to 2050, the demand for foodstuffs should increase at the rate of 1.5 %/year to reach, in 2050, 1.8 times the production today. Can this demand be met? The target seems accessible (Mazoyer 2008) thanks to the combination of increased farmed surfaces and increased mean per ha yield by a factor 0.75 %/year, extrapolating past and present progress of agricultural yields. This implies an approximate 34 % increase of farmed surfaces and of mean yields within 40 years.

It is difficult to reply to the question with definite arguments because of the large uncertainties, in particular concerning the impact of global warming on agricultural output. We must certainly not exclude the possibility that the supply–demand balance in 2050 could be much tighter than today and that it could be necessary:

- To revise the share of food production versus that of animal feed, as this appears to be an important adjustment factor (See Part 1)
- To implement stricter handling of food crops in order to avoid waste and loss of resources

Finally, a progressive stabilization of the world population, whether natural or encouraged, can be hoped for, and thought possible by the 2050s.

Question B: Can agriculture (agrofood and/or agrofuel) play a significant role in the reduction by a factor of two of GHG emissions and in the supply of liquid fuels for transportation and nonstationary mechanization by 2050?

Regarding GHG emissions, the impact of agriculture on CO_2 emissions is quite small even if N_2O emissions are taken into account. The emissions avoided thanks to agrofuels can play only a marginal role in the reduction by a factor of two of overall worldwide GHG emissions. We have considered the production of 0.5 Gtoe of agrofuels in 2050. According to the energy mix proposed in Bouneau et al. (2009), an energy mix that meets the reduction by a factor of two of GHG emissions, the use of

CO_2 emitting fossil energies is limited to 4.2 Gtoe. In such a mix, the contribution of 0.5 Gtoe would introduce some leeway and partial but appreciable energy self-sufficiency for those countries with large agricultural production and no fossil resources.

Question C: Would a competition between alimentation and agrofuel productions be threatening?

As we have seen, the competition with A-F production would become threatening only if the price of petroleum were to climb to $150–$200 per barrel, a price that is higher than the cost of producing liquid fuel from coal. Another point tends to show that this competition does not threaten: only 15 % of the world agricultural production participates in international exchanges, implying that 85 % of the production is auto-consumed in the producing countries. As a consequence, it is hard to think probable a situation in which these countries would let this competition develop at the expense of food production for their own population (Americas, Europe, Pacific region, Asia, North Africa). For the food importing countries, the situation could be harder to anticipate but we should keep in mind the fact that total overall value of food imports amounts to less than 15 % of world production and that the volume of food imports is even less. Regarding the countries whose alimentary autonomy is insufficient, a suitable international organization should be able to fill the need without insurmountable difficulties.

Still, it is mandatory that the responsibility for the alimentation of populations clearly remains in the hands of the states and that an uncontrolled globalization of exchanges susceptible to threatening this responsibility peculiar to each state, and the United Nations on the international level, be prevented.

Question D: Hunger in the world; how extensive is it, what determinants, what remedies?

This issue is the object of rather alarmist communication that can jeopardize the setting up of efficient solutions to this problem, which pertains to ethics. In its communication (FAO 2009) stating that one billion human beings suffer from hunger, the FAO could have also mentioned that 600 million of these live in Southern Asia (China, India, etc.) a part of the world that is not, for the most part, within the sphere of international food supply solidarity and, indeed, does not ask for international assistance to cope with the problem. In fact, 300 million persons who live mostly in Africa are, de facto, within the sphere of this international solidarity. The quantity of cereals that has to be delivered to these 300 million persons can be evaluated at 30 million tons, representing a material cost (at $150/t) that amounts to 5 billion dollars. This can be compared with the French, US, and world GDPs (2,000, 15,000, and 63,000 billion dollars). These orders of magnitude demonstrate that this issue can be settled if the international community so desires.

From these remarks, it appears that the problem, today, can be quantitatively and economically solved at the international scale provided this aid is organized efficiently. To diminish the impact and the frequency of recurring famines, we mention again the recent recommendation issued by the FAO to focus aid to these countries on farming mechanization and the use of inputs in order to attain economically viable farming conditions and, as a consequence, to increase their production potential and alimentary self-sufficiency (FAO 2009).

Question E: Is the development of agricultural production compatible with the conservation of biodiversity and the environment?

Regarding the environmental impact of agriculture beyond GHG emissions, research work should be undertaken on the pollution due to nitrates and certain plant protection products. For nitrates from intensive livestock breeding, there is the beginning of a solution in some regions that consists in transporting the manure to intensive and large-scale farming areas, where it can be used as fertilizer at low doses on these large areas. A rational analysis should be made on the N_2O emissions coming from fertilizer taking into account the new way to distribute nitrogen fertilizer.

Regarding biodiversity, we hold that the safeguard of sufficient biodiversity compatible with human alimentation is possible. Without being an expert in this field, one can wonder if the zones dedicated to the creation of biodiversity would not profit from better organization rather than attempting, at all costs, to reserve small, dispersed, unchanged areas for this function. For the large geographic zones of Africa, the Americas, and Asia that prove to be sanctuaries for this role of conservation and development of biodiversity, it seems obvious that international solidarity must come into play. Financial compensation must be implemented in favor of the host countries concerned, a compensation dedicated to the protection of these sanctuaries, and to compensation for the loss of earnings as compared to the income that would have been obtained from farming in the same zones.

14.5 Conclusion

This analysis of the triple A issue (agriculture, alimentary needs, agrofuels) has allowed us to explicate, quantify, and rationalize our assessment of the stakes concerning human alimentation during the next four decades on a world scale.

- It appears that agriculture can, indeed, satisfy humanity's alimentation needs in spite of an expected worldwide increase of the demand by a factor 1.7–2, through the combined increase of farmed surfaces and mean yields.
- Agricultural production of foodstuffs and agrofuels has only a marginal impact on the mandatory reduction of GHG emissions. Some relative improvements can, however, be obtained.
- The chances that agrofuel production may become economically competitive are small in the short term and its competition with food production does not seem to really threaten. Still, the large countries that enjoy plentiful agricultural production could subsidize agrofuel production as a strategic improvement of their energy autonomy with respect to liquid fuels. The price of imported foods could then increase. Logical countermeasures would be to assist those countries that are regular food importers in improving their alimentary autonomy by increasing the productivity of their own agriculture and to implement market regulation by an international organization. We have considered that the production of a gross 0.5 Gtoe of agrofuels could be envisioned, covering 10 % of the surfaces farmed in 2050.

- The objectives concerning the conservation of biodiversity could be reasonably achieved, but it is true that the necessary increase of agricultural productions does not make the task any easier. It appears that the solution of the problem of biodiversity conservation will not be simple if the only action considered is to maintain the already existing zones.
- The volatility of agricultural product prices, in particular of those that are intended for human alimentation is a fatality only insofar as the international community continues to consider that the free market can "naturally" optimize the adaptation of supply to demand. On the contrary, supply is subject to various hazards, in particular related to the climate, so that it cannot be rapidly adjusted whereas the demand is always, or almost always, urgent. For agricultural foodstuffs, clearly, regulation under the responsibility of an international organism is mandatory.
- Aid focused on re-establishing alimentary self-sufficiency in countries that suffer periodically from hunger should be considered by such an international governance body and it should take action without delay. Similarly, an increased and formalized awareness of the responsibility of governments in providing an adequate food supply to their population should be enforced.

More generally, it can be hoped that the international community will remain vigilant as to the triple A issue, in particular regarding human alimentation. But there is no reason to dramatize and excessive alarm could well serve only to amplify the instabilities and thus encourage speculation. The triple A issue continues to be an important one from an ethical point of view for all humanity which is, in addition, confronted by new constraints whose economical weight is far beyond that of alimentation (reduction by a factor of two of GHG emissions, imperative energy mix transformation, doubling of global energy consumption, and regulation of the financial markets). Trade globalization is probably irreversible but it appears that agriculture and human alimentation cannot be managed in an efficient, responsible, and ethical manner without appropriate regulation.

How pertinent is it to build up agrofuel production in terms of a partial solution to energy self-sufficiency, in terms of the reduction of GHG emissions, in terms of its impact on meeting humanity's alimentation needs?

The development of the agrofuel sector as a partial response to energy self-sufficiency seems legitimate and the major large-scale farming countries do seem to be determined to encourage this approach, even if economic profitability is not proven. In terms of the reduction of GHG emissions, their impact seems too marginal for this to serve as a relevant argument. In terms of their impact on human alimentation, the fact that the economic profitability of first generation A-F is not really obvious for the coming decades should limit the risk. Moreover, it seems unlikely, even unthinkable, that sufficient human alimentation would not be guaranteed by those states that are self-sufficient or even export alimentary products, and by an international organization for the countries that would continue structurally to depend on imports.

The crux of the matter is—and remains—to see to it that human alimentation needs are satisfied from now to 2050. This will be possible only through the implementation of a strong international organization that ensures this priority is respected. There must be at least one exception to the generalization of the free market for international exchanges and that concerns those products that have to do with human alimentation.

We have attempted a new approach to this alarming issue, the satisfaction of human alimentation needs in a context of scarcity of easy-to-use energies, mandatory reduction of GHG emissions, potential competition with agrofuel production, running astray of generalized free-trade where the responsibility for the satisfaction of human alimentation needs could be relinquished by the governments of some states.

The present analysis of the agriculture, alimentation needs, and agrofuels has been conducted on a world scale on a quantitative basis; it aimed at a better understanding and appreciation of the stakes, at putting fears in perspective, and at considering solutions that are within the reach of the international community. But we should remain vigilant and continue "to arm ourselves in thought" (Büchner 1834) because the situation of human alimentation in the 2050s will probably be quantitatively tight in some states, and hard to manage on an international scale.

Acknowledgments This chapter is largely inspired by the work done in the GT3A working group whose members were M. Bruschi, Y. Faure Miller, J. M. Most, H. Peerhossaini, J. B. Saulnier, and myself. Many thanks for their contributions.

Many thanks also to E. Huffer for her help in writing the English version of this paper.

References

ADEME DIREM (2002) Report: Bilans énergétiques et gaz à effet de serre des filières de production des biocarburants en France (Energy balances and greenhouse gas emissions from biofuels production in France), Sept 2002

Ballerini D (2006) Les Bio-carburants" edited by IFP Institut Français du Pétrole: Etat des lieux, perspectives, et enjeux du développement (Biofuels, current situation, prospects and development issues). IFP Publications, 92852, Rueil Malmaison, Paris

Bouneau S, David S, Loiseaux JM, Meplan O (2009) Construction d'un monde énergétique en 2050 (Elaborating a worldwide energy mix for 2050). Annales de Physique 34(1), 1–101

Büchner G (1834) Le Messager des campagnes hessoises (The Hessian campaign Messenger), with Friedrich Ludwig Weidig. Der Hessische Landbote, in Georg Büchner (éd. Bernard Lortholary), *Œuvres complètes, inédits et lettres*, Seuil, coll. «Le don des langues», Paris, 1988 (ISBN 9782020100281)

Cereals International Council (2008) Data of 2008. FAO, Geneva. Retrieved July 2009 from http://www.fao.org/docrep/012/ak349e/ak349e00.pdf

Dronne Y, Forslund A (2009) INRA Rennes, Marchés internationaux des produits agricoles et biocarburants (International markets for agricultural products and biofuels). Paper presented at 9th Ini-FCA meeting. Centre Inra de Rennes Domaine de la Motte, BP, France, 2 Apr 2009

FAO World Food Program 2009 Site Web. Retrieved July 2010 from http://documents.wfp.org/stellent/groups/public/documents/communications/wfp229437.pdf

Jancovici JM (2010) Web Site Manicore URL. Retrieved Nov 2010 from http://www.manicore.com/documentation/serre/index.html#Gaz_a_effet_de_serre

Mazoyer M (2008) Réorganiser les échanges agricoles mondiaux (Rethinking world agricultural trade). La Jaune et la Rouge N° 639 Novembre 2008, pp 12–15

Index

J.B. Saulnier and M.D. Varella (eds.), *Global Change, Energy Issues and Regulation Policies*, Integrated Science & Technology Program 2, DOI 10.1007/978-94-007-6661-7,

D

E

MIX
Papier aus verantwortungsvollen Quellen
Paper from responsible sources
FSC® C105338

If you have any concerns about our products,
you can contact us on
ProductSafety@springernature.com

In case Publisher is established outside the EU,
the EU authorized representative is:
Springer Nature Customer Service Center GmbH
Europaplatz 3, 69115 Heidelberg, Germany

Printed by Libri Plureos GmbH
in Hamburg, Germany